Aquaculture Economics and Financing
Management and Analysis

Aquaculture Economics and Financing

Management and Analysis

Carole R. Engle

Aquaculture/Fisheries Center
University of Arkansas at Pine Bluff

A John Wiley & Sons, Ltd., Publication

Edition first published 2010
© 2010 Carole R. Engle

Blackwell Publishing was acquired by John Wiley & Sons in February 2007. Blackwell's publishing program has been merged with Wiley's global Scientific, Technical, and Medical business to form Wiley-Blackwell.

Editorial Office
2121 State Avenue, Ames, Iowa 50014–8300, USA

For details of our global editorial offices, for customer services, and for information about how to apply for permission to reuse the copyright material in this book, please see our Website at www.wiley.com/wiley-blackwell.

Library of Congress Cataloging-in-Publication Data

Engle, Carole Ruth, 1952–
    Aquaculture economics and financing : management and analysis / Carole R. Engle. – 1st ed.
        p. cm.
    Includes bibliographical references and index.
    ISBN 978-0-8138-1301-1 (pbk. : alk. paper)   1. Aquaculture industry.   2. Aquaculture–Economic aspects.
3. Aquaculture–Finance.   I. Title.
    HD9450.5.E53 2010
    338.3–dc22

                                                                                                    2010013917

A catalog record for this book is available from the U.S. Library of Congress.

Set in 9.5/11.5 pt Times by Aptara® Inc., New Delhi, India

1   2010

This book is dedicated to my parents, Mildred Evelyn Orris Engle Wambold, Glenn Wambold, and Morris Engle, Jr.; my husband, Nathan Stone; our children Reina, Eric, and Cody; and to the fish farmers of Arkansas.

Contents

Preface ix

Acknowledgments xi

I Managing Aquaculture Businesses **1**

1 Starting an Aquaculture Business 3

2 Marketing Aquaculture Products 13

3 Developing a Business Plan for Aquaculture 23

4 Monitoring Economic and Financial Performance of Aquaculture Businesses 41

5 Financing an Aquaculture Business 57

6 Managing Cash Flow 67

7 Managing Capital Assets in Aquaculture Businesses 81

8 Managing Risk in Aquaculture Businesses 93

9 Managing Labor 105

II Economic and Financial Analysis of Aquaculture Businesses **115**

10 The Enterprise Budget and Partial Budgeting in Aquaculture 117

11 Financial Statements: Balance Sheet and Income Statement in Aquaculture 131

12 Cash Flow Analyses in Aquaculture 143

13 Investment Analysis (Capital Budgeting) in Aquaculture 159

14 Lending in Aquaculture 173

III   Research Techniques to Analyze Farm-Level Decision-Making   **183**

15   Use and Misuse of Enterprise and Partial Budgets   185

16   Risk Analysis in Production Aquaculture Research   197

17   Whole-Farm Modeling of Aquaculture   207

18   Managing Government Policies and Regulation in Aquaculture Businesses   219

Bibliography   231

Webliography   235

Glossary   241

Index   255

Preface

The volume of aquaculture production worldwide has grown at a rate of approximately 8% per year over the last decade. Continued growth is expected due to increases in world population and the apparent leveling off of the capture of many commercial fisheries species. Farmed salmon and shrimp production have grown to dominate their respective world markets over the last several decades. More recently, new global markets have emerged for farmed species such as tilapia (*Oreochromis* sp.), channel catfish (*Ictalurus punctatus*) and the basa/tra species (*Pangasius* sp.). Continued growth, however, depends not just on demand but also on the economic and financial viability of the businesses developed.

Aquaculture production presents some unique challenges for economic analysis. While there are many books that address the theory and methodology of economic and financial analysis, there are few that present clear details on applications to aquaculture businesses. The few that do are quite general in nature and rely on hypothetical examples that omit the often-messy details of the real world. The difficulties posed by aquaculture businesses are rarely discussed or addressed. The problem is exacerbated by the lack of practical knowledge and experience in aquaculture on the part of many economists. Simplifying assumptions make analyses more tractable, but too often obscure the problems and challenges faced by those attempting to make a living from aquaculture.

As a result, researchers who wish to add an economics component to a production aquaculture trial have little guidance. The unfortunate result is that key costs are too often ignored, invalid assumptions are made, and analytical tools are applied incorrectly. These errors degrade the quality of the work and may lead to erroneous or misleading conclusions.

The intent of this book is to provide a detailed and specific set of guidelines for both aquaculture businessmen and women and researchers related to the use of economic and financial analysis of aquaculture. The goal of the book is to remove the mystery or "voodoo" from economic analysis as it is applied to aquaculture and to provide a guide for its accurate application.

This book discusses key issues related to both financing and planning for aquaculture businesses, how to monitor and evaluate economic and financial progress, and how to manage the capital, labor, and risk in the business. The book works through the specific application of farm management and financial analysis tools for aquaculture. Particular attention is paid to those line items and valuation methods that are most often confused or in error in aquaculture. A section on use and misuse of budgeting techniques in research should assist aquaculture researchers to avoid common mistakes. Additional chapters on risk analysis and whole-farm modeling provide a sense of more advanced techniques and their applicability. Finally, a chapter on managing government regulations provides guidance for adjusting to an increased number of regulatory activities.

The book is based entirely on aquaculture examples and literature with an emphasis on farm-level data and analysis. It is written in terminology that aquaculture researchers and business persons will readily follow and understand. The section on the application of economic analysis in aquaculture research is unique; no other book outlines how to value parameters measured in aquaculture field trials.

The book includes a specific, detailed example of a practical application in each chapter. A section on other applications in aquaculture is included to paint a broad picture of the economics of aquaculture around the world while providing comprehensive guidance on each particular topic.

The three principal audiences for this book are: (1) aquaculture business owners and managers; (2) those who conduct research on aquaculture production systems, strategies, equipment, or management practices;

and (3) students preparing for careers either in the industry or in aquaculture research. The book should appeal to practitioners in a number of different countries, but especially those with aquaculture industries.

Aquaculture business owners and managers will likely be most interested in Section I: Managing Aquaculture Businesses. The chapters in this section are designed for those who are likely to hire an accountant to develop the analyses, but need to know what questions to ask and how to interpret the answers. This section works through the key questions related to starting an aquaculture business, some basic marketing considerations, business planning, understanding how to interpret the financial statements prepared by accountants for businesses, cash flow, financing, and management of capital assets, labor, and risk. The key focus in this section is on the use of information to make management decisions. Those aspects of aquaculture businesses that are unusual or different from other types of businesses are emphasized from the perspective of managing the business effectively for profit. Chapter 18 on managing government policies and regulations will also be of interest to aquaculture business owners.

Section II, Economic and Financial Analysis of Aquaculture Businesses, is for students and those who wish to understand the details of how to develop and complete the various types of economic analyses that are commonly used in the economic analysis of aquaculture. Each chapter presents in detail the mechanics and methodology for developing enterprise budgets, partial budgets, balance sheets, income statements, cash flow budgets, and investment analyses. Challenges and common pitfalls associated with use of each of these methodologies are discussed in each chapter.

Section III, Research Techniques to Analyze Farm-Level Decision-Making, is written especially for those who conduct research on aquaculture production systems, strategies, equipment, or management practices. Misleading assumptions, omitted costs, overestimating revenues, and misapplication of research data can be common in economic analysis based on aquaculture research. This section reviews these challenges and describes detailed approaches to developing accurate economic analyses with production research data.

The book includes an annotated bibliography and a webliography of resources. Software products available for economic and financial analyses are listed and described. It is my hope that you will find this book useful and that it will help aquaculture businesses to be efficient, viable, and profitable.

Acknowledgments

There are many people who contributed either directly or indirectly to the content of this book. Some of this material was drawn from earlier training programs developed with the assistance of Diego Valderrama, Steeve Pomerleau, and Ivano Neira. Ganesh Kumar has provided invaluable assistance throughout. Insightful and useful review comments were provided by Diego Valderrama, Anita Kelly, Madan Dey, and George Selden. Umesh Bastola, Pratikshya Sapkota, and Abed Rabbani also provided suggestions. Finally, fish farmers throughout Arkansas and other states provided the continuous ground truthing of the economics and financing of aquaculture.

Section I
Managing Aquaculture Businesses

INTRODUCTION TO SECTION I: MANAGING AQUACULTURE BUSINESSES

This section focuses on the use and application of economic tools and interpretation of their results. It is designed primarily for owners and managers who hire others to prepare financial statements and analyses. At the same time, those who are trained to conduct economic and financial analysis but who are not familiar with aquaculture should also find this section useful.

Aquaculture is a management-intensive business. The need for intensive and skilled management stems from the high level of capital invested in the facilities, and in the high levels of operating capital required to operate a competitive and profitable business. Throughout aquaculture, undercapitalization (not having enough capital to make payments and survive the sometimes lengthy startup periods) has been a consistent problem.

Individual companies must answer a series of questions that involve pricing, output, and market positioning. Key questions that the manager must answer include: (1) how much should be produced; (2) how much input should be used; (3) what is the optimal size of the business; (4) how should cash flow be managed; (5) how should risk be managed; (6) how will the business be financed; and (7) how can business performance be optimized? Thus, it is the manager who must develop the business plan, monitor economic and financial performance of the business, and manage cash, capital, labor, and risk. Each of the following chapters discuss specific aspects of the types of management functions and decisions that need to be made by the manager.

1
Starting an Aquaculture Business

Aquaculture has grown rapidly in volume and in complexity around the world in the last several decades. While aquaculture has a centuries-long history as a source of food for households in Asia and Africa, the most dramatic change in more recent years has been the development of aquaculture businesses into complex industries. These industries operate on national and international levels.

Development of efficient and viable businesses requires careful evaluation and thorough planning for the new business. This book presents details on the process of business planning (see Chapter 3) as well as on how to prepare the various types of financial statements needed for thorough planning (see Chapters 10 through 14). Chapter 1 begins by outlining steps to be considered before starting the business.

The new business owner must think carefully about what will set his or her business apart, both from other existing businesses and from other future businesses. It is critical to identify the strengths, abilities, and skills owned and available that will help the farm owner to be successful. This chapter discusses the motivation and goals for starting an aquaculture business, and the capital- and management-intensive nature of aquaculture. Marketing challenges and trade-offs associated with various organizational structures, financing, and the availability of resources are contrasted. It concludes with a discussion of permits, regulations, and sources of assistance. Figure 1.1 illustrates the various steps that will be needed to start a successful aquaculture business.

MOTIVATION AND GOALS

The first step to starting an aquaculture business is to carefully consider one's goals and motivation. An individual interested in starting an aquaculture business must fully understand why he or she wants to do this. Some individuals enjoy working outdoors with fish and dislike office work and paperwork. These individuals may do an admirable job like raising fish on the farm. However, inadequate attention to the business aspects of the aquaculture business will result in financial failure. If the owner spends all his or her time caring for the fish, who will take care of the permits, regulations, financial statements, and economic performance of the business?

Others who wish to start an aquaculture business may view it as a way to make a great deal of money. There certainly are success stories of aquaculture businesses that have become profitable businesses. However, aquaculture businesses are intensive businesses that require management committed to working long hours under often difficult conditions. Who will provide that level of management?

Still others view aquaculture as the wave of the future and want to get in on the ground floor. However, businesses developed to raise the latest "hot" species with the newest production technology frequently are beset with substantial levels of financial, price, and yield risk. Aquaculture entrepreneurs must be prepared to manage the degree of risk associated with their business model and plan.

It is important to develop clear and specific goals for the business from the beginning. For example, what is adequate revenue for one individual may not be sufficient to entice another to invest the necessary time and money. The effects of starting a new business on the farmer's family must be considered carefully. Will family members be supportive and helpful or will they resent the time that must be invested in the business? The early years of an aquaculture business may generate minimal revenue, and the family may be required to live for a time on reduced income.

Aquaculture Economics and Financing: Management and Analysis, Carole R. Engle, © 2010 Carole R. Engle.

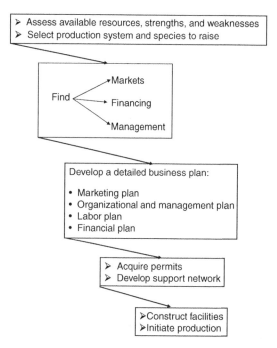

➤ Assess available resources, strengths, and weaknesses
➤ Select production system and species to raise

Find
→ Markets
→ Financing
↗ Management

Develop a detailed business plan:

• Marketing plan
• Organizational and management plan
• Labor plan
• Financial plan

➤ Acquire permits
➤ Develop support network

➤ Construct facilities
➤ Initiate production

Figure 1.1. Starting an aquaculture business.

AQUACULTURE IS A CAPITAL-INTENSIVE BUSINESS

The majority of aquaculture businesses require substantial amounts of both operating and investment capital. One of the largest problems encountered in starting an aquaculture business often is to acquire sufficient capital. Undercapitalized farms and processing plants rarely survive. Careful thought and planning need to go into determining the amount of capital needed to operate at an efficient level and to identifying sources for the needed capital.

Capital requirements begin with the investment capital needed to purchase land, build production facilities, and purchase equipment. Depending on the specific location, new roads may need to be constructed, electric power lines may need to be installed, or there may be additional infrastructure required that will increase the total amount of investment capital needed. Capital required for marketing facilities must be included in the planning. Is a shed needed to hold, grade, and sell the fish? If so, a water supply system to fill hauling trucks will also be required. Perhaps an ice machine is required, depending on the form of the products sold. If the farm will do its own hauling and

transportation, then the trucks, tanks, oxygen systems, and loading equipment will also be needed. In all, investment capital for an aquaculture business typically will be several thousand dollars an acre of production for pond systems, and can range from $0.30 to $7.00/lb across the many different types of species and production systems. More information on investment capital requirements can be found in Chapter 10, and techniques to measure the profitability of such an investment can be found in Chapter 13.

The high level of investment capital required for an aquaculture business results in high levels of annual fixed costs (see Chapter 10 for more details on what constitute annual fixed costs). The best way to reduce the fixed cost portion of the cost of producing fish is to produce at an intensive level with high yields. High yields spread the annual fixed costs over a greater level of production and lower the cost per pound of production.

Operating capital requirements often are as substantial as investment capital requirements for aquaculture businesses. Frequently, this is because high yields are needed to lower the per-pound annual fixed costs and keep production costs at a competitive level. Achieving high yields requires high numbers of fingerlings, large amounts of feed, greater electricity for aeration, and corresponding amounts of other inputs such as labor, repairs and maintenance, and fuel. Operating costs frequently can be $2,000–$5,000/acre for pond systems and $33,000–$150,000/acre for intensive systems such as raceways and indoor systems.

Operating cost requirements are compounded by the fact that some types of fish raised do not reach market size in one growing season. Thus, the prospective fish farmer often must prepare to operate the business for more than a year without receiving revenue from the business. Careful financial planning and good communication with one's banker are keys to having access to sufficient amounts of capital with which to build the business until it reaches its full production capacity.

The high levels of capital required for many aquaculture businesses result in substantial amounts of financial risk (see Chapters 8 and 16). The profit potential is often accompanied by a variety of risks, and the large sums of money invested in an aquaculture business can be lost quickly. The best method to prevent such losses is adequate and thorough planning, monitoring, and assessment of the economics and finances of the aquaculture business throughout its life. If the owner is not willing to spend the time to monitor the business' financial performance, then it is essential to hire or retain an expert to keep a constant and close eye on its economic aspects. Otherwise, the likelihood of failure and severe financial losses is high.

AQUACULTURE IS A MANAGEMENT-INTENSIVE BUSINESS

The high levels of investment and operating capital required in aquaculture businesses, along with the intensive nature of production of aquatic animals requires a high degree of management. When aquatic animals are confined in a production unit, constant attention is needed to the quality of the growing conditions for the fish. Maintaining adequate levels of oxygen and other critical water quality parameters, and preventing problems associated with the breakdown of waste products in the system, takes careful and constant monitoring. Diseases spread rapidly when animals are maintained in close confinement. Thus, attention must be paid to monitoring the health of the animals and taking necessary actions when there are indications of disease.

Marketing and sales of aquatic products can represent management challenges depending upon the nature of the target market. Farms engaged in direct sales will require a great deal of management attention to marketing functions and activities. Even farms that sell directly to a processing plant must have managers who pay close attention to the requirements of processors. These requirements include quality standards and delivery requirements, among others, to minimize dockages from fish that do not meet specifications. Top managers ensure that their farm is considered to be a preferred supplier, one that consistently delivers quality fish, within specification, and on time.

Management must keep a close eye on costs and production efficiencies throughout the production process. This includes monitoring the efficiency of labor, feed usage, use of electricity and fuel, and use and care of equipment. Feed, for example, is frequently the largest single component of the cost of raising an aquaculture crop. Feeding carefully and appropriately will ensure a better feed conversion ratio and will result in more pounds of fish produced per pound of feed applied. Similarly, judicious use of aeration can provide adequate oxygen levels by turning aerators on sequentially, as needed, rather than turning on more aerators than are needed at one time. Taking time to keep equipment well maintained and to ensure that it is operated correctly will reduce the costs of repairs and will extend its life. This will reduce the cost of equipment as a percentage of the total cost of production.

Moreover, the manager must be able to think and plan strategically. Preparing to stay ahead of future challenges requires examination of the business's strategic plan from a variety of different perspectives.

MARKETING CHALLENGES

Many individuals who wish to start an aquaculture business are captivated by the animals and plants that they wish to raise and will spend many hours talking about their biology and growing requirements. However, the marketing challenges of starting a new aquaculture business often are greater than the production challenges and ultimately more important. Successful aquaculture businesses are managed and owned by individuals who spend as much time exploring marketing options and trends as they do working on production efficiencies. Chapter 2 of this book discusses marketing issues and strategies related to aquaculture products in greater detail and Chapter 3 outlines steps in the development of marketing plans as part of an overall business plan. This chapter discusses some general concepts.

There are a number of overarching trends and challenges that prospective aquaculture business owners should consider. Most species that are being aquacultured at one time were primarily a wild-caught species. Many of these species have existing markets and demand that were based originally on supply from capture fisheries. Preferences for wild-caught as compared to farmed fish vary by region. Care must be taken to understand these preferences in the market targeted for the business's product. As farmed product becomes available in the market, it frequently must compete with wild-caught product that is already well established in the market. However, the cost of producing farmed fish, especially in the early years of startup businesses, requires a price that often is higher than that of wild-caught fish. To establish a new product in the market often requires differentiating it from wild-caught and other similar products to capture a price that will cover production costs.

The seafood market has undergone dramatic changes in the last several decades. The possible effects of current and emerging trends must be considered carefully in planning for successful marketing programs. Seafood in earlier decades was primarily a locally sourced, fresh product supplied by either fishermen or specialized jobbers and small-scale wholesalers. Changes in packaging and freezing technologies have opened the door for global trade in seafood that has continued to increase dramatically. The increase in global trade in seafood has resulted in a number of conflicts and competition with fish raised domestically. All types of fish and seafood are now shipped around the world to satisfy various markets.

Dynamic markets like those for seafood, while challenging, can also offer opportunities for entrepreneurs. For example, the shrimp and salmon industries worldwide have benefited from the increase in global trade and technology. These segments of aquaculture have grown to dominate shrimp and salmon markets worldwide.

Food marketing in general has undergone dramatic changes in recent decades that have resulted in changes along the supply chain. The driving force has been the emergence of strong market power at the level of the large hypermarket discount retail sector, exemplified by companies like Wal-Mart. In response to this concentration of market power, wholesalers and food service distributors have also become more concentrated. This has increased pressure on growers to either consolidate by integrating vertically to capture market power, or to form cooperative or other forms of organizations to be able to compete.

Startup aquaculture businesses must identify the specific market that the business plans to target. The overall marketing plan (see also Chapter 3) must also identify the competition and the unique position the company's product will occupy in the market. The product must be defined well and must match the way it is positioned in the market for the targeted customers. Careful attention should be paid to the size of the market, long-term price trends, and distribution patterns of similar products.

The marketing plan must lay out an effective promotion and advertising plan. Even the smallest-scale aquaculture farms must have a plan to spread the word about their products. Promotion is a way to transmit information about the attributes of the product, the price, and why the consumer should purchase it.

Appropriate and effective market channels must be developed. Is the farmer planning to transport all the fish produced to the various markets? The amount of time needed to transport fish to markets must be determined and adequate personnel included in the business plan. The length of round trips that can be undertaken feasibly can be an important factor. If the farmer does not intend to transport his or her own fish, relationships and agreements will be needed with a wholesaler or distributor.

ORGANIZATIONAL STRUCTURE FOR THE AQUACULTURE BUSINESS

Most farms in the United States have a single owner and are classified as sole proprietorships. In a sole proprietorship, the farmer is self-employed and has legal title to the property. This is the simplest form of business structure, but it also entails the greatest risk. Risk results from the liability for any debt obligations or accidents that falls entirely on the owner in a single proprietorship. Moreover, the liability is not limited to what the farmer has invested in the business. The farmer can lose his or her land and home as a result of severe adverse situations.

Some farmers form partnerships with family members or others to gain access to additional resources such as land, equipment, labor, or management. Partnerships can be either general or limited. Partners share in all ownership, management, and liability in a general partnership. Limited partners share in the profits and losses of the business but not in the management. In this way, the limited partner provides resources such as capital, but management decisions are under the control of the principal owner.

Some segments of aquaculture have integrated vertically and have developed into corporations. In a corporation, capital is provided by shares of stock, and the management and control are provided by the stockholders, the board of directors, and the officers. The board sets policies, and the officers manage the daily activities of the company. Stockholders, while owners of the company, are not personally liable for actions of the corporation. Their liability is limited to their investment in stocks.

There are also subchapter C and subchapter S corporations. With C corporations, dividends received by stockholders are taxed as income, while S corporations are taxed as limited partnerships. The officers are paid before the remaining profits are distributed.

AVAILABLE RESOURCES

An important step in starting an aquaculture business is to develop a frank assessment of the resources available for the business. New businesses fail more often than they succeed, often due to the lack of adequate resources. The assessment of available resources begins with the individual. The owner must be innovative, persistent, resourceful, and determined to find solutions to the many problems that will arise. The assessment should extend to physical resources available that include land, existing ponds, wells or other types of water supply, farm equipment, and buildings. The assessment must be thorough and detailed. For example, the individual may have an adequate quantity of land available, but current or impending zoning regulations may prohibit its use for aquaculture. Water and soil analyses should be done to check the suitability for the species to be considered. Some freshwaters have enough salinity to consider some crops like marine shrimp that can tolerate low levels of salinity. There must also be adequate backup equipment in the event of breakdowns, generators for power outages and backup aeration equipment. Resources also include the availability of adequate quantities of seedstock.

The availability of labor resources can be an important factor in the success or failure of a new business. The assessment of labor availability should include any family labor that is available to assist with the farm. A realistic assessment must be made of the local labor supply and the availability of adequately trained labor that can be hired for the aquaculture business. The type of labor is also an important consideration. Aquaculture businesses often require more skill than some other types of agriculture, and the ability of workers to handle the new responsibilities must be evaluated carefully. For example, workers who cannot swim or who are afraid of the water may have difficulty adjusting to working around it constantly. Much aquaculture requires long and irregular hours during the main growing season. Workers may or may not be willing to work such hours. The degree of equipment on the farm requires a great deal of maintenance. An aquaculture business requires either a mechanic hired on the farm or the business must be prepared to have higher repair and maintenance costs.

The availability of management resources must be assessed. The level of expertise and skill of the owner to manage the production, marketing, and financing of the aquaculture business must be evaluated frankly. If the owner has excellent aquaculture skills but is weak in financial analysis, an appropriate accountant or financial analyst will need to be hired, contracted, or retained. Similarly, if the owner has good business skills in marketing and financing, but lacks experience in culturing aquatic animals, hiring a production manager with adequate aquaculture skill and expertise will be essential.

Sufficient capital resources must be available as well. Both investment and operating capital are required in necessary quantities to be received at ap-

propriate times. The operating line of credit must be structured to continue the business throughout the entire startup period during which the business begins to generate returns. Depending on the type of business, this may be a period of 2–3 years before substantial revenue can be generated from the aquaculture business. The investment capital must be available in sufficient quantities to provide facilities to minimize risk. This includes sufficient redundancy in equipment to cover power outages, breakdowns, and unanticipated extended periods of adverse weather conditions. Maintenance requirements must be accounted for in financial planning. This includes the capital to be able to replace all equipment and facilities when necessary.

The availability of adequate credit will depend in part on the ability of the owner to finance the operation through equity or to have the credit capacity to borrow the necessary amounts of capital. This in turn will depend upon the individual's balance sheet, availability of collateral, and overall credit worthiness.

The particular species selected and their product forms are critical decisions. These must match the projected price point and the quantity demanded for that product form for that species. In selecting the species to be raised, thought must be given to whether there is competition from imported species or capture fisheries, or both. Diversifying farm production with several species also serves to spread the market risk of price downturns for one specific species.

The key to starting a successful aquaculture business is to match the species to be produced, the production system to be used, and the scale and scope of operation with the available markets and resources (labor, land, capital, and management). Mismatches are likely to result in business failure. For example, a particular species and production system may exhibit strong economies of scale. If the owner is unlikely to be able to acquire sufficient capital to construct and operate a farm of a large size, it is better to rethink the business plan to develop one that is workable with the capital resources available. Undercapitalization is often a major reason for failure of aquaculture farms and processing plants. Mismatches between projected and actual capital requirements result in financial failure.

FINANCING

Adequate financial resources are essential to a successful business, and the ability to acquire sufficient capital is a key factor. One of the first steps is to identify the sources of capital that are available. Venture capital can be difficult for aquaculture and often follows certain patterns and trends that may not always favor financing aquaculture businesses. Private capital from partners, whether active or silent, can be considered in establishing the business. However, private lenders finance most aquaculture businesses. Many lenders may be skeptical about aquaculture and view it as a risky business. Perceptions of high risk in aquaculture may lead to less favorable terms of lending, requirements for greater owner equity in the business, higher interest rates, or refusal to consider loans for aquaculture ventures.

Financing from private lenders can also be complicated by the fact that many lenders may not have substantial experience with aquaculture. Lack of familiarity with the business can result in unwillingness to assess business loan proposals; the loan officer may not be comfortable with the estimates of yields, costs, or efficiency measures that form the basis of the proposal. It may be necessary to spend a great deal of time working with a lender to help them understand the basics of aquaculture, introduce them to people who are knowledgeable about successful aquaculture businesses and the keys to their success, and to keep them informed of the most recent trends in aquaculture. It is important to plan for adequate capital to provide for the family through the very difficult early years of the business.

HARVESTING AND PROCESSING

Decisions must be made early on in the development of an aquaculture business on how the fish or other animals will be harvested and whether they will be processed, stored, and transported by the farm business. If not, these services will need to be contracted. Serious thought needs to be given to the implications of these decisions. In areas with little aquaculture production, these services may not be available. If the farm owner must hire a seining crew, process the product, and store and transport it, the owner likely will need to operate on a relatively large scale. There are some examples of small-scale aquaculture businesses that perform these functions, but typically these will require a larger scale of business. Processing in particular has substantial economies of scale that must be considered before proposing this type of component to the business.

Product handling throughout transportation and processing will affect the end quality of the product. If proper conditions are not maintained during harvest and transport, the quality of the fillet may suffer. Sim-

ilarly, if processed fillets are not stored or packaged properly, the result will be a poor quality product. Adequate planning for these functions is necessary.

PERMITS AND REGULATIONS

Part of a careful assessment for a startup business includes identifying the permits and regulations that will affect the new business both currently and in the future. Chapter 18 discusses the role of regulations and preparing to manage them in greater detail.

There are a wide variety of types of permits that are required in different states, provinces, and countries. These permits may refer to the site, the business, access to water supplies, discharges, predator control, or processing.

All legal and regulatory statutes relevant to the business must be understood and planned for. Some types of permits may require lengthy application periods that may delay startup of the business.

SOURCES OF HELP

There are a number of sources of help and technical assistance available to the individual considering a startup aquaculture business. Table 1.1 summarizes several types of assistance available. It is advisable to develop an excellent relationship with these groups. Universities, extension agents, trade associations, and diagnostic laboratories are all essential sources of support, technical assistance, and help. Joining the relevant trade association and inclusion on the mailing list of the local extension office will ensure that the new farmer receives the latest updates on permits, regulations, issues, and research.

Plans must include developing contacts with the local diagnostic laboratories, pathologists, and technicians. Understanding the best way to submit samples for diagnosis and training workers in the procedures required will reduce the time to initiate appropriate treatments.

International sources of help include international networks that promote aquaculture such as the Network of Aquaculture Centres in Asia-Pacific (NACA). A network in Eastern Europe, the Network of Aquaculture Centres of Central-Eastern Europe (NACEE), similarly promotes aquaculture and provides information to industry.

In the United States, available help includes personnel of United States Department of Agriculture-

Table 1.1. Sources of Help and Assistance for New Aquaculture Businesses.

Type of organization	Type of assistance
Extension services	Research-based information
Disease and water quality diagnostics laboratories	Diagnosis of disease and water quality problems
Industry trade associations	Updates on issues
State	Political action
National	Trade journals
International	News
Species specific	Updates on issues, meetings
Multispecies	Updates on issues, meetings
Government agencies	Information on permits and programs
State	Information on permits and programs
Federal	Information on permits and programs
Related industry segments	Information on trends, costs, products
Equipment suppliers	Information on trends, costs, equipment
Feed manufacturers	Information on trends, costs, products
Supply company representatives	Information on trends, costs, products, permits
Bank	Information on financial position and trends
Local government entities	Information on trends and permits
Chambers of commerce	News and local events
Economic development offices	Business and financial assistance and new programs

Animal Plant and Health Inspection Service/Wildlife Services. Permits are required in the United States to control fish-eating birds. Severe fines and penalties can be levied on farmers who have not obtained the necessary permits. Wildlife Services personnel have a variety of programs to provide assistance in the control of fish-eating birds and in the process of obtaining the necessary permits.

Local and state aquaculture associations can be of great help. Subscriptions to aquaculture journals, magazines, and newsletters help to keep abreast of current news and impending legislation.

Extension professionals are some of the best sources of information. These are trained scientists who are skilled in techniques to disseminate information effectively. They also have the latest research results at their fingertips and may offer opportunities for farmers to cooperate in on-farm or verification trials.

Some states and provinces have government offices that will assist aquaculture growers. Equipment suppliers, feed manufacturers, supply company representatives, and restaurant owners can be good sources of information on trends, costs, and market data. Local chambers of commerce, economic development offices, and banks can also provide relevant and useful information.

RECORD-KEEPING FOR AQUACULTURE BUSINESSES

The intensive nature of successful businesses requires managers to maintain detailed records. Those contemplating starting an aquaculture business should prepare to spend time to maintain records and to analyze them periodically throughout the year. This level of management can make the difference between success and failure.

Records required will include complete records on input purchases, use, and inventory. Labor and sales records that indicate the quantity sold, the price received, and any dockages incurred with each sale must be maintained. The ability to sort records into feed amounts and fish sales by pond or other fish culture unit will provide the manager with a means to evaluate pond-level performance and relate this back to management changes in that pond. Reports from diagnostic laboratories on disease incidence by pond will enable the manager to search for ways to minimize losses due to disease. Financial records must be maintained for each loan along with depreciation schedules for all equipment in the business. All paperwork related to permits and compliance with regulations must be maintained over time along with all chemical use on the farm.

Each of the following chapters presents detailed suggestions on how to organize and use records to monitor and evaluate farm performance relevant to the topic discussed in each chapter. Management decisions made from detailed farm records will be more effective and have greater positive results over time if based on detailed historical performance records of the farm business.

PRACTICAL APPLICATION

Throughout this book, each chapter includes an example of an application of the material presented in each chapter to the case of a fish farm. To start such a farm, the owner will need to begin to address the critical issues related to acquiring the necessary management skills and capital. If the owner is not skilled and experienced in managing a fish farm, it will be necessary to recruit and hire a skilled manager. Careful thought as to how to obtain the capital that will be needed must be given from the very beginning. Preliminary contacts with lenders will be necessary to identify those more likely to loan to fish farmers and to identify the levels of lending that each bank can provide.

Decisions related to the overall structure of the business can affect the supply of capital available for the fish farm. Developing a partnership or joint venture with a friend or family member can provide a source of capital. The overall financial plan needs to detail the capital that will be available from the owner and any partners and how much will need to be borrowed.

Marketing decisions also will need to be made early on in the planning process for the business. Overviews of the market for that particular business and analysis of its position in the market can be important. The rest of this book provides details of each component and analysis that is required to start and maintain a successful aquaculture business.

OTHER APPLICATIONS IN AQUACULTURE

Engle and Valderrama (2001) developed a training manual designed to assist shrimp growers to begin to develop business plans. The Engle and Valderrama (2001) document emphasizes the financial statements needed for a comprehensive business plan for shrimp farming in Honduras. A CD is provided with spreadsheet templates to assist those who wish to de-velop comprehensive business plans for their shrimp operations. Self-guided tutorials and exercises are included.

SUMMARY

Aquaculture businesses should be entered into only after considerable thought and analysis. Greater capital, more intensive labor, and high levels of management are required to be successful in aquaculture regardless whether the business is large or small.

Comprehensive business and marketing planning is necessary. However, other considerations such as effects on the family, personal motivations, and the availability of adequate resources must also be analyzed carefully. The remaining chapters in this book present detailed information on the steps needed both to start up and to maintain a successful aquaculture business.

REVIEW QUESTIONS

1. What types of specific goals must be set when starting an aquaculture business?

2. Why is it important to assess one's motivation to enter into an aquaculture business?

3. Why is aquaculture considered to be a capital-intensive business? Identify some specific examples of the capital requirements for various aquaculture businesses.

4. Where does financial risk come from?

5. Why is aquaculture considered to be a management-intensive business? Give some specific examples for various types of aquaculture production.

6. What are some of the marketing challenges involved in starting up aquaculture business? Give some specific examples.

7. What types of resources must be available for different types of aquaculture businesses? Pick two different aquaculture production systems and contrast the differences in resource availability that would be required.

8. List three types of organizational structures that can be used for aquaculture businesses and compare the advantages and disadvantages.

9. What are some of the key considerations related to financing new aquaculture businesses?

10. What are some sources of help and assistance for new aquaculture businesses?

REFERENCE

Engle, Carole R. and Diego Valderrama. 2001. Economics and management of shrimp farms training manual. In: M.C. Haws and C.E. Boyd (eds). *Methods for Improving Shrimp Culture in Central America*. Managua, Nicaragua: Editorial-imprenta, Universidad Centroamericana. pp. 231–261. (in English and Spanish)

2
Marketing Aquaculture Products

The focus of this book is on the economics and finance of aquaculture businesses. However, an aquaculture business will not be successful without a marketing program and plan that is appropriate and workable for that particular business. This chapter summarizes critical factors involved in identifying and developing markets for aquaculture products. For a complete discussion and presentation of marketing aquaculture products, readers are referred to Engle and Quagrainie (2006).

THE ESSENCE OF SUCCESSFUL MARKETING

The first essential point to understand about marketing is that marketing is not synonymous with "sales." A salesman may or may not be a marketer. Successful marketing results in sales, but product sales do not always mean that there is a successful marketing program in place. Aquaculture businesses will not succeed without a successful marketing program in today's food market.

Successful marketing involves development of a complete plan that is cohesive and meshes seamlessly with a number of factors. A successful business must identify which specific set of customers the business is seeking to attract and what that group of customers wants to buy. Moreover, the business must find a way to meet those customers' wants and needs better than any other business. The price charged needs to match the customers' expectations in such a way that they believe they receive enough value to justify what they pay for the product. This type of price/quality position for the product must be at a price point that results in a profit for the business. The business' promotional strategy must effectively communicate how the prod-

uct uniquely meets the preferences of the targeted consumer groups. Lastly, the aquaculture business must identify the most convenient locations for the exchange of product to occur for their targeted consumer group. These factors must all mesh and function together as a single company concept to be successful.

THE IMPORTANCE OF DEMAND CHARACTERISTICS

Quantitatively measured characteristics of consumer demand can shed some light on how some of these factors interact. Economists measure demand quantitatively by regressing the price of a product against independent variables that often include the quantity sold of the product, incomes, and tastes and preferences, among others. Economists then use the estimated demand relationships to calculate elasticity, the proportional change in quantity demanded in response to a change in price. Elasticity is important to a discussion of markets because, if demand is elastic, an increase in price will result in a relatively larger decrease in the quantity demanded, such that total revenue to the producer will go down. However, with inelastic demand, the price increase may still result in a lower quantity demanded, but the quantity response is proportionately less than the change in price. When this occurs, the total revenue to the grower will increase as prices increase. Development of specialty markets typically depends upon inelastic demand to successfully command the higher prices common in specialty markets.

One characteristic of products with inelastic demand is that there are few close substitutes for the product. After all, if all shrimp sold (imported, domestic, farmed, wild-caught) are identical, why would a buyer pay a higher price for one particular farm's shrimp? Thus, one key to developing a product for a specialty market is to differentiate it from the competition.

Aquaculture Economics and Financing: Management and Analysis, Carole R. Engle, © 2010 Carole R. Engle.

PRODUCT DIFFERENTIATION AND POSITIONING

Efforts to develop a differentiated product must carefully consider where the product needs to be positioned in the marketplace. This decision must consider the costs of production, the competition, consumer perceptions of the product, and closely competing products. A product-space map (Figure 2.1) can be used to help make this decision. Products can be positioned as a high-quality and high-priced product or, at the other end of the spectrum, a low-quality and low-priced product. The key is to match consumers' expectations with the price. Certain market segments are willing and able to pay high prices for a product that is expected to be of high quality.

Product positioning reflects the combination of the species, product form, packaging, and its price. Any attribute that adds value, such as a spice package, marinade, or the package size, may affect the way a product can be positioned and how it can be differentiated from other products.

Aquaculture growers, particularly smaller-scale growers, often prefer to target specialty markets because prices frequently are higher. Specialty marketing is a choice to produce a high-quality product to capture a high price. To be successful, the marketer must have a clear understanding of which attributes will entice a consumer to pay a higher price. Equally important is how the specific product embodies those attributes to a greater extent than the competition. Why should someone pay more for your product? What is it specifically that uniquely presents the value that the consumer is searching for? Products must be positioned to be differentiated from other products in such a way that the value of the product is worth what the consumer must pay for it.

Successful products attract competition over time. Other companies will learn to produce similar products and perhaps extend the product concept and line. When this happens, the product will no longer be unique. The business needs to be prepared for competition that will come if the new products are successful. Each product that becomes established in the market goes through a type of life cycle (Figure 2.2). In the early stages following the introduction of product, sales grow rapidly as customers become aware of the product and begin to experiment with it. However, at some point, the product reaches a stage of maturity when it attracts competition. The rate of growth of product sales begins to slow as the sales volume reaches a maximum. Product sales then begin to decline. Companies with products in the maturity stage must begin to either search out new markets (either new geographic markets or new market segments) or begin to develop new products.

The price point of the product as the business has positioned it in the market must be adequate to cover not just production costs, but marketing costs as well.

Figure 2.1. Generalized example of a product-space map with various types of seafood species. The exact position of a product will reflect not only the species, but product form, size, and handling.

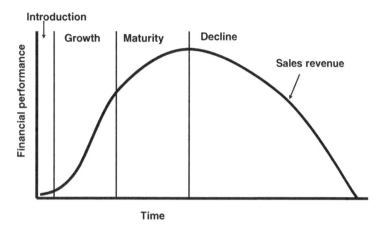

Figure 2.2. A theoretical diagram of a product life cycle indicating its various stages.

A high price is meaningless if total costs of production and marketing are higher still. The business will not be profitable. If the business plan requires the producer to carry out all marketing functions and activities, these costs must be included in the financial analyses of the plan. Marketing costs can be considerable and can frequently include the costs of hauling live fish or transporting processed fish, off-flavor checking, marketing equipment such as baskets, dip nets, and oxygen systems, supplies like ice and bags, and communication systems that include telephone, fax, and Internet charges. Unfortunately, farmers often ignore such marketing costs. An often-overlooked cost is the fish or shrimp that are produced but not sold (Engle and Stone 2008). Few farmers sell all that they produce. This is particularly true of newer farms and those that target higher-risk markets like specialty markets.

Successful market development must also place the product where it is easy for the targeted consumers to purchase it. Successful marketing managers understand where and when their customers prefer to shop. Product is then supplied to those physical locations with the greatest likelihood of being available to the target consumers. For example, it would be unwise to attempt to sell a high-valued specialty product in a discount store.

The last essential component of successful marketing is how products will be promoted. Promotion is the effort by the company to provide information on the product. The customer has to understand the benefits of the product, appreciate its value, and realize why they need to purchase it.

SUCCESSFUL MARKETING REQUIRES DETAILED PLANNING

Marketing plans must evolve constantly to stay ahead of lifestyle and consumer trends. A complete marketing strategy involves development of the product, careful pricing, identifying the most promising locations for sales, and developing an effective and targeted promotion program. Customer segments should be selected on the basis of whether the firm can address their needs effectively. The business's image and logo must reflect and communicate the unique and specific attributes of its products to not only justify but also validate its higher price. No one business can service everyone. The key is to identify which customer segments work best for that particular company and its product line. Long-term success depends upon how well the company understands its customers. Why should someone buy that farms' product? Is it the freshest? Is it locally grown? Is it the highest quality? Is it because that farm treats the animals in the most humane manner? How does the farm create an "experience" that will bring people back?

Table 2.1 outlines what needs to be included in a thorough and comprehensive marketing plan. It is a long outline, but shortcutting the various components listed is a good way to ensure business failure. The market summary requires research and careful assessment of current demographics such as the size of the geographic area, age groups, family structure, gender, income, education, lifestyle factors, and spending habits. Out of this assessment must come a clear understanding of what products are needed, how they need

Table 2.1. Outline for a Marketing Plan.

I. Executive summary	5. Market trends
II. Overall market situation analysis	a. Supply
A. Market summary	b. Packaging
1. Consumer demographics	c. Health consciousness
a. Geographic area	6. Market growth
b. Age groups	B. Analysis of strengths and weaknesses of business
c. Family structure	1. Strengths
d. Gender	2. Weaknesses
e. Income	3. Opportunities
f. Education	4. Threats
g. Lifestyle factors	C. Competition
h. Spending habits	D. Product offering
2. Supermarket demographics	E. Keys to success
a. Geographic areas	F. Critical issues
b. Age groups	III. Marketing strategy
c. Family structure	A. Mission
d. Gender	B. Marketing objectives
e. Income	C. Financial objectives
f. Education	D. Target markets
g. Lifestyle factors	E. Distribution channels
h. Spending habits of customers	F. Positioning
3. Restaurant demographics	G. Strategies
a. Geographic areas	H. Marketing mix
b. Age groups	I. Marketing research
c. Family structure	IV. Financial analysis
d. Gender	A. Planned expenses
e. Income	1. Sales force requirements
f. Education	2. Advertising expenditures
g. Lifestyle factors	B. Sales forecast
h. Spending habits of customers	C. Break-even analysis
4. Market needs	V. Controls
a. Product(s)	A. Implementation
b. Convenience/service	B. Marketing organization
c. Pricing	C. Contingency planning

to be packaged and supplied, and how they need to be priced.

The strengths and weaknesses of the business must be identified and described objectively along with opportunities and threats facing it. The business must position itself to produce those products identified in the market assessment for which it has a unique advantage and for which it can do a better job than the competition. The specific factors that will have the greatest effect on the success or failure must be identified.

The steps outlined above should have resulted in identification of the specific target markets, products and packaging, and pricing that mesh well with the company's unique strengths and advantages. The next step is to lay out the specific marketing, sales, and

financial objectives for the business. The detailed financial analysis develops detailed projections of production, costs, sales force requirements, advertising expenditures, sales forecasts, and a break-even analysis. Chapter 3 provides additional detail on business planning.

PRICING

Pricing strategies used by the company are clearly important determinants of the success or failure of the business. Prices received must cover costs of production for the business to survive. As obvious as this is, there are many examples of aquaculture businesses dumping product on the market at low prices when cash flow deficits require immediate receipts. This type of behavior reflects poor business planning and analysis, inadequate liquidity, and perhaps inadequate capitalization. Dumping product is not a pricing strategy but rather a symptom of a company with a poor financial structure. In this book, Chapter 3 discusses proper and thorough business planning, Chapter 4 presents detailed measures of how to monitor and control financial performance, and Chapters 11 and 12 illustrate what constitutes adequate liquidity and capitalization.

Engle and Quagrainie (2006) discuss various pricing systems and strategies and factors that affect market prices. What is most important is to identify a price level that reflects the value to be derived by the consumer. Consumers must perceive that what they spend to acquire the product is worth that amount of money. The market planning process, if done well, should result in clear understanding of what consumers want, how the company's product uniquely meets those particular needs and solves those consumers' problems, and how much value consumers will receive from purchasing that product. That value is what the price needs to be. If the price is not high enough to cover production costs, the business is not feasible and additional planning is needed to identify a product that can be produced at a cost that also provides adequate value to the consumer.

MARKETING ISSUES SPECIFIC TO VARIOUS SIZES OF FARMS

LARGE-SCALE GROWERS

Large-scale growers clearly need to identify a high-volume market outlet such as a processing plant. It is essential that a potential grower invest the time to meet with the processing plant buyers before construction of production facilities.

There are a number of important considerations taken into account when considering sales to a processing plant. Information on historical prices paid to fish farmers for fish from this plant should be compared to historical prices paid to fish farmers from other plants. The plant's policies on dockage rates (poundage or percentage deducted from the total delivery amount for trash fish, out-of-size fish, turtles, or other reasons) should also be compared across processing plants. Some plants pay based on dressout yield. Some plants provide harvesting and transportation services while others expect the farmer to make arrangements to harvest and transport fish. There is a great deal of variation on contracting arrangements between processors and fish farmers. Details of the agreement should be examined carefully. The farmer should ask specific questions related to delivery volume requirements and seasonality issues that may affect delivery to the plant. Size requirements can be important. For example, some plants will not pay for fish that are smaller or larger than a specified size range while other plants may dock a percentage off the entire load if the proportion of out-of-size fish is higher than the plant's standard. The grower must fully understand the processor's expectations with regard to quality standards, flavor sales, and meat quality.

Payment frequency to growers and typical length of time between the time of delivery of fish and receipt of payment should be compared across plants. Some states require that bonds be posted by processing companies to ensure that farmers are paid for fish delivered. The processor's record of payment should be reviewed.

SMALL-SCALE GROWERS

Small-scale producers will need to identify higher-priced alternative marketing outlets to maintain a profitable operation. This is because economies of scale result in higher costs of production on smaller as compared to larger farms. The prices paid to farmers by processors frequently fall below the break-even prices on smaller-scale farms. Alternative market outlets for small-scale growers can include the following: live sales with custom processing, fee-fishing or pay lake operations, or sales to local grocery stores and restaurants, and sales to live haulers.

In areas with populations exhibiting regular fish consumption, sales of live fish can be a means of achieving higher prices. The capability to process fish according to preferences of the customer may attract a broader clientele. State and local health codes, permits, and Hazard Analysis of Critical Control Points (HACCP) plans must be considered before developing this type of marketing plan.

Fee-fishing or pay lake operations essentially sell a recreational opportunity to their customers. If located within 30–50 miles of a major population center, fee-fishing may offer a viable market outlet for farm-raised fish. Pay lake operators also purchase fish for stocking. Increased interest in urban and community fishing programs managed by state game and fish agencies may provide marketing opportunities for fish farmers. Sales to pay lakes may require arrangements with a livehauler to transport fish. Livehaulers are firms or individuals who truck live fish to pay lakes or live sale markets. Larger fish frequently are required for pay lake outlets. For catfish, a larger (2–3 lb) fish typically is required.

Sales to local grocery stores and restaurants require on-site processing unless restaurant personnel clean the fish. Typically, only managers of very exclusive seafood restaurants will purchase whole fish to be cleaned by their personnel. State and local health codes, permits, and HACCP plans must be considered before developing this type of marketing outlet.

Targeting any of these sales outlets requires careful estimates of volumes, size preferences, and costs associated with the sales. Moreover, state and local regulations must be evaluated carefully, and implications for costs of the business must be assessed.

HOW DOES A GROWER BEGIN TO ASSESS THE MARKET?

Fish and shellfish markets are dynamic. Each market segment has its own buying patterns, quantities purchased, product forms, pricing schemes, and delivery requirements. The specific needs of buyers must be determined in detail. To do so, it is important to talk to as many different buyers as possible. Prospective growers can also talk to growers already in the business and to buyers in regions where the product is already being sold. This will provide a good idea of how the product is positioned, priced, and promoted in that area. Paying attention to the current level of competition will also provide insight into whether it is time to begin to develop new products.

It is more difficult to assess the market potential for an innovative new product not currently sold. One approach is to gather information on substitute products sold locally and inquire about the market for these substitutes. For example, someone interested in a market for freshwater prawns could evaluate the market for shrimp in a particular area and attempt to find a niche within the overall line of shrimp products sold that is not currently being met. These efforts should result in an overall view of the types of needs that buyers have.

The next step is to quantitatively estimate the potential size and volume of the market. Secondary data can be collected on the overall population demographics. Consumer census data and business or economic development data from local chambers of commerce can be used to estimate the number of potential buyers and the total expected sales in the targeted market area. With the market assessment data, the aquaculture business can then set specific market objectives. These objectives would include specifying target sales goals by market segment.

As an example, a targeted population might be couples without children in an age range of 25–40 years with income levels above $40,000. The target market area includes 20,000 households that can be classified within this demographic segment. The aquaculture business believes that 10%, or 2,000, of these households will purchase their product each year. If these households are expected to purchase product once a month, there will be 24,000 purchases a year. If customers are expected to purchase 2 lb of fish at each purchase, the business will project sales of 48,000 lb of product/year.

A key component of the marketing plan is to include strategies to adapt to changing market conditions. Prices, consumer preferences, product contamination, and safety issues can have drastic effects on a farm business, and the farm must be prepared to react quickly to these.

MARKET POWER AND WHAT IT CAN MEAN TO AQUACULTURE BUSINESSES

One of the striking trends in food marketing has been the trend toward increased concentration in the retail and wholesale sectors. The emergence of very large supermarket chains such as Wal-Mart, Tesco, and Carrefour has been accompanied by consolidation in the food service distribution sector as well. This increased consolidation has led to increased market power on the part of these large companies.

Market power is a term used to refer to the ability of an individual company to affect exchanges in the market. A single large buyer often is in a position to strongly influence price of the product, quality standards, and how products flow to consumers. The ability to affect market exchanges constitutes power in the marketplace.

Many aquaculture businesses are small relative to buyers like Wal-Mart, Tesco, and Sysco. Thus, in negotiations related to price, products, and volume, the large buyer typically will be in a position of strength and will be better able to negotiate for favorable terms.

There are three broad strategies that can be used by farming businesses to position themselves more favorably in this type of environment. The best-known strategy is to integrate vertically. A vertically integrated company is one that controls various stages in the supply chain or the stages that a product moves through to reach the end consumer. Examples in aquaculture include Marine Harvest (salmon) and Clear Springs, Inc. (trout). Vertically integrated companies often control both production and processing stages, and sometimes also control feed manufacturing or other activities. As a result, vertically integrated companies are large enough to be able to negotiate more effectively on price, product, and volume because they control many of the various components and are large enough to control a substantial volume of production.

A second strategy that can be used by farming businesses when confronted by a high degree of concentration in their markets is to integrate horizontally. There are many examples of horizontal integration in agriculture. By definition, horizontal integration refers to the formation of business relationships with other companies that operate at the same level in the supply chain. Horizontal integration can take many forms that can include cooperatives, bargaining associations, marketing orders, marketing agreements, and others. In integrating horizontally, care must be taken not to violate laws designed to prevent unfair pricing practices. Often, these take the form of antitrust legislation within a particular country, or can constitute provisions of international treaties such as the World Trade Organization.

Farmers who organize to integrate horizontally to gain parity in the market (when faced with disproportionately greater market power by buyers) may be exempted from charges of antitrust violations. In the United States, such an exemption is provided on a limited basis by the Capper-Volstead Act of 1922. The Fishermen's Collective Marketing Act of 1934 extended the provisions of the Capper-Volstead Act to aquaculture. Specific provisions of the Capper-Volstead Act include: (1) associations must be operated for the mutual benefit of the members; (2) all members (voters) must be agricultural producers; (3) each member must have one vote or there must be a cap on stock dividends of 8% per year; and (4) the value of business conducted with members must exceed that with nonmembers. Forming a cooperative or creating a marketing agreement or order allows for that organization to control a greater amount of supply that can lead to greater influence over price, products, and volumes when negotiating with larger buyers.

The third strategy, which can be used by an individual farm when there is a high degree of concentration in the market, is to form an alliance with other companies across the supply chain that are allied with a particular buyer. This creates a chain of specific suppliers at each stage that are typically allied through a series of contracts. These types of supply chain agreements have been used by growers of specialty products to meet growing consumer interest in product diversification. Supplying large customers with many different types of products has become more complex and supply chain agreements have emerged as one solution that has been used by some companies. These alliances can be beneficial to small growers. Clearly specifying prices, volumes, and quality standards and grades can provide stability for the grower, facilitates farm and financial planning, and reassures lenders of repayment capacity. However, a grower who commits all their production to one buyer also faces market risk associated with the buyer's ability to remain viable.

RECORD-KEEPING

All data compiled throughout the process of developing the marketing plan should be carefully organized and filed. Marketing conditions change over time. The marketing plan, and the data that underlie its assumptions and targets, will need to be modified from year to year. Maintaining careful records that provide the foundation for each decision will facilitate adjusting the plan as conditions change.

Some of the key pieces of data to be maintained will include the long-term series of product prices over time in different markets. Information that includes those segments that demonstrate growth and those that demonstrate declines in numbers and income will be important to compare over time. By maintaining these data over time, key trends can be forecast to facilitate timely changes in the overall marketing plans for the business.

PRACTICAL APPLICATION

This practical application will be a sample market plan for a 256-acre catfish farm, for example. This hypothetical farm is located in northeast Louisiana. It is owned and operated by a family who also hires one additional worker. The loss of the major processing plant in this area has required that the family develop a new, well-organized and well-thought-out marketing direction.

EXECUTIVE SUMMARY

This family-owned and operated fish farm has been in operation for over 30 years. The family has a strong work ethic, has kept up with the latest research in catfish farming, and is known as a reliable and dependable producer. The family plans to diversify their markets by beginning to sell some larger fish to the fee fishing market.

Vision

This business is based on a commitment to be a preferred supplier of high-quality live catfish. The farm plans to provide value to its customers by providing precision-graded fish of optimal sizes to supply to processing plants and to the livehaul market.

Overall Market Situation Analysis

Market Summary

The U.S. catfish industry supplies a traditional market for catfish in the states along the Mississippi river and in the southeast and southwest portions of the United States. Newer markets have been developed on both east and west coasts of the United States. U.S. farm-raised catfish is positioned as a mid-priced white fillet, freshwater fish product. While still recognized primarily for its southern-style fried method of preparation, its versatility lends itself to a variety of preparation styles and cuisines. Increased volumes of low-priced imports in recent years have gained market share and made it difficult to pass increased input prices on to the end consumers.

The principal market channels for U.S. farm-raised catfish are for farms to sell to processing plants. Processors then sell to food service distributors, supermarkets, and other outlets. The increased competition from imports has created pressure on processing plants to find new ways to increase their productivity. The primary means of enhancing the productivity of the plant is to seek to gain control over the distribution of fish sizes delivered to the plant.

Analysis of Strengths and Weaknesses of the Business

The farm family has 30 years experience raising channel catfish, has a strong work ethic, and has a reputation for good management and being an efficient farm. The family owns their land and ponds and there are no financial claims on them. The family has traditionally relied on family labor and contracted with a custom harvesting company to seine their ponds.

Thus, the family is limited in its ability to harvest more frequently.

The key to success of their new venture will be to acquire an in-pond grader to be able to guarantee uniform sizes of fish to deliver to the processing plant, to acquire the seining equipment to be able to harvest and seine their own ponds, and to hire additional labor over the summer to harvest fish for sale to pay lakes. Critical issues will include acquiring the capital to purchase the additional equipment, the availability of summer labor, and convincing the custom harvester to grade fish with the in-pond grader.

Marketing Strategy

The mission of this farm business is to become a preferred source of high-quality live catfish of the size most desired by processing plants and pay lake operators. The marketing objectives are as follows:

1. Sell 10,000 lb/week for 20 weeks to pay lakes in other states, at a price of at least $0.90/lb. The farm will harvest these fish themselves from May through the end of September.
2. Sell 952,000 lb of catfish of the size preferred by catfish processors. The farm will continue to use the services of the custom harvesting company, but plans to work with the custom harvester to grade fish using the in-pond grader.
3. Implement a grading program on the farm to supply a consistent and uniform size of fish to processors.

Financial objectives are:

1. Obtain financing and acquire an in-pond grader and seining equipment.
2. Increase cash flow and revenue by 5% in Year 1.
3. Repay loans for grader and other seining equipment in 5 years.

Target markets will include contracts with two new processing plants and pay lakes (publicly and privately owned). Targeted states will include those with the largest pay lake programs. The initial target is to select five different states and become the preferred supplier for those five. The key characteristics to promote in the development of the new markets will be the availability of preferred sizes, guaranteed health status with laboratory certification, guaranteed delivery volumes, and low percentages of trash fish, debris, or other extraneous materials.

Planned expenses begin with the historic production expenses. The costs associated with financing the grading and seining equipment, and hiring additional summer labor will be added to the cost analysis.

OTHER APPLICATIONS IN AQUACULTURE

Markets for seafood are dynamic and aquaculture growers must constantly keep attuned to changes in the marketplace. Andersen et al. (2009) analyzed the rapid changes occurring in the Russian food market. Russian consumers have begun to adopt new seafood products, as well as other new food products, at a rapid pace. The Russian seafood market includes traditional wild-caught species like herring and new aquaculture products like pangasius from Vietnam and salmon from Norway. This analysis showed that, even in a country like Russia with strong traditions and preferences for herring, white fish, such as tilapia and pangasius can compete with wild-caught traditional pelagic species. If incomes grow in Russia, the market for high-value aquaculture products like farmed Atlantic salmon would be likely to grow.

SUMMARY

This chapter provides a brief discussion of several key factors that are critical for successful marketing. Without a targeted and well-conceived marketing plan, few businesses will survive, even those whose fundamental plan is to sell to a processing plant. The key to successful marketing is to identify the products that can be produced successfully by the company that uniquely meet consumers' problems and needs and are valued by consumers at a level that exceeds the costs of production.

REVIEW QUESTIONS

1. Explain the difference between marketing and sales.

2. Why is it important to know the elasticity of the product the farmer is raising? What are the implications of elasticity for marketing strategy?

3. Draw a product life cycle curve and describe the characteristics of the different stages of the product's life cycle.

4. Explain how a product-space map can help a business manager to make decisions related to product positioning.

5. What are the major components of a comprehensive marketing plan?

6. What are some of the key considerations in developing a pricing strategy?

7. List some marketing issues that differ for large-scale and small-scale farms.

8. How should a farmer begin to assess the market?

9. What is market power? What are the implications of increasing concentration in the retail sector for marketing actions of fish farmers?

10. Compare and contrast vertical and horizontal integration.

REFERENCES

Andersen, Trude B., Kristin Lien, Ragnar Tveterås, and Sigbjorn Tveterås. 2009. The Russian seafood revolution: shifting consumption towards aquaculture products. *Aquaculture Economics and Management* 13(3):191–212.

Engle, Carole R. and Kwamena Quagrainie. 2006. *Marketing Aquaculture Products*. Ames, IA: Blackwell Publishing.

Engle, Carole R. and Nathan Stone. 2008. The hidden costs of small-scale fish farming. *Aquaculture Magazine* 33(5):24–27.

3
Developing a Business Plan for Aquaculture

INTRODUCTION

Most modern aquaculture businesses are capital-intensive. The magnitude of the capital resources necessary to construct, equip, and operate a modern aquaculture operation requires intensive management that begins with a complete business plan. A thorough business plan constitutes a road map for the business and indicates the strengths of the business, where and when the potential problems are likely to occur, and analyzes alternative strategies for overcoming the obstacles. This chapter reviews the structure and detail of thorough business plans with guidance for compiling, evaluating, and implementing them.

THE BUSINESS PLAN

A written business plan, modified annually, provides a structure for continual analysis and evaluation of the business over time. It must integrate marketing, production, and financial components of the business in an in-depth fashion. The plan must be consistent across all its various component analyses. Moreover, the plan needs to include short-term (about a year) as well as long-term goals (for the next 5–10 years). Integrated with the goals are both short- and long-term strategies to achieve them. With a written plan in place, it is easier to identify the source of problems and potential solutions to be prepared for times when something goes awry. Adequate planning for the business may help avoid mistakes and can also minimize risks associated with financing, producing, and marketing products of the business.

Aquaculture Economics and Financing: Management and Analysis, Carole R. Engle, © 2010 Carole R. Engle.

A good business plan takes months to develop. It is an organized and structured document that analyzes market potential for the products produced, examines production technologies selected, and evaluates financial performance of the business.

Business plans and business loan proposals are sometimes confused. However, all businesses should have a thorough business plan regardless of whether the capital in the business is borrowed or equity (provided by the owner). However, for a prospective fish farmer, who will need to borrow money, the business plan constitutes a major portion of the loan application.

Many prospective and existing fish farmers indicate that they have experienced difficulties in obtaining financing for aquaculture businesses. In many cases, lenders are not familiar with aquaculture practices and market potential for aquaculture products. In other instances, lenders are concerned with the uncertainty and risk associated with aquaculture ventures. These concerns frequently lead to rejection of loan applications. However, even if a lender understands aquaculture and its potential, an inadequately prepared business loan proposal will still result in rejection by the lender. Lenders complain about prospective aquaculture borrowers submitting requests for multimillion dollar loans with a few numbers "scratched on a napkin." A successful application for a loan for an aquaculture operation requires a high level of documentation.

The end result of a business plan is an in-depth understanding of whether the proposed business is likely to be successful or not. It will indicate where the major problems lie and what strategies are the likeliest to provide a way to overcome challenges.

The business plan addresses several fundamental questions. How will you produce the product? Is the proposed business economically feasible? Is it financially feasible? A good business plan will include the following components:

1. Executive summary
2. Background information
   a. Analysis of farm's industry
   b. Analysis of farm's position within industry
   c. History of farm
3. Strategic goals and objectives
   a. External opportunities and threats
   b. Internal opportunities and threats
   c. Short-term business goals and objectives
   d. Long-term business goals and objectives
4. General description and characterization of business
   a. Description of production system
   b. Resources available to the farmer
5. Marketing plan
6. Production plan
   a. Products selected (foodfish, fingerlings, eggs)
   b. Technology selected (ponds, cages, intensive, semi-intensive)
   c. Targeted harvest size, stocking rates, stocking sizes
7. Financial plan
   a. Estimated cost and returns
   b. Estimate of required financing
   c. Current appraisal of farm
   d. Balance sheet
   e. Income statement
   f. Cash flow budget
8. Staffing plan
9. Personal financial statement
10. Brief resume of borrower

The two components of the business plan that require the most investigation and analysis are the marketing plan and the financial analysis. Taken together, a well-conceived marketing plan and a carefully documented financial analysis will result in more favorable consideration by a lender.

EXECUTIVE SUMMARY

The executive summary should summarize the goals and key strategies for the business. It should include a concise description of plans to achieve goals, and describe resources needed to carry out the plan.

BACKGROUND GOALS AND OBJECTIVES

Background information on specific permits that will be required, procedures, and probable time frame to obtain required permits should be presented.

Analysis of Farm's Industry

Aquaculture operations proposed for areas without a history of aquaculture may work with lenders who are unlikely to have knowledge of aquaculture practices and potential. In this situation, it is useful to present information on the size of the industry, current trends, and overall growth potential.

Analysis of Farm's Position within Industry

A concise analysis of the farm's position within the industry is needed. Is the farm a trendsetter, one that serves as a leader in adopting new technologies? Is it one of the largest that is able to influence prices and policies? Is it a small business that has little influence on the overall industry? Especially important is to understand whether the farm is a high- or low-cost producer.

History of Farm

Historical changes in the farm business can shed light on its growth and development over time. The business proposal must begin with a description of the site where the aquaculture operation is to be established. Suitability of the site in terms of soil characteristics, environmental conditions, and water supply must be presented.

STRATEGIC GOALS AND OBJECTIVES

The farmer must be constantly aware of external opportunities and threats to the business and explicitly use these as a basis for forecasting changes in the business environment over the upcoming planning period. By forecasting price and business conditions in the future, farms can attempt to adjust production and financial plans to be prepared when the external events come to pass. For example, the price of fish feed increased dramatically in 2008. The high feed prices caught many farmers unaware and unable to adjust before falling into serious cash flow and financial difficulty. Similarly, imports of basa or tra from Vietnam and channel catfish from China have substituted for catfish in the U.S. market.

Economic downturns often decrease demand for seafood and can cause prices to decline. Changes in the marketing chain, such as consolidation and mergers of processing plants can impose challenges on the farm business. The possible effects of these external threats on the business must be assessed. External changes can also create new opportunities. New gas drilling initiatives to extract gas from shale formations create opportunities to treat the discharge water in lined ponds. While fish cannot be raised in these units, the ponds can be used profitably in this way to generate supplemental income.

External threats can include new regulations that affect aquaculture businesses. Regulations can increase costs, restrict market access, or otherwise have detrimental effects on the business. New regulatory initiatives since 2005 include proposals for restrictions on the shipment of live fish, on the discharge of effluents, on the use of nonnative species, on the welfare of fish, shrimp, and shellfish, and cap-and-trade programs. Careful planning each year, with an eye on the external threats likeliest to affect the business, provides an opportunity to make changes and adjustments to minimize adverse effects. Attention to the changing environment also may provide insights into new opportunities.

The internal strengths and weaknesses of the business also need to be evaluated. A business may have strengths in skilled management but have a weakness in terms of its capacity to borrow additional capital. These strengths and weaknesses must be assessed each year in terms of the effect on the overall plan. Will the weaknesses constrain the production system to the point where production levels will fall short of targeted goals? If so, can current markets be sustained at those levels? What will production shortfalls mean for the cash flow and financial position of the business? Analysis of these types of questions should be a part of the annual planning and evaluation process.

Internal strengths should be assessed as well as weaknesses. For example, a small company with a higher cost of production may be in a better position to develop higher-valued niche markets. A farmer with expertise in spawning a type of fish that is difficult to spawn may have an advantage as a hatchery supplier while another business with access to large amounts of land may concentrate on growout.

Internal weaknesses may include assets that are out of date, such as ponds that are old, have not been renovated, and may have become shallow. Aging staff not able to provide the physical labor required, or too many people to operate profitably can represent a weakness of the business. A business's strength may lie in detailed knowledge of markets, excellent engineering and maintenance skills, or skill in financial analysis.

The internal analysis must include careful consideration of the financial resources available for market research and any new investment or operating capital requirements. New directions may require reallocation of company resources, and the company must have a thorough understanding of what those implications will be.

Analysis of internal strengths and weaknesses should also include: (1) relationships (with buyers, suppliers, people who work in the business, and with other businesses); (2) reputation; (3) innovation; and (4) strategic assets. Relationships are the key to success of any business. Establishing and maintaining good relationships with buyers will give a business an advantage over the competition (Palfreman 1999). Special relationships with suppliers and repeated transactions may enable a business to benefit from improved services, short-term credit, improved quality, or possibly better prices. Within the business, a higher degree of commitment or team spirit may result in greater productivity or efficiency.

Both short- and long-term goals and objectives must be set for the plan. Short-term goals typically are for the upcoming year while long-term goals are for periods greater than a year, up to 5–10 years. The best goals are those that are very specific and quantifiable.

The most useful goals and objectives are those expressed in a specific and measurable way. Businesses often focus on short-term financial goals such as: (1) earn a profit (increase net farm income by 10% to become positive); (2) reduce liabilities and debt outstanding of the business (reduce the debt/equity ratio from 50 to 30% over the next 5 years); and (3) increase sales by 5%. Other longer-term objectives may relate more to indirect effects on the financial position of the business, such as: (1) increase market share to have greater influence on price of the product; (2) integrate either vertically or horizontally to have greater influence on the price of the product and to stabilize prices; and (3) increase net worth by 2%.

However, a farmer may have other, noneconomic, objectives. He or she may wish to increase the value of the land or business to be able to sell it for retirement in 5–10 years. The farmer may wish to increase time for hunting or for family activities, for example. All goals should be stated as explicitly as possible.

GENERAL DESCRIPTION AND CHARACTERIZATION OF BUSINESS

The business proposal should include a thorough discussion of the proposed production system. Stocking rates, fingerling sources, anticipated feed rates, and aeration strategy need to be presented clearly and consistently. Internal consistency checks should focus on whether the feed rate is appropriate for the stocking rate, if the number of aerators planned is appropriate for the stocking rate, and whether assumed production levels are realistic given the stocking size, stocking rate, and the feeding rate.

Harvesting methods should be included. Capital investment requirements differ for farmers who plan to harvest their own ponds and for those hiring custom harvesters. If the business plan calls for custom harvesting, then a listing of those serving the area where the farm will be located should be included. Information on fees charged, volume requirements, and scheduling constraints should be presented. Possible production problems such as off-flavor and its effect on inventory management should be discussed. This information demonstrates to the lender an awareness of the problem and also alerts the lender to potential cash flow or debt repayment problems.

Resources Available

The proximity of the proposed farm to processing facilities, feed mills, aquaculture supply firms, equipment repair services, disease and diagnostics laboratories, and the Extension office for technical assistance should be made clear. This demonstrates to the lender both the distance the farm will be from these services and that the individual knows where to find these services.

MARKETING PLAN

For detailed information on aquaculture marketing and developing marketing plans for aquaculture, see Engle and Quagrainie (2005). Each business should have a marketing plan of action and strategy that addresses product, price, promotion, and place. These are factors that are controlled by the business. For example, the farmer decides what species and size of fish to raise as the product to sell. Place refers to the geographic market, or where the farmer will sell the fish. The "place" decision involves deciding whether to sell on the farm, haul to a processing plant, or sell to other farms. The farmer decides what types of advertising to use to promote the product. The farmer also decides the price at which to sell his or her product. Whether or not the product is then sold depends on market conditions.

When to sell the product can also be important. For example, assume that a baitfish farmer has borrowed money from the bank to produce his or her crop, and does so successfully. He or she plans to sell the crop in the fall to generate revenue to make a bank payment. This farm will be in serious financial difficulty because

the main crop of baitfish is sold in the spring, not the fall. The business will fail because it did not develop an adequate marketing plan at the outset.

The marketing plan and strategy must identify the specific market segments to be targeted by the business. Diversifying production, for example, is likely to increase costs. Thus, careful analysis is required to identify the most profitable market segments, those with the greatest overall potential for achieving the company's objectives. The company must then target expenditures on the development, production, inventory, and promotion of products most likely to be successful in that market segment. A segment must be of sufficient size, with potential for further growth, not overoccupied by competition and have an identified need that the company can satisfy uniquely.

The selection of products and product lines must be developed concurrently with the selection of target markets in the company's market plan. A product with a high cost of production will need to be of sufficient quality to charge a price high enough to be profitable. Clearly, the target market for such a product would be one in which consumers not only value the particular attributes of that product, but also have high enough income levels to be able to pay the price level required. The market segment selected must contain enough consumers to offer the volume of sales required to provide an adequate return on the investment in the necessary product development.

The identification and selection of products and product lines for the business is an essential component of a successful business and market strategy. Product lines are a series of closely related but somewhat differentiated products. For example, several catfish processing companies have a marinated fillet product line that may include lemon pepper, Cajun, or preparations with other seasonings and flavors. The marinated fillet product line is distinct from the nugget, steak, and whole-dressed product lines. Companies with single product lines may have lower costs of production due to production efficiencies, but may also have higher market risk. Differentiated product lines and multiple product lines allow a company to spread the risk associated with changing market and economic conditions. For some companies, the move to more extensive and varied product lines may fit the company's business plan whereas such an investment in sales force and processing and packaging infrastructure would not be feasible for others. Over 20,000 new products are introduced to U.S. grocery stores each year. Over 90% do not last more than 3 years. Thus, careful market analysis and testing are required to successfully introduce new products.

A business that plans to develop a new aquaculture product should carefully evaluate the product life cycle effects. For more information on product life cycles, see Engle and Quagrainie (2005). Successful marketing strategies differ depending upon the stage of the product in its product life cycle. The introductory stage requires a market penetration strategy while the maturity stage requires a product differentiation strategy.

Businesses must make critical decisions related to prices and positioning of their product(s) in the market place. Consumers' willingness to purchase a product is related to how closely its price matches their perception of its quality. Consumers will pay very high prices for seafood that they view as of the highest quality. This clearly holds true only for markets that include consumers with income levels that allow them to pay these prices. Conversely, they will refuse to pay high prices for a product they view as low quality.

Positioning a product as the highest quality may not always be a successful strategy. The quantity demanded for the highest level of quality might not be sufficiently high for the company to meet its revenue requirements. High quality products frequently require additional costs related to providing and guaranteeing that level of quality. If such costs are higher than what consumers are willing to pay for that particular product, then the product is not feasible even if it meets high quality standards. If the financial analysis completed in the marketing plan shows that the costs of guaranteeing the highest quality exceed what consumers are willing to pay, an alternative strategy might be to target a higher-volume, but lower-priced market for which quality standards are not quite as rigid.

Marketing plans and strategies for fish species with existing demand must be different from those for new species to be successful. The difference is that consumers have already developed attitudes related to the quality and price they are willing to pay for a fish with an existing market. For high-valued species, like turbot, the existing market price for wild-caught species may be high enough to provide for profitable sales of aquaculture products. However, the history of the salmon industry shows that, if aquaculture expands dramatically, the market price may decline as aquaculture supplies increase. On the other hand, if the wild-caught product is considered to be low quality as is the case for carp in the United States, farmers will have to overcome that image and differentiate their farm-raised product.

Farmers who raise species for a market in which buyers have no previous experience will have to create and develop the market. While this can be a long and sometimes expensive process, it is easier to develop a market for an unknown species than to overcome negative perceptions associated with a species. The companies that export tilapia fillets to the United States have successfully introduced an entirely new species into the U.S. seafood market. New products offer opportunities for price skimming and price penetration. The marketing plan should clearly delineate whether the marketing goal is to sell product as a commodity in an industrial market, as an input into a longer supply chain, or to develop niche markets.

Nevertheless, niche markets are commonly viewed as low-volume, high-priced, specialty markets, although some experts argue that all markets are niche markets. Niche markets typically consist of a small segment of a large market. Sales frequently are lower in niche markets but the strategy is to sell fewer products at a higher price. Smaller companies that successfully identify niche-marketing opportunities may have less competition from larger firms. Typically, a niche market is developed through a specific contact, and the grower uniquely supplies a custom product to that one particular market. The grower will need to provide full support in terms of providing material for taste tests, sampling, and point-of-sale materials, as well as to guarantee consistent product quality.

Niche markets in aquaculture typically have consisted of direct sales from the grower to the end consumer. Thus, the fish farmer performs wholesaling, distribution, and retail functions of the supply chain. In return, the grower captures the profit margins of each of these phases. However, each of these functions also entails costs, sometimes in the form of the time of the grower, in addition to costs related to holding or processing facilities, utilities, labor, advertising, transportation, and packaging (Morris 1994).

The marketing plan must also lay out the preferred form of business structure. Many fish farmers are sole proprietorships or partnerships, but others are vertically integrated companies with control over several stages of the market channel or supply chain. For example, a shrimp company that owns its own farm, hatchery, and packing plant is vertically integrated. It controls its own supply chain and, thus, is in a position to be more flexible in terms of meeting customer demand throughout the supply chain. Still other companies integrate horizontally to gain market power through joint marketing efforts.

The marketing plan begins with a summary of the current market. The market demographics are described and include geographic information, potential numbers of customers by outlet types (supermarkets, restaurants), age, gender, education levels, household income, or lifestyle segment. Consumer needs, likes and dislikes, and buying trends by geographical area are important information. Substitute products sold locally should be identified and market inquiries made.

The marketing plan continues with a description of the product situation. The recent history of sales and revenue for current products, the size, goal, market share, product quality, and marketing strategies of competing firms already in the markets of interest are described and assessed. Important competitive attributes may include: price, product form, product quality, species availability, sources of competing supply, and buyer preferences. The existing distribution situation, in terms of sales through brokers, wholesalers, and retailers, should be described in detail.

If the target market is a processing plant, it is still important to visit the plant and identify delivery requirements. Some important types of information to obtain from a processor include: historical prices paid, dockage rates and policies, transportation charges, if any, frequency of payment to growers, seasonality trends as these affect fish deliveries at the plant, delivery volume requirements, fish size requirements, quality standards and quality control procedures, delivery quotas and scheduling, contracts, and bonding requirements. Transportation requirements and plans should be included in the marketing plan.

The final section of the marketing plan is to forecast sales and set annual sales goals. Consumer census data and business or economic development data can then be used to estimate the number of potential buyers in the targeted market area.

RESOURCES AVAILABLE TO THE FARMER

It is important to develop a careful listing of the resources available to the farmer. These need to be categorized as those that are already available, those that could be acquired readily, and those that would be difficult to acquire. Land is essential, of course. How many acres of land are available for the project? What kind of land is it? Are there any limitations to its use? For example, if the land available is prone to flooding, extra expense may be incurred due to construction of adequate facilities to prevent damage to the business during flooding events. Severe risk of flooding may

prohibit development of the business on that site all together. Is the land appropriate for the proposed business? Weather and temperature can be critical components. Raising a tropical animal like tilapia in a temperate climate will result in higher costs of production due to the necessity of maintaining stocks indoors in heated facilities over the winter.

The availability of adequate water resources is critical. The total volume of water needed must be calculated to assess whether the available water resources are adequate. The quality of the water must also be assessed carefully. What is the salinity level? Is it appropriate for the species planned to be cultivated? Along with the availability of adequate supplies of water, the adequacy of the site in terms of discharges must be assessed. While pond facilities discharge water only rarely, there needs to be a plan even for occasional discharge events.

Capital is a major resource needed to begin an aquaculture business. Capital requirements tend to be high in aquaculture. The assessment of access to adequate levels of capital must be done carefully, objectively, and realistically. The availability of labor resources must also be assessed.

DETERMINE OPTIMAL LEVELS OF INPUT AND OUTPUT FOR THE PRODUCTS TO BE PRODUCED

A projected whole-farm plan should be developed based upon the product and input prices projected. Chapter 17 presents details of how to develop whole-farm plans.

THE FINANCIAL ANALYSIS

There are many excellent books on the preparation of a business loan proposal, on farm management, and on the financial analysis of agricultural businesses. It is beyond the scope of this chapter to present definitions of the terms and concepts used in development of the financial statements discussed below. Chapters 10–14 provide details of the terms, concepts, and calculations. However, a list of reference materials is included at the end of this chapter for additional background and detail.

A table of annual cost and returns, an enterprise budget, should be estimated for the proposed production system. This statement indicates whether or not the proposed production system is profitable. Table 3.1 in this chapter provides an example. Chapter 10

Table 3.1. Enterprise Budget for a 60-acre Catfish Farm (Stocking 5,690 4- to 6-inch Fingerlings/acre; Feed Fed at 4.78 tons/acre/year; Yield of 4,500 lb/acre; Fingerlings Purchased Off Farm; Ponds Owned by Farmers).

Item	Description	Unit	Quantity	Price/cost	Total
Gross receipts	Catfish foodfish	lb	270,000	0.70	*189,000*
Variable costs					
Feed	32% protein floating	Ton	286.80	385	110,418
Fingerlings	5-inch	Inch	1,707,000	0.010	17,070
Labor	Part-time	FTE*	0.4	16,608.00	6,643
Plankton control	Empirical average[†]	Acre	60	14.40	864
Gas and diesel	Empirical average[†]	Acre	60	130	7,800
Electricity	Empirical average[†]	Acre	60	289	17,340
Repairs and maintenance	Empirical average[†]	Acre	60	97	5,820
Bird depredation supplies		Acre	60	6.25	375
Seining and hauling	Catfish foodfish	lb	270,000	0.05	13,500
Telephone	Empirical average[†]	Acre	60	17	1,020
Office supplies	Empirical average[†]	Acre	60	11	660
Interest on operating capital		$	151,258[‡]	0.10	15,126
Total variable costs	Per farm				*196,636*
	Per acre				3,277
Income above variable costs					*−7,636*
Fixed costs					
Farm insurance	Empirical average[†]	Acre	60	43.6	2,616
Legal/accounting	Empirical average[†]	Acre	60	18.80	1,128
Investment					
Land	Empirical average[†]	$	49,320[§]	0.10	4,932
Wells	Empirical average[†]	$	14,000[¶]	0.10	1,400
Pond construction	Empirical average[†]	$	83,880[**]	0.10	8,388
Equipment	Empirical average[†]	$	138,100	0.10	13,810
Annual depreciation					
Equipment	Empirical average	Acre	1	14,110	14,110
Total fixed costs	Per farm				*46,384*
	Per acre				773
Total costs	Per farm				*243,020*
	Per acre				4,050
Net returns to operator's labor, management, and risk per farm					*−$54,020*
	Per acre				*−$900*
Opportunity costs					
Operator's labor	Family	Total	1	9,965	9,965
Operator's management	Family	Total	1	2,610	2,610
Total opportunity costs of family labor and management					12,575
Total costs					*255,595*
Net returns to operator's risk per farm					*−66,595*
	Per acre				*−1,110*

(Continued)

Table 3.1. (*Continued*)

Item	Description	Unit	Quantity	Price/cost	Total
Noncash costs					42,640
Net returns above cash (to operator's risk) per farm					*−23,955*
	Per acre				*−399*
Breakeven price	Above variable costs				*0.73*
	Above total costs				*0.95*
Breakeven yield	Above variable costs	Per farm			*280,908*
		Per acre			4,682
	Above total costs	Per farm			*365,135*
		Per acre			6,086

* FTE, full-time equivalent. One person working one 10-hour day is 1 FTE. Two people working 5-hour days is 1 FTE.
† From survey data reported in Engle (2007).
‡ Operating capital was assumed to be used for 10 months of the year.
§ Land values = $822/acre.
¶ Two wells at $7,000 each.
** Pond construction costs = $1,398/acre.

presents details on developing an enterprise budget. The calculated net returns from the statement indicate the general level of profitability. Furthermore, the annual cost and returns statement indicates breakeven costs and breakeven yield.

The business proposal must clearly summarize financing requirements for the fish farm. Required financing should be divided into the following loan categories: operating, equipment, and real estate. Chapter 14 provides detail on lending to aquaculture businesses.

The amount of capital for an operating loan is based on the amount of variable cost required. Equipment loans cover the purchase of any new or additional equipment necessary, while a real estate loan covers the cost of constructing ponds, buildings, or other relatively permanent structures. Repayment schedules should be specified to demonstrate how revenues will cover debt payments. See Chapter 12 for details on repayment ability as determined by cash flow budgets.

The owner may wish to schedule payments in such a way as to either defer payment or only pay interest the first year. Construction of ponds or other facilities and weather delays may cause revenues to be delayed the first year. The borrower will need to present a sound plan to demonstrate how interest will be paid during the construction phase and throughout the first year's production season.

In many cases, it may be 18–24 months before income is realized. For operating lines of credit, a lien on the fish crop and a first mortgage collateral position will be required by most lenders as a minimum standard.

A lender will require a current appraisal that reflects the value before and after ponds and facilities are constructed. This will be used to calculate a loan-to-appraisal value ratio. Most lenders will want a loan-to-appraisal value ratio of 50–125%, depending upon the borrower's financial strength. Some lenders may require a Farm Service Agency or other type of guarantee for those with loan-to-appraisal values above 50%.

The balance sheet lists the assets and liabilities (debts) that would be for the entire business including the new aquaculture operation. From the balance sheet, net worth (owner equity) can be calculated as well as the following financial ratios: equity/asset ratio (owner equity), debt/asset ratio, debt/equity ratio, and current ratio. Table 3.2 provides an example. Chapter 11 provides detail on development of the balance sheet.

Minimum standards used by lenders to evaluate the current ratio range from 1.3 to 1.5 with the higher level being preferred. See Chapter 4 for details of ratio analysis. The actual dollar amount of working capital is also compared with the value of farm production.

Table 3.2. Balance Sheet for a 60-acre Catfish Farm, December 31.

Item	Total value
Assets	
1. *Current assets*	
Cash on deposit	$3,715
Fish inventory*	$62,370
Total current assets	*$66,085*
2. *Noncurrent assets*	
Equipment	$138,100
Ponds	$83,880
Wells	$14,000
Land	$49,320
Total noncurrent assets	*$285,300*
3. *Total assets*	*$351,385*
Liabilities	
4. *Current liabilities*	
Payments on debt due and payable over next year	
Operating loan interest	$768
Operating loan principal	$41,046
Total current liabilities	*$41,814*
5. *Noncurrent liabilities*	
Equipment loan	$0
Real estate loan	$0
Total noncurrent liabilities	*$0*
6. *Total liabilities*	*$41,814*
7. *Net worth (3–6)*	*$309,571*

* 4,500 submarketable fish/acre at 0.33 lb each @ 0.70/lb.

Table 3.3. Income Statement for a 60-acre Catfish Farm, December 31.

Item	Total value
Catfish farm revenue	
Cash catfish sales	$189,000
Total catfish farm revenue	**$189,000**
Catfish farm expenses	
Cash operating expenses	
Feed	$110,418
Fingerlings	$17,070
Labor	$6,643
Plankton control	$864
Gas, fuel, and oil	$7,800
Electricity	$17,340
Repairs and maintenance	$5,820
Bird depredation supplies	$375
Seining and hauling	$13,500
Telephone	$1,020
Office supplies	$660
Legal/accounting	$1,128
Insurance	$2,616
Total cash farm expenses	$185,254
Depreciation	*$14,110*
Total operating expenses	*$199,364*
Cash interest paid	
Interest on operating line of credit	$10,744
Interest paid on long-term loans	
Land	$4,932
Wells	$1,400
Pond construction	$8,388
Equipment	$13,810
Total interest paid	$39,274
Total expenses	**$238,638**
Net farm income from operations	**−$49,638**
Family labor	**$12,575**
Net returns to operator's risk	**−$62,213**

The income statement itemizes anticipated farm income and expenses after the second full year of operation. The income statement is also called the profit and loss statement. Net farm income, return to capital, return to labor and management, and return to equity are calculated from the income statement. Table 3.3 provides an example. See Chapter 11 for details of developing income statements.

A cash flow budget shows cash receipts and cash expenses by month, quarter, or year. It includes only cash expenses and provides an indication of when cash will be available for loan repayment. Table 3.4 provides an example. See Chapter 12 for details on development of cash flow budgets.

The cash borrowing requirements projected by the cash flow budget must be examined carefully. Given the borrowing capacity of the farmer, can these borrowing requirements be met feasibly? Are there ways to restructure purchases or production to alleviate cash flow problems? An applicant needs to provide 3 years of tax records as well as 3 years (including a current statement) of financial statements.

Operating capacity and management skills will be critical to the success of a fish farming business. If the owner does not have these skills, the business proposal must include funds to hire a manager or demonstrate that the owner has either taken or will be taking courses in fish production. The résumé should note these courses along with experience raising fish that includes acreage managed, yields, stocking, and feeding rates, and hatchery experience.

Table 3.4. Monthly Cash Flow Budget for a 60-acre Catfish Farm without Financing.

Item	January	February	March	April	May	June	July	August	September	October	November	December	Total
Beginning cash	0	3,889	−976	−6,136	16,110	24,826	19,588	5,727	−5,603	−16,453	1,824	8,612	
Pounds of catfish sold, lb	10,800	10,800	10,800	48,600	27,000	27,000	16,200	27,000	27,000	45,900	18,900	0	270,000
Receipts from catfish sold, $	7,560	7,560	7,560	34,020	18,900	18,900	11,340	18,900	18,900	32,130	13,230	0	189,000
Total cash inflow	7,560	11,449	6,584	27,884	35,010	43,726	30,928	24,627	13,297	15,677	15,054	8,612	459,000
Operating cash expenses													
Feed	1,105	2,209	2,209	7,731	5,522	17,669	17,669	22,086	22,086	6,626	3,297	2,209	11,0418
Fingerlings	0	8,535	8,535	0	0	0	0	0	0	0	0	0	17,070
Labor, seasonal, part-time	0	0	0	0	996	1,661	1,661	1,661	664	0	0	0	6,643
Plankton control	0	0	0	0	216	0	216	0	432	0	0	0	864
Gas, diesel fuel, and oil	468	234	234	234	390	624	1,014	1,014	1,248	1,248	546	546	7,800
Electricity	347	347	520	694	867	1,734	2,601	2,948	2,948	2,948	867	520	17,341
Repairs and maintenance	698	116	233	233	291	587	757	757	582	291	291	989	5,820
Bird depredation	56	56	56	38	19	0	0	0	0	38	56	56	375
Seining and hauling	540	540	540	2,430	1,350	1,350	810	1,350	1,350	2,295	945	0	13,500
Telephone	82	82	61	102	102	82	82	82	82	82	82	102	1,020
Office supplies	53	7	33	13	132	132	66	33	33	26	33	99	660
Farm insurance	209	209	209	209	209	209	235	209	235	209	235	235	2,616
Legal/accounting	113	90	90	90	90	90	90	90	90	90	90	113	1,128
Total operating expenses	3,671	12,425	12,720	11,774	10,184	24,138	25,201	30,230	29,750	13,853	6,442	4,869	185,255

Debt servicing													
Real estate													
Interest	0	0	0	0	0	0	0	0	0	0	0	0	0
Principal	0	0	0	0	0	0	0	0	0	0	0	0	0
Subtotal	0	0	0	0	0	0	0	0	0	0	0	0	0
Equipment													
Interest	0	0	0	0	0	0	0	0	0	0	0	0	0
Principal	0	0	0	0	0	0	0	0	0	0	0	0	0
Subtotal	0	0	0	0	0	0	0	0	0	0	0	0	0
Operating													
Interest	0	0	0	0	0	0	0	0	0	0	0	0	0
Principal	0	0	0	0	0	0	0	0	0	0	0	0	0
Subtotal	0	0	0	0	0	0	0	0	0	0	0	0	0
Total debt servicing	3,671	12,425	12,720	11,774	1,0184	24,138	25,201	30,230	29,750	13,853	6,442	4,869	18,5255
Total cash outflow	3,889	−976	−6,136	16,110	24,826	19,588	5,727	−5,603	−16,453	1,824	8,612	3,743	273,745
Cash available	0	0	0	0	0	0	0	0	0	0	0	0	0
New borrowing	0	0	0	0	0	0	0	0	0	0	0	0	0
Ending cash balance	3,889	−976	−6,136	16,110	24,826	19,588	5,727	−5,603	−16,453	1,824	8,612	3,743	273,745
Summary of debt outstanding													
Real estate													
Equipment													
Operating													

REVISIONS TO THE BUSINESS PLAN

The plan must also include a detailed methodology for monitoring and evaluating the company's performance once the plan is put into place. The business plan typically will need to be revised following the evaluation of the marketing and production plans developed. Problems detected in cash flow, in financial strength, borrowing capacity, or overall feasibility require that the proposed business be modified to overcome these problems. Once modifications are proposed, then the marketing and production plans and financial statements must also be revised and reanalyzed to see if the modifications made have resolved the problems.

Ideally, the best would be to analyze all alternative strategies with potential to achieve the goals and objectives of the business. In reality, few aquaculture businesses have the capability to do this. Thus, the iterative process suggested above may be more practical for many farmers.

EVALUATION OF A LOAN PROPOSAL

In evaluating a business proposal and loan application, lenders will take into consideration several factors. The overall character and honesty of the individual is considered on the basis of his or her history of paying other bills and character references. Owner equity, the current ratio (from the balance sheet), the loan-to-appraisal value, and the value of farm production are key indicators for many lenders. Earnings will be examined in great detail along with repayment capacity. These will be viewed in terms of sustaining production over a 3-year price cycle. Collateral and capital of the individual operator will also affect the level of the lender's decision. The lender will look at the financial condition of the processor to make certain that the producer will receive payment within 2–3 weeks of fish delivery. A comprehensive and thorough proposal will greatly facilitate the loan application process.

RECORD-KEEPING

A business plan requires detailed records on every phase of the business, from production and marketing data to complete financial information. The most useful data are longer-term series of price and cost data. These provide for trend analyses and indicate where prices and costs are headed for better planning.

Detailed market information is also valuable. Maintaining files of market reports and changing conditions is a good basis for developing the marketing plan portion of the business plan.

PRACTICAL APPLICATION

Tables 3.1–3.4 present the financial analyses for a 60-acre catfish farm in the United States. This farm plans to sell food-sized fish to a processing plant. The farm's production technology is the same as that of the 256-acre farm for which financial statements are presented in Chapters 10–12. For comparison between the two farm sizes, Table 3.5 summarizes key financial indicators for a 256-acre farm, demonstrating its financial feasibility.

The 60-acre farm is facing a different set of circumstances from those of the 256-acre farm. Feed and energy prices have escalated rapidly but fish prices have not. Thus, Tables 3.1 and 3.3 shows that this farm is not profitable. Its balance sheet (Table 3.2) shows a high net worth. This is primarily because the farm used its own capital, borrowing the operating capital necessary for cash flow. However, the cash flow budget (Table 3.4) shows that there are cash shortfalls in the months of February, March, August, and September. Thus, new borrowing will be needed in these months. The new borrowing results in adequate cash flow (Table 3.6), but incurs an operating loan of $48,068 and an outstanding operating loan of $7,022 at the end of the year. Thus, with the operating loan, this farm can make its payments in the short term.

However, few farms can provide all the investment capital needed. If the farm has to borrow just 30% of the capital needed to purchase the equipment, buy the land, and build the ponds, the cash flow changes substantially. Table 3.7 shows that less of the operating loan was paid off throughout the year, leaving an operating loan balance of $22,030 that will need to be carried over until the next year. Thus, this farm has problems of profitability as well as cash flow problems if even just 30% of the long-term capital is borrowed.

This farm will need to make some adjustments. Breakeven price of catfish sold above total costs from the enterprise budget is $0.95. The farm will need to explore some alternative markets to see if developing a live market can result in achieving a higher price. Careful analysis will be needed to determine the volume of fish likely to be sold as a live product. In some areas, live catfish are sold for as much as $1.00/lb, but volumes tend to be only a few thousand pounds a week. Alternatively, the farm may need to look at the feasibility of either expanding its size to capture

Table 3.5. Summary of Results of Financial Analyses of 256-acre Catfish Farm Example.

Analysis	Unit	Farm value	Per-acre value	Source (table in this volume)
Enterprise budget				
Net returns to operator's risk	$	−$18,502	−$72	Table 10.2
Net returns to operator's labor, management, and risk	$	$2,792	$11	Table 10.2
Breakeven price above variable costs	$/lb	$0.57	$0.57	Table 10.2
Breakeven price above total costs	$/lb	$0.72	$0.72	Table 10.2
Breakeven yield above variable costs	lb/acre	932,104	3,641	Table 10.2
Breakeven yield above total costs	lb/acre	1,178,431	4,603	Table 10.2
Balance sheet				
Total current assets	$	$281,961	1,101	Table 11.1
Total noncurrent assets	$	$1,003,890	3,921	Table 11.1
Total assets	$	$1,285,851	5,023	Table 11.1
Total current liabilities	$	$110,584	432	Table 11.1
Total noncurrent liabilities	$	$423,630	1,655	Table 11.1
Total liabilities	$	$534,214	2,087	Table 11.1
Net worth	$	$751,637	2,936	Table 11.1
Income statement				
Net farm income from operations	$	$2,792	10.91	Table 11.2
Cash flow budget				
Cash available	$	Increasing across months	n.a.*	Table 12.1
New borrowing	$	0	n.a.*	Table 12.1
Debt outstanding	$	0	n.a.*	Table 12.1

* Not applicable to a cash flow budget.

economies of scale or to decrease its size to focus on producing the volumes that local live markets can bear. Complete financial analyses are needed for each new scenario to judge its feasibility.

OTHER APPLICATIONS IN AQUACULTURE

Engle and Neira (2005) developed a series of pro forma financial statements for tilapia businesses in Kenya. Data were collected from tilapia farms in Kenya from which sample enterprise budgets were developed. The base enterprise budget for a 1-ha tilapia farm in Kenya was based on a stocking rate of 3.0 tilapia/m². Fish were assumed to be fed a pelleted diet. The enterprise was profitable, with positive net returns. The income statement also demonstrated profitability with positive net farm income. The balance sheet showed positive owner equity, indicating a strong financial position. The cash flow budget showed a positive cash flow balance, with an ending cash balance greater than

the beginning cash balance. Overall, the business plan showed that the overall plan is feasible.

SUMMARY

This chapter presents an outline of the content of a thorough business plan. Aquaculture businesses are sophisticated and capital-intensive businesses. Preparation of a comprehensive plan can help the owner to avoid major problems or redirect the business in a manner that is likelier to succeed.

REVIEW QUESTIONS

1. Why is a business plan important?

2. What is the difference between a business plan and a business loan proposal?

3. What are the major components of a business plan?

Table 3.6. Monthly Cash Flow Budget for a 60-acre Catfish Farm with New Borrowing from an Operating Loan.

Item	January	February	March	April	May	June	July	August	September	October	November	December	Total
Beginning cash	0	10,000	10,000	10,000	15,832	24,548	19,310	10,000	10,000	10,000	10,000	10,000	
Pounds of catfish sold, lb	10,800	10,800	10,800	48,600	27,000	27,000	16,200	27,000	27,000	45,900	18,900	0	270,000
Receipts from catfish sold, $	7,560	7,560	7,560	34,020	18,900	18,900	11,340	18,900	18,900	32,130	13,230	0	189,000
Total cash inflow	7,560	17,560	17,560	44,020	34,732	43,448	30,650	28,900	28,900	42,130	23,230	10,000	459,000
Operating cash expenses													
Feed	1,105	2,209	2,209	7,731	5,522	17,669	17,669	22,086	22,086	6,626	3,297	2,209	110,418
Fingerlings	0	8,535	8,535	0	0	0	0	0	0	0	0	0	17,070
Labor, seasonal, part-time	0	0	0	0	996	1,661	1,661	1,661	664	0	0	0	6,643
Plankton control	0	0	0	0	216	0	216	0	432	0	0	0	864
Gas, diesel fuel, and oil	468	234	234	234	390	624	1,014	1,014	1,248	1,248	546	546	7,800
Electricity	347	347	520	694	867	1734	2,601	2,948	2,948	2,948	867	520	17,341
Repairs and maintenance	698	116	233	233	291	587	757	757	582	291	291	989	5,820
Bird depredation	56	56	56	38	19	0	0	0	0	38	56	56	375
Seining and hauling	540	540	540	2,430	1,350	1,350	810	1,350	1,350	2,295	945	0	13,500
Telephone	82	82	61	102	102	82	82	82	82	82	82	102	1,020
Office supplies	53	7	33	13	132	132	66	33	33	26	33	99	660
Farm insurance	209	209	209	209	209	209	235	209	235	209	235	235	2,616
Legal/accounting	113	90	90	90	90	90	90	90	90	90	90	113	1,128
Total operating expenses	3,671	12,425	12,720	11,774	10,184	24,138	25,201	30,230	29,750	13,853	6,442	4,869	185,255

Debt servicing													
Real estate													
Interest	0	0	0	0	0	0	0	0	0	0	0	0	0
Principal	0	0	0	0	0	0	0	0	0	0	0	0	0
Subtotal	0	0	0	0	0	0	0	0	0	0	0	0	0
Equipment													
Interest	0	0	0	0	0	0	0	0	0	0	0	0	0
Principal	0	0	0	0	0	0	0	0	0	0	0	0	0
Subtotal	0	0	0	0	0	0	0	0	0	0	0	0	0
Operating													
Interest	0	51	92	136	0	0	38	0	133	224	74	18	765
Principal	0	0	0	16,279	0	0	0	0	0	18,053	6,714	0	41,046
Subtotal	0	51	92	16,415	0	0	38	0	133	18,277	6,788	18	41,811
Total debt servicing	0	51	92	16,415	0	0	38	0	133	18,277	6,788	18	41,811
Total cash outflow	3,671	12,476	12,812	28,189	10,184	24,138	25,201	30,268	29,883	32,130	13,230	4,887	227,065.8
Cash available	3,889	5,084	4,748	15,832	24,548	19,310	5,449	−1,368	−983	10,000	10,000	5,113	231,934
New borrowing	6,111	4,916	5,252	0	0	0	4,551	11,368	10,983	0	0	4,887	48,068
Ending cash balance	10,000	10,000	10,000	15,832	24,548	19,310	10,000	10,000	10,000	10,000	10,000	10,000	280,002
Summary of debt outstanding													
Real estate													
Equipment													
Operating	6,111	11,027	16,279	0	0	0	4,551	15,919	26,902	8,849	2,135	7,022	

Table 3.7. Monthly Cash Flow Budget for a 60-acre Catfish Farm with New Borrowing from an Operating Loan and 30% Borrowing for Equipment and Real Estate Capital.

Item	January	February	March	April	May	June	July	August	September	October	November	December	Total
Beginning cash	0	10,000	10,000	10,000	15,832	24,548	19,310	10,000	10,000	10,000	10,000	10,000	10,000
Pounds of catfish sold, lb	10,800	10,800	10,800	48,600	27,000	27,000	16,200	27,000	27,000	45,900	18,900	0	270,000
Receipts from catfish sold, $	7,560	7,560	7,560	34,020	18,900	18,900	11,340	18,900	18,900	32,130	13,230	0	189,000
Total cash inflow	7,560	17,560	17,560	44,020	34,732	43,448	30,650	28,900	28,900	42,130	23,230	10,000	459,000
Operating cash expenses													
Feed	1,105	2,209	2,209	7,731	5,522	17,669	17,669	22,086	22,086	6,626	3,297	2,209	110,418
Fingerlings	0	8,535	8,535	0	0	0	0	0	0	0	0	0	17,070
Labor, seasonal, part-time	0	0	0	0	996	1,661	1,661	1,661	664	0	0	0	6,643
Plankton control	0				216	0	216	0	432	0	0	0	864
Gas, diesel fuel, and oil	468	234	234	234	390	624	1,014	1,014	1,248	1,248	546	546	7,800
Electricity	347	347	520	694	867	1,734	2,601	2,948	2,948	2,948	867	520	17,340
Repairs and maintenance	698	116	233	233	291	587	757	757	582	291	291	989	5,820
Bird depredation	56	56	56	38	19	0	0	0	0	38	56	56	375
Seining and hauling	540	540	540	2,430	1,350	1,350	810	1,350	1,350	2,295	945	0	13,500
Telephone	82	82	61	102	102	82	82	82	82	82	82	102	1,020
Office supplies	53	7	33	13	132	132	66	33	33	26	33	99	660
Farm insurance	209	209	209	209	209	209	235	209	235	209	235	235	2,616
Legal/ accounting	113	90	90	90	90	90	90	90	90	90	90	113	1,128
Total operating expenses	3,671	12,425	12,720	11,774	10,184	24,138	25,201	30,230	29,750	13,853	6,442	4,869	185,254

	1	2	3	4	5	6	7	8	9	10	11	12	Total
Debt servicing													
Real estate													
Interest	0	0	0	0	0	0	0	0	0	922	0	0	922
Principal	0	0	0	0	0	0	0	0	0	3,943	0	0	3,943
Subtotal	0	0	0	0	0	0	0	0	0	4,865	0	0	4,865
Equipment													
Interest	0	0	0	0	0	0	0	0	1,650	0	0	0	1,650
Principal	0	0	0	0	0	0	0	0	8,286	0	0	0	8,286
Subtotal	0	0	0	0	0	0	0	0	9,936	0	0	0	9,936
Operating													
Interest	0	51	92	136	0	38	133	0	224	157	142	0	972
Principal	0	0	0	16,279	0	0	0	0	8,117	1,766	35	0	26,197
Subtotal	0	51	92	16,415	0	38	133	0	8,341	1,923	177	0	27,169
Total debt servicing	0	51	92	16,415	0	38	133	0	18,277	6,788	177	0	41,970
Total cash outflow	3,671	12,476	12,812	28,189	10,184	24,138	25,201	30,268	29,883	32,130	13,230	5,046	227,223.6
Cash available	3,889	5,084	4,748	15,832	24,548	5,449	−1,368	−983	19,310		4,954		231,776
New borrowing	6,111	4,916	5,252	0	0	4,551	11,368	10,983	0	0	5,046	0	48,227
Ending cash balance	10,000	10,000	10,000	15,832	24,548	10,000	10,000	10,000	19,310		10,000		280,003
Summary of debt outstanding													
Real estate	48,576	48,576	48,576	48,576	48,576	48,576	48,576	48,576	48,576	44,633	44,633	44,633	
Equipment	41,430	41,430	41,430	41,430	41,430	41,430	41,430	41,430	33,144	33,144	33,144	33,144	
Operating	6,111	11,027	16,279	0	0	4,551	15,919	26,902	18,785	17,019	22,030	22,030	

4. What is the difference between a commodity and a niche market?

5. Why does business planning take so much time?

6. Describe several examples of the types of external and internal opportunities that can affect an aquaculture business.

7. Describe several examples of the types of external and internal threats that can affect an aquaculture business.

8. What are the major components of the marketing plan for the business?

9. Outline the content of the production plan that must be included in a complete business plan.

10. What types of resources are available for developing aquaculture business plans?

REFERENCES

Engle, Carole R. 2007. *Arkansas Catfish Production Budgets*. Arkansas Cooperative Extension Program MP466. Pine Bluff, AR: University of Arkansas.

Engle, Carole R. and Ivano Neira. 2005. *Tilapia Farm Business Management and Economics*. Corvallis, OR: Oregon State University.

Engle, Carole R. and Kwamena Quagrainie. 2006. *The Aquaculture Marketing Handbook*. Ames, IA: Blackwell Publishing.

Morris, J.E. 1994. *Niche Marketing Your Aquaculture Products*. Publication #TB-107. Ames, IA: Iowa State University Press.

Palfreman, Andrew. 1999. *Fish Business Management*. Oxford: Blackwell Science.

SUGGESTED REFERENCES

Barry, Peter J., John A. Hopkin, and C.B. Baker. *Financial Management in Agriculture*. Danville, IL: The Interstate Printers & Publishers.

Furlong, Carla B. 1993. *Marketing for Keeps: Building Your Business by Retaining Your Customers*. New York: John Wiley & Sons.

Kay, Ronald D., William M. Edwards, and Patricia A. Duffy. 2008. *Farm Management*, 6th ed. New York: McGraw-Hill.

Lee, Warren F., Michael D. Boehlje, Aaron G. Nelson, William G. Murray. 1988. *Agricultural Finance*, 8th ed. Ames, IA: Iowa State University Press.

Libbin, James D., Lowell B. Catlett, and Michael L. Jones. 1994. *Cash Flow Planning in Agriculture*. Ames, IA: Iowa State University Press.

O'Hara, Patrick D. 1989. *SBA Loans*. New York: John Wiley & Sons.

Olson, Kent D. 2004. *Farm Management Principles and Strategies*. Ames, IA: Iowa State University Press.

Rachlin, Robert and Allen H.W. Sweeny. 1993. *Handbook of Budgeting*, 3rd ed. New York: John Wiley & Sons.

Warren, Martyn. 1998. *Financial Management for Farmers and Rural Managers*, 4th ed. Ames, IA: Blackwell Science.

4

Monitoring Economic and Financial Performance of Aquaculture Businesses

Efficient management of a farm can make the difference between profits and losses especially in years with unfavorable prices and costs. However, farm management involves more than just taking care of the biological processes involved; it includes paying close attention to economic and financial measures of the farm business. Few farmers enjoy spending time on financial analysis, but doing so is essential to the success of the business. Even if the business retains an accounting firm to generate the analyses, spending time to use the results to plan to make improvements is necessary.

Chapter 3 described the process of developing a comprehensive plan for the aquaculture business. Having the plan developed is a first step, but the owner/manager must follow up on the plan each year with equally careful analysis and assessment of the performance of the business over the past year. This analysis should go through the entire plan to assess whether the short-term goals for that year were met or not. There are a series of biological, economic, and financial ratios and indicators that can be used to identify why goals were not met and (or if they were) to further evaluate the business's performance over the year. From this evaluation, new goals for the upcoming year must be developed and the overall plan modified as appropriate, to keep the business on track to meet its longer-term goals.

Solvency and liquidity are important financial measures of the overall well being of a business. Solvency refers to the value of the assets owned by the business as compared to the amount of liabilities. Assets refer to the value of anything owned by the business whereas liabilities refer to any debt obligations that the business has outstanding. Liquidity refers to the ability of a business to meet cash flow obligations. Liquidity is critical for the financial transactions of the business to run smoothly. The balance sheet and income statements are used as the bases for measuring and monitoring solvency and liquidity in the farm business. For details on developing balance sheets and income statements, see Chapter 11 of this book. This chapter reviews and discusses various types of economic and financial indicators that can be used to measure performance of the aquaculture business.

PRODUCTION AND INPUT USE EFFICIENCY

PRODUCTION EFFICIENCY

Production efficiency refers to the biological performance of the farm. Monitoring the efficiency of the aquaculture business begins with evaluating these biological factors. The first such measure to monitor is the total production from the farm, or gross yield. Gross yield is measured in weight of the production per unit area (lb/acre or kg/ha for pond production and lb/cubic foot or kg/cubic meters for cages). Net yield is the gross yield minus the weight of the fingerlings or postlarvae stocked. In Table 4.1, the gross yield of catfish was 4,500 lb/acre while net yield, after subtracting out the weight of fingerlings stocked was 4,318 lb/acre. Over time, increases in gross and net yield will result in reduced production costs and greater productivity.

Aquaculture Economics and Financing: Management and Analysis, Carole R. Engle, © 2010 Carole R. Engle.

Table 4.1. Production Efficiency Measures, 256-acre Catfish Farm.

Measure	Calculation	Unit	Farm value
Gross yield	Weight of fish sold ÷ number of acres	lb/acre	4,500
Net yield	(Weight of fish sold − weight (lb) of fish stocked) ÷ number of acres	lb/acre	4,318
Survival	(Number of fish sold ÷ number of fish stocked) × 100	%	79%
Average size of fish harvested	Weight of fish harvested ÷ number of fish harvested	lb	1.5
Growth rate	(Average size of fish harvested − average size of fish stocked) ÷ number of days of production	g/day	2.2

Net yield measures the gain in production over time and is the more accurate measure of biological efficiency. Gross yield is the weight sold and is used to calculate the total revenue for the business. Monitoring yields following management changes on the farm will provide a basis for understanding the effects of the change.

Measuring and monitoring survival from year to year provides an indication of the effectiveness of the measures undertaken to control and minimize mortalities. These may include disease control and predator control measures. Table 4.1 shows a survival of 74% for the 256-acre catfish farm. Improving survival will increase profits and will improve other financial indicators as discussed below.

Records maintained on the farm must also look at the average size of fish harvested. Changes in the average size of fish harvested each year can provide an indication of effects of other management changes on the farm, such as changes in the stocking density. Size thresholds may have implications for prices paid for various size groups of fish. Table 4.1 shows an average size of fish harvested of 1.5 lb. Decreases in the average weight of fish sold might indicate inefficient grading and sales of larger numbers of submarket sized fish while increases in the average weight might indicate inefficient harvesting (growing numbers of larger fish that had previously escaped the seine).

Growth rates are an important indicator also, but can be difficult or impossible to calculate for farms that are in continuous production. Increasing growth rates will increase turnover rates, can improve cash flow, and are likely to increase profitability.

INPUT USE EFFICIENCY

Measures of the efficiency of input use should also be used to evaluate and monitor farm efficiency. The feed conversion ratio (FCR) is the most important measure of input use efficiency, but similar measures can be calculated for labor, utility use, and other inputs. FCR is the most important for many aquaculture businesses because feed is the largest single cost on many aquaculture farms. Efficient use of feed inputs will assist the farmer to keep costs per unit of fish produced as low as possible. Table 4.2 shows a FCR of 2.12 for the 256-acre catfish farm. If this farm could reduce its FCR across the farm to 2.0, its total costs would be reduced by 2.2%, but net farm income would increase sevenfold, to $20,450 from $2,792.

It can be instructive to also calculate similar measures for other inputs, particularly for businesses in which an input other than feed is the most important. These can include fish production per worker (efficiency of labor), fish production per dollar of working capital (efficiency of operating capital), and fish production per 1,000 fingerlings stocked (efficiency of fingerlings stocked). For the example farm, the input use efficiency measures were: 791 (fingerling efficiency), 6.72 (operating capital efficiency), and 1,800 (labor efficiency) (Table 4.2). Year-to-year monitoring of these efficiency measures will demonstrate whether these measures show improvement or decline. For example, purchasing an improved strain of fish would be expected to result in improved fingerling efficiency, while hiring new workers may result in improved labor efficiency. Conversely, if fingerling efficiency declines after purchase of a more expensive, improved strain, it would be advantageous to discontinue expenditures on it.

The farm manager should review the measures of input use efficiency at least once a year and compare them with previous years. The remainder of this chapter concentrates on economic and financial measures that are less commonly used by aquaculture farm managers.

Table 4.2. Measures of Input Use Efficiency, 256-acre Catfish Farm.

Measure	Calculation	Unit	Farm value
Feed conversion ratio*	Weight of feed fed ÷ weight of fish sold	No unit	2.12
Fish production per 1,000 fingerlings stocked (fingerling efficiency)	Weight of fish sold ÷ (number of fingerlings stocked ÷1,000)	lb/1,000 fingerlings	791
Fish production per dollar of working capital (operating capital efficiency)	Weight of fish sold ÷ working capital†	lb/$	6.72
Fish production per worker (labor efficiency)	(Weight of fish sold ÷ number of acres) ÷ FTE‡ of labor	lb/acre/ worker	1,800
Cost of production	Total costs ÷ weight of fish sold	$/lb	$0.72
Stocker turnover ratio	(Total biomass on farm ÷ sales) × 365 days	Days of stock	570

*Marketable feed conversion.

†Working capital is determined from the balance sheet (total assets − total liabilities).

‡FTE, full-time equivalent. Determined by adding up the total months of labor and dividing by 12. Combines all seasonal and year-round labor into one value.

COST OF PRODUCTION

The cost of production measures the cost of producing a single unit of product. It is also referred to as the breakeven price and is typically calculated in units of $/lb or $/kg. The cost of production can be calculated from the enterprise budget (see Chapter 10 for more detail) or from the income statement (see Chapter 11 for more detail). It is a rapid way to view the profitability of the business because a quick comparison between the breakeven price and the current market price indicates whether there is a profit margin.

STOCK TURNOVER RATIO

The turnover ratio provides a measure of how quickly the inventory of fish produced is sold, or "turned over" into cash revenue. It compares the value of the inventory of fish or shrimp on the farm to the value of the sales, and then multiplies that ratio by 365 days. The answer calculated is expressed in the number of days of inventory on hand. Lower values demonstrate more rapid turnover and are preferable. Higher values mean that fish inventories on the farm take longer to reach market size. Because the fish stocks in inventory on the farm must be fed regularly, longer turnover periods tie up working capital that could otherwise be used in more productive ways. Table 4.2 shows a stock turnover ratio of 570 days. This is a long period and demonstrates a rather inflexible production system. Management changes that result in faster turnover, perhaps by switching to a faster-growing strain or re-

ducing stocking density to obtain better growth, will reduce the stock turnover ratio. A lower stock turnover ratio improves cash flow, reduces financial risk, and may improve financial performance.

PROFITABILITY

A business that is both solvent and liquid will not necessarily be profitable. Profitability is calculated generally by subtracting total costs from total revenue. It is measured from the income statement. However, net farm income can be further partitioned into returns or profits attributable to each of the four primary factors of production: land, labor, capital, and management. Returns to capital can be further partitioned into returns to equity capital (capital owned by the farmer) and returns to debt capital (borrowed capital).

PROFIT MARGINS

Two profitability ratios commonly calculated include the gross profit margin and the net profit margin. The gross profit margin is calculated by dividing the gross profit by the sales and multiplying by 100. The gross profit margin measures the sales and production performance, can be tracked over time, and can be used to compare performance of other similar aquaculture businesses. The net profit margin divides net profits (gross profit minus operating costs) by sales and multiplies by 100.

NET FARM INCOME

The primary measure of farm profitability is net farm income. Net farm income measures the return to operator's equity, capital, unpaid labor, and management. It is measured from the income statement. Net farm income is measured as follows:

> Total revenue − total expenses
>   = net farm income from operations
>   ± the gain/loss on the sale of capital assets
>   = net farm income

For the example farm, net farm income from the income statement for the 256-acre catfish farm is $2,792 (Table 4.3; Chapter 11). This farm will need to improve its low net farm income over time to provide adequate compensation for the owners.

RATE OF RETURN ON FARM ASSETS

The rate of return on farm assets (ROA) measures the profits obtained from the use of all capital (debt and equity capital) invested in the business and can be compared to rates of return on other long-term investments. It has several names including rate of return on assets (ROA), return on assets (ROA), return to capital, and return on investment (ROI). It can be viewed as an interest rate earned on the total investment of capital in the farm. It is a measure of profitability because it compares the profits (returns minus expenses) to the value of the assets of the business. If the farm maintains a beginning and ending balance sheet for each year, the value of the total assets is averaged between the beginning and end-of-year balance sheet. The ROA is not related to the way that assets are financed.

The ROA is independent of the type and amount of financing. It can be compared to other similar farms, returns from other investments, opportunity costs of the farm's capital, and past ROA values for the farm to measure profitability.

ROA is calculated as follows:

$$ROA = \frac{return\ to\ assets}{average\ asset\ value} \times 100 \qquad (4.1)$$

Return to assets is calculated as follows:

> Adjusted net farm income
>   (= net farm income from operations plus interest expenses)
>   − opportunity cost of unpaid labor
>   − opportunity cost of management
>   = return to assets

Interest is added back to net farm income (in the calculation of adjusted net farm income) because interest expenses were subtracted out in the income statement. This is because the ROA measures the return to all capital used; if interest has been charged out for the use of the capital, then the returns are to the equity alone, not the total amount of capital used. In other words, the use of the capital would have been charged out twice and would give a false measure of the profits generated by the capital used in the business.

In calculating the ROA, adjustments are also made for unpaid labor and management that is provided by the farm owner and family. If not subtracted out, then the calculation of net farm income would include the earnings produced by the farm labor and management. Because wages were not paid directly to the family labor and management, opportunity costs are used to assign a value to these resources. The ROA is an average return and should not be used to make investment decisions. Chapter 13 provides details on the appropriate tools for assessing investment decisions.

In Table 4.3, the ROA for the example farm is 10%. Rates of return in agriculture are frequently lower than for other types of businesses. An ROA of 10% is competitive for many agriculture enterprises, particularly during times when interest rates are low.

OPERATING PROFIT MARGIN RATIO

The operating profit margin ratio (OPMR) measures the proportion of gross revenues left after paying expenses by calculating operating profit as a percent of total revenue. The indicator measures the operating efficiency of the business. It is calculated as:

$$OPMR = \frac{return\ to\ farm\ assets}{gross\ revenue\ of\ farm} \qquad (4.2)$$

The higher the value, the more profit the business has generated per dollar of revenue. Farms with large investments in fixed assets such as land and few operating expenses will show a higher OPMR. A high OPMR results from a business that maintains expenses in proportion to the value of the farm's production. Farms with more rented assets will have a higher ROA but a lower OPMR. Problems of profitability arise when both the ROA and OPMR are below average. A low OPMR indicates that more attention must be paid to improving this ratio before looking at expanding production levels. Low OPMRs can result from inefficient production, low prices, or high operating expenses.

Table 4.3. Profitability Measures, 256-acre Catfish Farm.

Measure	Calculation	Unit	Farm value
Net farm income	Total revenue − total expenses	$	2,792
	= net farm income from operations		
	± gain or loss on sale of capital assets		
Rate of return on assets (%)	Net farm income	$	2,792
	+ interest expense	$	150,579
	= adjusted net farm income	$	153,371
	− operating cost of unpaid labor	$	10,140
	− operating cost of management	$	11,154
	= return to assets	$	132,077
	ROA = return to assets ÷ average farm asset value[*] × 100	%	10%
Rate of return on equity (%)	Net farm income	$	2,792
	− operating cost of unpaid labor	$	10,140
	− operating cost of management	$	11,154
	= return on equity	$	−18,502
	ROE = return on equity ÷ average equity[†] × 100%		−2%
Operating profit margin ratio	Net farm income	$	2,792
	+ interest	$	150,579
	− operating cost of unpaid labor	$	10,140
	− operating cost of management	$	11,154
	= operating profit	$	132,077
	Operating profit margin ratio	$	132,077
	= operating profit ÷ total revenue × 100%	%	16%
Financial leverage ratio	ROE ÷ ROA	%	−0.2%
Return to labor and management		$	24,786 =
	Net farm income from operations	$	2,792
	+ interest expenses	$	150,579
	= adjusted net farm income	$	153,371
	− operating cost of all capital (10%)	$	128,585
Return to labor		$	$13,632 =
	Return to labor and management	$	$24,786
	− operating cost of management		−$11,154
Return to management		$	$14,646 =
	Return to labor and management	$	$24,786
	− operating cost of labor		−$10,140

[*]Average farm asset value is the average of the beginning and ending value of total asset values from the farm's balance sheet. In this example of the 256-acre catfish farm, the total asset value is $1,285,851 found on Table 11.1 on the balance sheet of the farm. For this example, in the absence of beginning (January 1) and ending (December 31) balance sheets, the simplifying assumption is that this value did not change over the year.

[†]Average farm equity is the average of the beginning and ending value of net worth, from the farm's balance sheet. In this example of the 256-acre catfish farm, the net worth (owner's equity) is $751,637 found on Table 11.1 on the balance sheet of the farm. For this example, in the absence of separate beginning (January 1) and ending (December 31) balance sheets, the simplifying assumption used is that this value did not change over the year.

The OPMR for the example 256-acre catfish farm is 16%. This indicates that, for every dollar of revenue, $0.16 remains as profit after paying the operating expenses needed to generate that dollar.

RATE OF RETURN ON FARM EQUITY

The rate of return on farm equity (ROE) measures the returns on the owner's share of capital (equity) used in the business. It can be viewed as an interest rate earned on the business' average equity or net worth. It is indicative of the farm's financial progress because, if the business should fail, the equity capital will be what the farmer has left after paying off all liabilities. It measures the percent return to owner's net worth or equity. If the farm has no debt, the ROE is equal to the ROA. It is calculated as:

$$\text{ROE} = \frac{\text{return on equity}}{\text{average equity}} \times 100 \qquad (4.3)$$

The ROE is calculated as follows:

Net farm income from operations
    − opportunity cost of unpaid labor
    − opportunity cost of management
    = return on equity

In contrast to the calculation of the ROA, interest charges are not added back in to net farm income from operations to calculate the return on equity. This is because interest is a charge on borrowed capital. Because ROE measures only the return to equity capital, it is correct to have charged out the interest cost on borrowed capital prior to calculating ROE. The opportunity costs of unpaid family labor and management must still be charged out to appropriately account for their contribution to the returns generated from the farm.

The rate of return on equity in Table 4.3 is −2%. This low rate reflects the negative returns on equity for the farm. The low returns on equity for the 256-acre catfish farm result from the low net farm income generated from the farmer's equity in the business. The majority of capital in the business is equity capital that is earning a low return. The negative value results after charging out the opportunity costs of unpaid labor and management.

FINANCIAL LEVERAGE RATIO

The financial leverage ratio compares debt with equity capital by dividing the ROE by the ROA as follows:

$$\text{Financial leverage ratio} = \frac{\text{ROE}}{\text{ROA}} \qquad (4.4)$$

If the financial leverage ratio is greater than 1 (because ROE is greater than ROA), then debt capital is being used effectively in the business. If the financial leverage ratio is less than 1 (ROE is less than ROA), the cost of debt capital is higher than the return to it. Thus, the financial leverage ratio indicates whether debt capital is being used effectively. Table 4.3 shows a financial leverage ratio of −0.2%. This value reflects the negative value for ROE and indicates that debt capital is not used effectively.

RETURN TO LABOR AND MANAGEMENT

Net farm income can be distributed among the four principal factors of production: land, labor, capital, and management. This is accomplished by charging out the costs associated with the other three factors of production. What remains is the income that can be attributed to that one factor of production.

The return to labor and management is what remains from net farm income after charging out returns for the use of all capital. Some businesses have more assets or borrow more money than others and will have differing costs of capital. Subtracting out the cost of capital for the particular business being analyzed allows the manager to see what part of income has been produced by the labor and management.

Return to labor and management is calculated as follows:

Net farm income from operations
    + interest expenses
    = adjusted net farm income
    − opportunity cost of capital

Return to labor and management for the 256-acre catfish farm was $24,786 (Table 4.3).

RETURN TO LABOR

The returns to labor and management can be further partitioned into returns to either labor or returns to management by subtracting out the cost of the other production factor. These measures indicate whether net farm income was sufficient to provide a return at least equal to the opportunity costs of labor and management. Return to labor is calculated as:

Return to labor and management
    − opportunity cost of management
    = Return to labor.

Returns to labor for the 256-acre catfish farm in Table 4.3 are $13,632, a positive return to the labor resources used.

RETURN TO MANAGEMENT

Return to management is that portion of adjusted net farm income remaining after opportunity costs of both labor and capital have been subtracted. It represents a residual return to the owner for the management input, or the decisions made by the manager. Negative returns to management are not unusual, but positive net returns should be the goal. Returns to management can be highly variable from year to year. Returns to management are calculated as:

Return to labor and management
− opportunity cost of labor
= return to management.

Returns to management for the 256-acre catfish farm are $14,646 (Table 4.3). These are positive returns to the management resource used.

SOLVENCY

Solvency refers to the value of assets owned by the business (equity) compared to the amount of liabilities owed. Measures of solvency show how the assets, liabilities, and equity of the business are related. A positive measure of solvency indicates that if the business were to be sold, there would be enough capital to pay off all debts. In other words, it measures whether the value of assets is greater than the value of the liabilities of the business. If the value of the liabilities is greater than the value of the assets, then the business is deemed to be insolvent.

Insolvent businesses are at risk of being forced to declare bankruptcy. Thus, financial risk is associated more with its financial structure and solvency than with liquidity. A low net worth provides little margin of reserves with which to react to emergencies or financial downturns.

CHANGE IN NET WORTH, CHANGE IN EQUITY

A change in net worth indicates business growth, additional capital investment, and a greater borrowing capacity. This is because net worth measures the difference between total assets and total liabilities; thus, net worth gives a measure of the cushion above financial obligations. Equity increases primarily from increases in the value of assets or from decreases in the liabilities. These changes frequently reflect earnings from net farm income that have been retained in the business to pay down debt or to acquire additional assets.

Net worth = total assets − total liabilities

The owner would want to see this number increase because the objective of a business is to increase net worth over time, to accumulate capital. The 256-acre catfish farm showed a decrease in net worth of −$245,994 over the previous year (Table 4.4). The decrease in net worth is due to decreasing prices of catfish (decreased current assets), falling land prices (decreased noncurrent assets), and increased borrowing of operating capital (increased current liabilities) to cover cash flow shortfalls.

NET CAPITAL RATIO

The net capital ratio also measures solvency of the business. It is calculated with the following equation:

$$\text{Net capital ratio} = \frac{\text{total assets}}{\text{total liabilities}} \quad (4.5)$$

If total assets are equal to total liabilities, then the net capital ratio will be 1.0. The net capital ratio is the inverse of the debt/asset ratio. For the same reason, a value of 1.0 indicates that there is no equity in the business. Higher values are preferred. For the catfish farm

Table 4.4. Solvency Measures, 256-acre Catfish Farm.

Measure	Calculation	Calculations	Farm value
Change in net worth	Net worth (Year 2) − net worth (Year 1)	$505,643 − $751,637	−$245,994
Net capital ratio	Total assets ÷ total liabilities	$1,285,851 ÷ $534,214	2.41
Debt/asset ratio	Total liabilities ÷ total assets	$534,214 ÷ $1,285,851	0.42
Equity/asset ratio	Net worth ÷ total assets	$751,214 ÷ $1,285,851	0.58
Debt/equity ratio	Total liabilities ÷ net worth	$534,214 ÷ $751,214	0.71
Debt structure ratio	Current liabilities ÷ total liabilities	$110,584 ÷ $534,214	0.21
Interest coverage ratio	Returns on assets ÷ interest payments	$21,294 ÷ $150,579	0.14

example, the net capital ratio was 2.41, demonstrating a solvent business.

DEBT/ASSET RATIO

The debt/asset ratio is a common measure of business solvency. It measures the amount of debt as compared with the total value of assets of the business. It shows the extent to which assets were acquired with debt versus equity capital. It is calculated by dividing total farm liabilities by total farm assets using current market values for each.

$$\text{Debt/asset ratio} = \frac{\text{total farm liabilities}}{\text{total farm assets}} \quad (4.6)$$

Thus, the debt/asset ratio measures how much of the total farm assets are obligated to a banker or other lender.

Smaller values are preferred to larger ones. Smaller values indicate a better chance of maintaining the solvency of the business should it be faced with a period of adverse economic conditions. Low debt/asset ratios may also indicate that a manager is reluctant to use debt capital to take advantage of profitable investment opportunities. Higher values indicate that a greater amount of the farm's assets are obligated to the firm's creditors.

A business that has borrowed all the capital needed for the business would have a debt/asset ratio of 1.0. This is because the total amount of liabilities will exactly equal the total amount of assets; in such a case, there is no equity because all the capital used has been borrowed. A debt/asset ratio greater than 1 indicates an insolvent business because the value of the liabilities is greater than the value of the assets; if the business would be sold, the amount of cash obtained from sale of the assets would not be enough to pay off the debts.

In Table 4.4, the debt/asset ratio for the example catfish farm is 0.42. This value is less than 1, an acceptable value, and indicates a solvent business. If this value is decreasing, however slightly, this means that assets are growing faster than debt, which would be a positive sign of improving solvency. This indicator should decrease as equity in the business grows.

EQUITY/ASSET RATIO

The equity/asset ratio is calculated in a similar manner, but compares only the equity that the owners have in the business to the value of the assets. It indicates the proportion of total assets that have been provided through the owner's equity capital. The fol-

lowing equation is used to calculate the equity/asset ratio:

$$\text{Equity/asset ratio} = \frac{\text{total equity}}{\text{total assets}} \quad (4.7)$$

The value of the equity/asset ratio cannot exceed 1.0 because the farmer cannot own more than the total value of the assets in the business. A value of 1.0 also indicates that there are no liabilities; all capital for the business has been provided by the owner. In general, higher values are preferred because the overall goal of most businesses is to accumulate capital and value over time in the form of growing equity. As equity in the business increases, assuming that the total value of the assets remains the same, the equity/asset ratio will grow in proportion to the growth in equity. Higher values indicate that the farmer owns a larger share of the farm's assets relative to the debt capital. A negative equity/asset ratio can occur if equity is negative and is an indicator of insolvency.

The equity/asset ratio for the example catfish farm in Table 4.4 is 0.58. This is an acceptable value because it is positive; a negative value would indicate an insolvent business. If the equity/asset ratio increases from year to year, this indicates that equity is growing in relation to asset levels. In this example, the equity/asset ratio of 0.58 indicates high equity in the business with little debt load. Over time, as the loans are paid off, equity increases in relation to the level of assets.

DEBT/EQUITY RATIO

The debt/equity ratio (also called the leverage ratio) compares the level of borrowing with the level of capital provided by the owner. It is calculated with the following equation:

$$\text{Debt/equity ratio} = \frac{\text{total liabilities}}{\text{total equity}} \quad (4.8)$$

Businesses must pay down their debt over time and seek to increase equity in the business. Thus, values for the debt/equity ratio that are lower and decreasing over time are preferable because this shows that the business is successfully paying down the principals on loans, and equity is increasing as a result. If half of the business' capital is provided by the owner as equity and half is borrowed, then the value of the total liabilities and the total equity are the same. In such a case, the debt/equity ratio will be 1.0. The debt/equity ratio will approach zero as liabilities approach zero. Very large values result from very low equity, which means an increasing chance of insolvency. A rule of thumb for the debt/equity ratio is a value of 1.0. This means that the owner is providing as much capital as it borrows. This standard equates to a net capital ratio of 2.0 and a debt/asset ratio of 0.50.

The debt/equity ratio for the example catfish farm in Table 4.4 is 0.71. In this example farm, there is very high equity because the majority of the capital is owned, indicating high solvency. Thus, this low value indicates a relatively low level of financial risk in the business. New businesses tend to have lower levels of equity which results in high debt/equity levels. The debt/equity ratio would be expected to be higher for newer farms that are highly leveraged. Newer farms with high initial debt/equity ratios would be expected to show decreasing debt/equity levels over time as debt is paid off and equity in the business increases.

DEBT STRUCTURE RATIO

The debt structure ratio demonstrates how much of the debt is due and payable in the next year. It calculates the percent that current liabilities compose of total liabilities and is calculated from the following equation:

$$\text{Debt structure ratio} = \frac{\text{current liabilities}}{\text{total liabilities}} \quad (4.9)$$

A business that has a high debt structure ratio is one that has most of its debt as an operating loan or other types of payments due the coming year. As long as the liquidity measures indicate that there are sufficient sources of cash to meet the financial obligations of the coming year, this is not a problem. However, if there are liquidity problems, then corrective actions must be taken as quickly as possible. Restructuring loans might be necessary to move current to noncurrent liabilities, for example. The debt structure ratio cannot exceed 1 because it is not possible for the current liabilities to exceed total liabilities. For the example catfish farm, the debt structure ratio is 0.21. This indicates that 21% of the debt consists of current liabilities.

INTEREST COVERAGE RATIO

The interest coverage ratio is considered a measure of solvency and of financial risk. It relates a firm's financial charges (interest) to the firm's ability to service debt. It also indicates how much of the firm's returns to assets are available for every dollar in interest commitment. It covers a period of time, rather than a point in time.

The interest coverage ratio is calculated as the net income after taxes plus interest paid and accrued, di-

vided by the interest. This is calculated from the income statement. The higher the interest coverage ratio, the less the burden that interest payments have on income.

It is calculated as follows:

$$\text{Interest coverage ratio} = \frac{R}{I} \quad (4.10)$$

where R is the business' returns to their assets and I is the amount of interest charges. However, the interest coverage ratio does not account for the principal payments that are equally important to the firm's ability to service its debt.

The interest coverage ratio calculates the operating profit as a ratio of interest charges (and is multiplied by 100). This ratio indicates the extent to which profit before interest will cover the interest charges to be accrued. For the example farm, the interest coverage ratio is 0.14 (Table 4.4).

LIQUIDITY

Liquidity is the ability of a business to meet cash flow obligations. Liquidity is important to keep financial transactions of the business running smoothly. Adequate liquidity indicates that the business will be able to pay its bills and meet all its financial obligations while the business proceeds in a normal fashion. In other words, no extraordinary measures (such as selling off needed equipment to quickly generate cash) are needed to make the payments necessary. Given that no one can forecast or project the future perfectly, it is critical to plan to have liquid reserves available to have the capacity to adjust to unforeseen adverse circumstances.

Liquidity is concerned with the short-run ability to meet financial obligations, typically, over the upcoming accounting period. Thus, the analysis of liquidity involves use of the current assets and current liabilities from the balance sheet as well as other measures from the cash flow budget. Current liabilities measure the need for cash over the next year while current assets measure the amount of cash that will become available over the next year.

CURRENT RATIO

The current ratio is a quick indicator of a firm's liquidity. Current assets are those that will be sold or turned into saleable products in the near future and will generate cash to pay debt obligations that come due. The current ratio is calculated as follows:

$$\text{Current ratio} = \frac{\text{current farm assets}}{\text{current farm liabilities}} \quad (4.11)$$

From this equation, it is clear that the current ratio compares the value of current farm assets with the value of current farm liabilities. A business that expects to have just enough cash available in the coming year to meet its financial obligations for that year will have a current ratio of 1, because current assets will exactly equal current liabilities. A current ratio of 1 indicates that the business is liquid for the upcoming year, but there is no margin for adverse circumstances that are all too common in aquaculture. Thus, the higher the value of the current ratio, the better, because the business is more liquid and better able to meet its financial obligations over the next year even if faced with adverse conditions. A current ratio of 2 is generally considered indicative of a favorable liquidity position.

The current ratio of 2.55 for the example catfish farm in Table 4.5 is an acceptable level. There are adequate current assets to cover current liabilities, with some safety margin. In future years, this value should increase because payments on debt will lower liability levels.

WORKING CAPITAL

Working capital is calculated as the difference between current assets and current liabilities. Thus, it is not a ratio but a dollar value. It represents excess dollars available from current assets after current liabilities have been paid. It is the money that the farm has available for use in the near future. Working capital provides a measure of the safety margin that would be available over the next year to meet financial obligations in the event of adverse circumstances. Working capital also measures whether funds would be available to make additional purchases for the business or family, after meeting the required financial obligations for the next year.

Working capital for the example farm in Table 4.5 is $171,377. This value represents the margin of safety for liquidity. Its positive value and its magnitude indicate a high degree of liquidity for the catfish farm. This value will increase as growth of the fish crop increases asset values.

CASH FLOW COVERAGE RATIO

The cash flow coverage ratio indicates the extent to which the excess cash generated by the business provides a cushion for or flexibility in covering debt-servicing requirements. It excludes payments on the operating loan balance because they are so flexible.

The cash flow coverage ratio is calculated by dividing the excess available cash by the cash required for interest and principal payments. It is calculated from the cash flow budget. Higher ratios indicate more favorable liquidity. For the example catfish farm, the cash flow coverage ratio is 5.63. This indicates a substantial amount of cash available for each dollar of interest and principal payments.

DEBT-SERVICING RATIO

The debt-servicing ratio is calculated by dividing the cash required for interest and principal payments by the total cash available. It is measured from the cash

Table 4.5. Liquidity Measures, 256-acre Catfish Farm.

Measure	Calculation	Calculations	Farm value
Current ratio[*]	Current assets	$281,961	2.55
	÷ current liabilities	÷ $110,584	
Working capital[†]	Current assets	$281,961	$171,377
	− current liabilities	− $110,584	
Cash flow coverage ratio[†]	Excess available cash	2,019,448	5.63
	÷ cash required for interest and principal payments	÷ 358,639	
Debt-servicing ratio	Cash required for interest and principal payments	$358,639	0.18
	÷ total cash available	÷ $2,019,448	

[*]From balance sheet.
[†]From cash flow budget.

flow statement. Higher ratios indicate that there is a greater burden of debt on cash flow and lower liquidity. The debt-servicing ratio for the example catfish farm is 0.18 (Table 4.5), indicating that the interest and principal payments represent only 18% of the total cash available.

FINANCIAL EFFICIENCY

OPERATING EXPENSE RATIO

The operating expense ratio provides a ratio of expenses to gross farm revenue. It is calculated as:

Operating expense ratio
= total farm operating expenses
− depreciation/amortization
÷ gross revenues

The operating expense ratio for the example catfish farm is 0.70 (Table 4.6).

DEPRECIATION EXPENSE RATIO

The depreciation expense ratio compares the depreciation expenses as a percent of gross revenues. It is calculated as:

Depreciation expense ratio
= depreciation/amortization
÷ gross revenues

The depreciation expense ratio for the example farm is 0.05.

INTEREST EXPENSE RATIO

The interest expense ratio compares the interest expenses as a percent of gross revenues.

Interest expense ratio = total farm interest expense
÷ gross revenues

The interest expense ratio provides a measure of the portion of expenses that are attributed to interest payments on borrowed capital. The more leveraged (greater the debt load) the business and the higher the interest rates, the more important is the interest expense ratio.

TURNOVER RATIO

The turnover ratio (also called the asset turnover ratio) is calculated by dividing the gross farm revenue by the average total farm assets. It is a measure of efficiency and measures how quickly the business' assets are turned into revenue.

It is calculated as follows:

$$\text{Asset turnover ratio} = \frac{\text{gross revenues}}{\text{average total farm assets}}$$

Higher values indicate greater efficiency of the farm business because higher values indicate that assets are turning over more rapidly. This means that there are greater opportunities for profits (assuming that profit margins are positive). Asset turnover ratios are commonly in the range of 40–50%. The asset turnover ratio for the example farm is 0.63 (Table 4.6).

DEBTOR TURNOVER PERIOD

The debtor turnover ratio calculates a ratio that compares the funds owed to the business from its sales and is multiplied by 365 days (as in the stock turnover ratio). A low debtor turnover ratio reflects a low amount of accounts receivable as compared to sales. A high ratio indicates that payments must be received more frequently. This value typically is preferred to be less than 1 month. The debtor turnover period is 42 days for the example catfish farm (Table 4.6).

Table 4.6. Financial Efficiency Measures, 256-acre Catfish Farm.

Measure	Calculation	Calculations	Farm value
Operating expense ratio	Operating expenses − depreciation ÷ gross revenues	$610,322 − $42,707 ÷ $806,400	0.70
Depreciation expense ratio	Depreciation ÷ gross revenues	$42,707 ÷ $806,400	0.05
Turnover ratio	Gross farm revenue ÷ average total farm assets	$806,400 ÷ $1,285,851	0.63
Debtor turnover period	(Cash owed to business ÷ sales) × 365 days	$100,000 ÷ 806,400 × 365	45.26
Percent owed	Debt capital ÷ total assets × 100	$534,214 ÷ $1,285,851 × 100	42
Acid test ratio	(Liquid assets ÷ current liabilities)	$15,849 ÷$110,584	0.14

PERCENTAGE OWNED

A simple measure of the overall stability of a business is to calculate the percentage owned. This is a quick measure of the proportion of the business owned by the owner. It is calculated by dividing the net capital by the total assets and multiplying by 100.

ACID TEST RATIO

The acid test ratio is calculated from the balance sheet and indicates the extent of a safety margin for the short term. It is similar to the current ratio but includes only cash assets. It is calculated by dividing the liquid assets by the current liabilities and multiplying by 100. The acid test ratio indicates the extent to which the business can survive if all debts are called in by the lenders. An acid-test ratio of 100% indicates that the business will be able to meet its short-term debt obligations. Businesses that carry large inventories of product may use the acid test ratio usefully because it considers only the cash assets. The acid test ratio for the example catfish farm is 0.14.

FLOW OF FUNDS ANALYSIS

In a flow of funds analysis, the flow of cash is traced through the business over the course of the year and monitors changes in the values of assets and liabilities. It divides cash flow into categories of the sources and uses of cash. This analysis shows the effects of decisions on long-term uses of cash. It ensures that there is sufficient cash available when needed.

REPAYMENT CAPACITY

The ability of the farm to repay its loans is critical to the long-term success and viability of the business. Repayment capacity must include the ability to repay both the interest and principal. The best way to analyze and assess repayment capacity is from the cash flow budget. This can be viewed by monitoring the line on the cash flow budget that presents the outstanding balance on the operating loan. High repayment capacity is indicated by rapid declines of the outstanding balance on the line of credit and if the operating loan is repaid that same year.

TERM DEBT AND CAPITAL LEASE COVERAGE RATIO

The term debt and capital lease coverage ratio is also known as the term debt coverage ratio. It measures whether the business will generate enough cash to meet its term debt payments. Term debt payments are the scheduled principal and interest payments on noncurrent liabilities such as equipment and real estate loans. It is calculated as follows:

> Term debt coverage ratio
> = net farm income from operations
> + total miscellaneous revenue/expense
> ÷ total nonfarm income
> + depreciation/amortization expense
> + interest on term debt
> + interest on capital leases
> − total income tax expense
> − owner withdrawals
> ÷ (annual scheduled principal and interest payments on term debt
> + annual scheduled principal payments on capital leases)

CAPITAL REPLACEMENT AND TERM DEBT REPAYMENT MARGIN

The capital replacement and term debt repayment margin is the dollar value that remains after subtracting out all expenses, taxes, family living expenses, and debt payments. It also represents the cash that is available to purchase additional equipment or other capital items. It is calculated as follows:

> Capital replacement and term debt repayment margin
> = net farm income from operations
> + total miscellaneous revenue/expense
> + total nonfarm income
> + depreciation/amortization expense
> − total income tax expense
> − total owner withdrawals
> − payment on unpaid operating debt from a prior period
> − principal payments on current portions of term debt
> − principal payments on current portions of capital leases
> − total annual payments on personal liabilities (if not included in withdrawals)

GROWTH RATIOS

Ratios can also be calculated for growth in values such as net capital, assets, and liabilities. These are calculated by subtracting the value for the previous year from that of the current year and multiplying by 100. These growth ratios give an overall indication of the trends in each parameter and, over time, can be useful to determine the direction of the business.

CASH FLOW RISK MEASURES

Table 4.7 lists several measures of cash flow risk, based on the practical application of the 256-acre catfish farm presented in Table 4.3. Operating cash flow as a proportion of current liabilities is 45%, slightly less than half of the liabilities for the current period. Thus, there is room for a similar increase in liabilities that can be covered by available operating cash flow. Operating cash flow as a percent of the largest short-term credit balance is 26%. Thus, there is a cushion that represents just 26% of the greatest amount of the outstanding operating loan. The ending cash balance is positive, and much (but not all) of the operating loan has been paid off over the course of the year.

RECONCILING FINANCIAL STATEMENTS

The financial statements developed each year must be reconciled with each other. This type of balance and check system will uncover errors in calculations and will allow the business owner or manager to examine the business in greater detail.

RECORD-KEEPING

The financial indicators discussed in this chapter are all calculated from information available on farm inventory records (discussed in Chapter 2), the balance sheet and income statements (records for these are discussed in Chapter 11), and the cash flow budget (records for these are discussed in Chapter 12). A spreadsheet that lists the indicators as calculated for each year should be maintained. Such a spreadsheet provides an easy method of tracking performance of the business over time from the various perspectives measured by each set of indicators. These can be used to develop graphs and trend lines that facilitate viewing changes over time.

PRACTICAL APPLICATION

Table 4.8 illustrates a practical application of the use of financial indicators over time to make decisions. Year 1 is the year demonstrated for the example farm of 256 acres. Values are taken from the balance sheet in Chapter 11 (Table 11.1) and from Table 4.4 of this chapter.

These tables were based on an average price of $0.70/lb, listed in the column labeled as Year 1. Year 2 represents a condition of economic downturn during which price falls to $0.60/lb, and land prices begin to decline also. The fall in catfish prices results in a decline in current assets on the balance sheet while the fall in land prices results in a decline in noncurrent assets. Thus, the value of total assets also falls. If the farmer's bank agrees to forego principal payments on the loan, in response to the economic downturn and falling prices, then noncurrent liabilities will remain the same. However, a farm business in this condition

Table 4.7. Cash Flow Risk Measures (Calculated From Tables 11.1, 11.2, and 12.3).

Measure	Calculation	Unit	Farm value
Cash flow risk	Operating cash flow* ÷ current liabilities ($49,499 ÷ $110,584)	$	0.45
	Operating cash flow ÷ total liabilities ($49,499 ÷ $534,214)	$	0.009
	Operating cash flow ÷ largest short-term credit balance ($49,499 ÷ $188,036)	$	0.26
Ending cash balance		$	1,000
Ending operating loan balance		$	25,830

*Operating cash flow = net income + depreciation − withdrawals − increase in accounts receivable − increase in inventories + increase in accounts payable + increase in accrued liabilities.

Table 4.8. Practical Application of Use of Financial Indicators, 256-acre Catfish Farm. Values Taken From Table 11.1 in This Volume.

Balance sheet category	Year 1	Year 2	Year 3
Assets			
Current	$281,961	$228,096	$228,096
Noncurrent	$1,003,890	$889,170	$889,170
Total	$1,285,851	$1,117,266	$1,117,266
Liabilities			
Current	$110,584	$187,993	$319,588
Noncurrent	$423,630	$423,630	$423,630
Total	$534,214	$611,623	$743,218
Net worth	$751,637	505,643	$374,048
Debt/asset ratio	0.42	0.55	0.67
Current ratio	2.55	1.21	0.71

likely will need to increase the amount of operating capital borrowed because the reduced revenue results in lower levels of cash inflow. The increased borrowing causes current liabilities to increase. The overall result is a decline in net worth (although net worth is still positive), an increase in the debt/asset ratio, and a decrease in the current ratio. A full assessment of these indicators shows that the business, while not profitable that year and with a decline in net worth, still has adequate liquidity (as evidenced by a current ratio greater than 1). Its debt/asset ratio is still acceptable.

If the economic downturn continues into Year 3, the financial condition of the business becomes much worse, even if catfish and land prices stabilize. In such a scenario, current, noncurrent, and total assets do not change in Year 3. However, the lower revenue levels continue to make payments on long-term debt difficult. Additional operating capital borrowed results in further declines in net worth, a continued increase in the debt/asset ratio, and further declines in the current ratio. Net worth remains positive; thus the farm continues to be solvent. The debt/asset ratio, while higher, remains below 1 and is still manageable. However, the current ratio has grown to exceed 1.0. Thus, this business will have liquidity problems in Year 3 that will have to be addressed. An increase in prices in Year 3 would begin to bring these values back into line.

OTHER APPLICATIONS IN AQUACULTURE

Sotorrío (2002) developed a model of the most relevant economic and financial indicators to analyze the economic performance of a finfish mariculture business.

This model was then used to develop a case study related to the Spanish market for the finfish produced.

Profitability was the indicator used for success. Balance sheet and income statement data were collected from 16 Spanish mariculture companies over a 3-year period. These companies included 10 firms that were in growout production only, 1 firm that was exclusively a hatchery, and 5 firms that had both growout facilities and hatcheries.

Results showed that production efficiency and the learning curve of beginning a business based on new technology explained the majority of the profits or losses in the businesses. The use of farm financial indicators showed the effect of technological efficiency improvements necessary for businesses to be profitable and measured the effect of adoption of new technologies. The analysis also showed the high risk in mariculture businesses that result from fluctuating supply, market prices, and farm revenue. The financial indicators also measured the improvement in efficiency over time due to the learning curve for mariculture companies.

The companies studied were rarely profitable in the first few years. The availability of financial indicators for each year demonstrated the number of years required for the firm, its management, and its workforce to learn, adapt, and become proficient with the technologies necessary for a successful mariculture business.

SUMMARY

This chapter presents a set of financial indicators that can be used to develop a detailed assessment of the performance of an aquaculture business. These indicators can be tedious to calculate and can also be confusing due to the number of different possible indicators. Yet, the size of many aquaculture businesses and the amount of capital invested requires the manager to pay this level of attention to detailed analysis to be successful.

REVIEW QUESTIONS

1. Describe some of the most critical indicators of production efficiency on an aquaculture farm.

2. Give three examples of indicators of input use efficiency and explain what each indicator measures.

3. Explain the difference between the ROA and the ROE.

4. Describe how to apportion net farm income among capital, labor, and management and what value these have for management decisions.

5. Define solvency and explain its importance in financial analysis of aquaculture businesses.

6. Explain how changes in net worth should be used to assess the financial condition of the business.

7. Contrast the debt/asset, equity/asset, and debt/equity ratios in terms of how they are calculated, how they are related, and what they indicate about the aquaculture business.

8. Explain the importance of liquidity, describe three common methods to calculate liquidity, and explain how each is interpreted.

9. What two financial statements can be used to evaluate liquidity?

10. Describe two measures of repayment capacity.

REFERENCE

Sotorrío, LL. 2002. Economic analysis of finfish mariculture operations in Spain. *Aquaculture Economics and Management* 6(1/2):65–79.

5
Financing an Aquaculture Business

INTRODUCTION

Financing has been difficult for many aquaculture businesses. This is primarily because aquaculture has been considered a risky enterprise by many lenders. As a consequence, interest rates on aquaculture loans have tended to be higher and terms of lending more stringent than for other types of loans. Conventional loans may not be available at all for aquaculture businesses in some areas because lenders are too uncomfortable with the business. A lender who does not understand the production system, expected patterns of cash flow and price fluctuations will tend to be unwilling to extend credit without a substantial level of financial security to back up the loan.

This chapter discusses lending to aquaculture businesses from the lender's perspective, financial characteristics of borrowers and their behavior, credit reserves, and credit capacity. Credit problems faced by small-scale aquaculture growers in developing countries are also discussed. Chapter 14 presents information related to the mechanics of structuring loan payments and discusses various types of loans and terms of lending.

LENDING FROM THE BANKERS' PERSPECTIVE

A bank is a private, commercial business with the goal of making a profit, just as an aquaculture business is developed to make a profit. Banks approve loans to earn money from the interest charged. Interest is the "cost" of money to the borrower, but to the banker, the interest rate is the "price" charged for the resource and service provided by making capital available to other businesses.

Aquaculture Economics and Financing: Management and Analysis, Carole R. Engle, © 2010 Carole R. Engle.

A bank obtains its profit over time as loans are paid back. The risk that the capital loaned out will not be returned in its entirety constitutes a substantial risk for the bank. A loan made to a risky business venture will put the bank's capital at risk also. Successful banks must put safeguards in place to minimize the risk of losing their capital. As a bank's capital is reduced by defaulted loans, the bank must draw upon its reserves to maintain adequate liquidity. Without adequate capital reserves, the bank will fail and go out of business.

While the borrower benefits directly from any profits obtained as a result of a loan, the lender does not. The lender's "profit" or returns are restricted to the interest payments that are specified in the loan agreement. The lender may benefit indirectly if the borrower's business grows over time and its financing needs increase.

A BANK'S BUSINESS

Banks, like aquaculture farms, are businesses. They operate for the purpose of generating profits to distribute to their stockholders. Just as aquaculture farm businesses must account for risk, a bank business must also manage its business with respect to risk and liquidity.

Banks manage their capital resources (financial assets) in a number of different ways. First, banks must maintain a certain amount of capital in the form of reserves. Secondly, capital is used to extend loans and to make investments. Additional capital is used to provide services to its customers. Capital is also used to pay dividends to stockholders and to pay income taxes.

A bank's loan portfolio can include a variety of types of loans that may include commercial and industrial, agricultural, consumer, and real estate loans. Extending loans results in greater profits, but lending also entails the greatest risk and reduces bank liquidity.

Commercial banks must maintain a high degree of liquidity in their asset structure. The loan portfolio is typically a major part of the asset structure of a bank.

EQUITY VERSUS DEBT

Not all capital for aquaculture businesses is obtained from traditional lenders. Some agricultural operations also obtain capital in the form of outside equity. Equity capital in agriculture has been obtained mostly from retained earnings from the business, inheritances, and gifts. However, equity capital can be obtained in other ways. Partnerships, joint ventures, and family corporations can be used to pool capital and provide a source of equity capital. Some equity capital can be available from outside investors seeking to shelter high nonfarm income from taxes.

Limited partnerships have been used as a way to provide equity capital in some types of agricultural operations, particularly cattle feedlot operations. However, such a source of capital can be withdrawn as easily as it is made available. In such an arrangement, the investor incurs risk associated with limited opportunities to sell out their financial interest in the farm business.

Because new equity must come from either the current owners or new investors, it is important to understand the costs involved. Current owners can provide equity capital through investing the earnings or profits from the business back into the business or by investing capital earned through other sources. Outside equity capital can be obtained by adding partners to the business.

Investing the profits from the business back into the business also represents a cost to the business, despite the fact that there is no definite repayment schedule. The cost of reinvesting retained earnings is a type of opportunity cost that has been referred to as the required rate of return to compensate for the investment options that are forfeited. In addition, if the risk associated with the aquaculture business is greater than that of other investment opportunities, the opportunity cost is greater.

The cost of obtaining outside capital for the farm is related to the rate of return that would be necessary to attract it. In other words, the rate of return from the aquaculture business will need to exceed that of other types of investment opportunities to attract investors.

Equity capital frequently ends up being more expensive (after taxes) than debt capital. This is particularly so when the cost of retained earnings is considered, as it should be in a complete analysis.

The costs of debt capital include the interest paid along with any loan settlement costs. Loan settlement costs can include points charged at the initiation of the loan, any transactions fees, fees for credit reports, and any insurance premiums that are required. This cost is the difference between the funds that are received by the borrower and the total funds used to close out the loan.

LENDER PREFERENCES

Different types of lenders will have different preferences for various types of lending terms. High-risk loans are often less preferred and will deplete credit rapidly. High-risk loans frequently have low loan limits to restrict the overall extent of losses. High management ability is also preferred by lenders. Highly skilled and experienced management increases the likelihood of repaying loans in a timely manner.

Lenders tend to prefer loans that are self-liquidating and that generate assets. Self-liquidating loans are those for which the loan is paid off with the income generated by the equipment or other uses of the capital before the new piece of equipment wears out. These are considered to be loans with lower levels of risk. If the loan matures earlier than can be paid from the income generated by the asset acquired by the capital, income from other sources may need to be used to make the payments. Crop inputs such as fingerlings and fertilizer typically mature when the foodfish are sold, but machinery and equipment may have loans that mature sooner than the number of years over which the piece of equipment will be used.

An asset-generating loan is one used to acquire capital goods such as equipment, ponds, or land that result in an increased value of assets in the aquaculture business. As the value of assets in the business increases, these assets can be used as collateral in the future.

Lenders consider loans that are both asset-generating and self-liquidating as most preferable. Those that are asset-generating but not self-liquidating and those that are self-liquidating but do not generate assets are of medium preference by lenders. Those that are neither self-liquidating nor asset-generating are not preferred by lenders and are much less likely to be approved.

The lender may prefer a loan length that is shorter than the economic life expectancy of the asset being financed. On the other hand, the borrower might prefer a longer length of the loan. Open discussion may be necessary to work out an acceptable loan agreement.

EXPECTATIONS OF BANK REGULATORS ON LENDING PRACTICES

Banks are regulated in different manners and by different agencies depending upon the type of charter under which they operate. State-chartered banks are regulated by individual state banking commissions whereas federally chartered banks are regulated by the Federal Deposit Insurance Corporation. Regulatory agencies and the laws that govern them were created to ensure a safe and stable financial sector. Bank regulations specify the extent of capital reserves, lending limits, and deposit insurance that each bank must maintain. The relevant regulatory agencies have authority to audit and conduct examinations of banks to ensure compliance with all relevant regulations. Those who conduct the audits are termed as examiners. Bank examiners have guidelines for the adequacy of capital, liquidity, and asset quality that are used each time the examiner inspects the bank.

Each bank is required to maintain certain levels of reserves to ensure that the bank is operating with adequate liquidity. While the level of reserves varies due to their charter and the relevant regulations, required reserves frequently are approximately 3% of transaction balances. The Federal Deposit Insurance Corporation Improvement Act of 1991 set standards for the adequacy of a bank's capital. This guideline differentiates between banks with different types of asset risk and divides a bank's capital into different categories. Required capital for each bank is based on percentages for each type of capital as well as total capital.

When a loan is not viable and the lender is forced to declare bankruptcy, the bank risks losing some portion of the principal that had been loaned out. Banks must cover this portion by transferring reserves to cover the loss. When reserves are transferred, the dollar amount that the bank can lend is also reduced. Thus, absorbing loan losses reduces the volume of business that can be conducted by the bank because it loses reserves as a consequence.

The amount that a bank can lend to any one individual varies with the amount of equity capital owned by the bank. Banks with national charters have loan limits that must be less than 15% of the bank's net worth unless the loan is secured by collateral that can be marketed readily. Loan limits of state-chartered banks are variable based on state regulations.

Many agricultural banks are small, rural banks. Their small size is reflected in lower amounts of reserves. Thus, agricultural banks frequently find it more difficult to grant large agricultural loans. Some small agricultural banks develop relationships with other lenders to meet the lending requirements of larger customers.

FINANCIAL CHARACTERISTICS OF BORROWERS AND SAVERS

The need to borrow capital is related to the level and rate of savings of the farm owner and manager. Different reasons exist for savings. Seasons of the year in which farm sales are prevalent result in positive cash flows that can be a source of savings. Savings can be used for investments that produce an adequate financial yield at an acceptable level of risk. The decision to save earnings from the farm can be used to develop cash reserves that will be available to the farm business during times of adverse conditions.

Some individuals who save do so to accumulate wealth. Accumulated wealth can be used to generate higher earnings in later years.

Borrowing by farmers has increased over time as the total investment per farm has increased. In particular, investment in farm cooperatives, real estate, and equipment has increased substantially. Farmers have increasingly turned to debt to finance business growth. Overall, the size of the loans and the size of the repayments required for farmers have increased over time.

For farmers to borrow funds, individuals who save funds must be willing to allow them to be used for farm loans. Financial intermediaries, such as traditional lenders and federal agencies have developed programs to provide some guarantees that reduce the risk exposure of savers' investment funds to the risks inherent in farm lending. These guarantees include those from the Federal Deposit Insurance Corporation, the Federal Housing Administration, Small Business Administration, Veterans Administration loans, and the Farm Credit System. Financial intermediaries can spread the risk of loans over broader areas and numbers of borrowers. The federal Farm Credit System, for example, can sell securities in national financial markets.

SOURCES OF LOAN FUNDS

There are a number of different sources of loan funds that can be used by aquaculture businesses. The major source of loan funds for agriculture is from commercial banks. Many of those are rural agricultural banks. Commercial banks are a major source of operating capital as well as some real estate loans. Smaller banks, without the reserves that enable them to grant loans to larger farms, can develop an arrangement for a correspondent bank to supply loans to larger customers.

The Farm Credit System has been a major source of capital for aquaculture in the United States. It obtains funds through the sale of bonds and notes. Four regional Farm Credit Banks and one Agricultural Credit Bank provide loan funds to local associations. These loans can be used for noncurrent loans for livestock, machinery, buildings, and land. Some operating capital loans are also made by the Farm Credit System.

The Farm Service Agency of the U.S. Department of Agriculture has been an important player in credit for U.S. aquaculture by providing guarantees to commercial banks for agricultural loans. These guarantees have encouraged private banks to grant loans to fish farmers. With a guaranteed loan, the FSA guarantees up to 95% repayment if the borrower defaults on the loan.

Farm supply companies have also been a source of operating capital for fish farms in the United States. Some of these types of operating loans are tied to purchases of feed or to contracts to supply fish to a processing plant.

ESTABLISHING AND DEVELOPING CREDIT

Why should a lender approve a loan application? A bank is not a charitable organization, but a for-profit business. The burden to prove that the borrower will repay the loan with interest in a timely manner is on the prospective borrower. Individuals that have a demonstrated history of timely repayment of loans will likely find it easier to obtain credit, borrow greater amounts of capital, and perhaps obtain more favorable terms of lending.

Key factors involved in establishing and developing credit include the character of the individual, demonstrated management ability, financial position as evidenced by the balance sheet, repayment capacity, the purpose of the loan, and the amount of collateral available to secure the loan. Banks weight these various factors differently, but most banks consider all the above factors to one degree or another.

The personal integrity of the prospective borrower is an important characteristic. A reputation as an honest businessperson will help to convince the lender that the estimated values in the business loan proposal are accurate.

Aquaculture businesses are management-intensive and require skilled management. Clear evidence of management experience of either the owner or the hired manager is key to a successful loan application. Established farmers will have records to demonstrate their management expertise.

The balance sheet indicates the financial strength or weakness of the borrower. Some lenders view the balance sheet as the most important financial statement because it shows the degree of leveraging of the business and the availability of collateral to use to secure new loans. The financial position of the business also indicates whether the net worth of the business has grown over time. Steady growth of the business reflects positively on the management and financial position of the business.

The lender is concerned primarily with repayment of the loan. Thus, the borrower must demonstrate adequate liquidity and a favorable cash position. A detailed cash flow budget will demonstrate clearly the ability of the business to repay new loans that are proposed.

Self-liquidating loans frequently are financed more easily than other types of loans. This is because the loan can be repaid from the sale of the crop or livestock financed by the loan. Loans used to acquire capital assets may require additional collateral because these assets are not typically sold themselves to generate revenue.

Land is one of the more frequently used types of collateral for aquaculture businesses. Buildings, equipment, and, occasionally inventory of livestock can also be used as collateral. Award of the loan should be based on clear evidence that the loan payments can be made from revenue from the farm rather than collateral. However, lenders ask for collateral to cover the risk of unexpected adverse conditions that may result in default of the loan. In aquaculture, personal assets such as a home have been used as collateral.

BORROWING BEHAVIOR

The capacity to borrow funds can be turned into borrowing behavior or it can be held as a reserve for future borrowing as needed. The decision as to whether to borrow the maximum that a lender's criteria will allow or to borrow some lower amount and reserve the capacity to borrow for some other purpose in the future should be based on the relative profitability of the alternatives. The level that constitutes the maximum an individual can borrow is based on the credit worthiness of the borrower.

The value to an individual of holding credit in reserve rather than borrowing the maximum will vary with the level of risk aversion of the individual and the strategies undertaken on the farm to manage risk (see Chapter 16 for a more complete discussion of risk and Chapter 8 for more discussion on managing risk on aquaculture farms).

CREDIT CAPACITY AND CREDIT RESERVES

Credit reserves refer to additional borrowing that a farmer likely would be able to obtain if desired. At any point in time, one or more lenders may be willing to loan additional funds to a firm or individual borrower to finance transactions. Credit reserves exist on farms that have not borrowed the maximum amount calculated by the lender when determining the credit worthiness of the individual. The credit worthiness will vary with the profitability of the business, the individual's balance sheet, net worth, and assets available for use as collateral, as well as the individual's demonstrated management skill. The maximum amount that can be borrowed will also fluctuate with overall trends of national and international economies.

The availability of credit reserves is particularly important when the farm faces adverse conditions. Economic downturns may result in lower prices and revenue for the farm or conditions may occur that result in increases in prices of key inputs such as feed. During these times of economic adversity, the cash flow of the farm business may suffer and credit reserves may be accessed to increase borrowing of operating capital. Alternatively, credit reserves may be used to refinance loans, defer loan payments, or carry over or extend loan maturity dates. Credit reserves can play an important role in managing the debt on the farm, particularly during times of adverse financial conditions.

Holding credit reserves means that the farm is not using all its potential for borrowing. Credit is an excellent source of liquidity and, if holding credit reserves results in levels of liquidity that are lower than optimal, the wise decision may be to increase liquidity by taking advantage of the credit reserves and borrowing additional funds. However, maintaining only low levels of credit reserves increases financial risk because there is little to draw upon during difficult economic times. Moreover, greater control of the business is exercised by lenders when the business has a high degree of debt.

DETERMINANTS OF CREDIT

The availability of credit is a complex phenomenon that is influenced by characteristics of the particular lender as well as trends in financial markets. Lenders seek profit and prefer higher rates of return earned over a shorter period because risks typically are less and more clear over shorter time periods. Lenders must also maintain a high degree of liquidity in their business. This may result in banks preferring shorter loans that may be less than the economic life of the asset.

Current trends in the macroeconomy will influence the availability of credit. Current national fiscal and monetary policies and international trade conditions will affect credit conditions. The financial condition of the marketplace will affect the availability and the cost of credit. Monetary and fiscal policies, inflation rates, the overall national money supply, and the aggregate national economic performance will have an effect on the availability of credit.

Financial conditions of the individual bank will affect how readily it will make loan funds available. The profitability of the bank, its loan loss rate, the rate of loan delinquency, and its size will all affect the amount of capital to be loaned. Each bank will also be affected by its size, costs, and the particular legal and regulatory environment within which it operates. The bank's personnel and board of directors will have an impact on the policies established by each bank.

Finally, financial conditions of the borrower's business will affect the extent to which loan funds will be directed toward it. Overall profitability of the industry, effects of international trade conditions, key input price trends, and other factors will affect the repayment capacity of the industry. The aggregate profit prospects for farm commodities as a whole, changes in the value of farmland and resources, international trade conditions that specifically affect the borrower's business, and how government farm programs are expected to affect the business will impact the availability of credit.

CREDIT EVALUATION

Lenders evaluate the credit level of prospective borrowers in various ways. Some are mostly subjective and informal while others have more formal credit scoring methods. The purpose of a credit evaluation is to identify the borrower's credit worthiness and by doing so to measure the credit risk. Credit evaluation is used to compare good and poor credit risks to decide which loans to approve and which to reject. In some cases, interest rates charged may depend upon the credit evaluation or if any specific monitoring system needs to be put in place.

The credit worthiness of the individual borrower is the lender's evaluation of the profitability and the risks associated with lending to an individual borrower. The evaluation of a borrower's credit worthiness is based on the evidence presented by the borrower to assure the lender that lending risks will be minimal and that debt servicing will meet the terms of the loan contract.

Credit scoring has been used commonly in consumer lending. To develop a credit score, key characteristics are identified and then assigned a weighting factor. The weighted averages of these characteristics are then summed to generate a credit score. The credit characteristics frequently include the estimated profitability of the business, the projected liquidity of the business (current ratio), solvency of the business (debt/equity ratio), the collateral position (ratio of the maximum loan balance to assets pledged as security), and loan repayment capacity (debt-servicing ratio). The credit score will change over time as the values of each characteristic change for the business. The assets (both quantity and quality) available for security for the loan, repayment and income expectations, and other financial management practices of the borrower will affect one's credit profile and score.

SOLVENCY

Solvency of a business is a key measure of the financial strength or weakness of the business. A solvent business is one that, if the business were to be sold, the value of the assets sold would exceed the value of the liabilities, and all debts would be paid in full.

The overall financial structure of the business affects the level of financial risk involved. There are various measures available to monitor the use of capital and credit in the farm business. Chapter 4 presents details of how to calculate a variety of different financial ratios that provide measures of the solvency of the business. One of the more important measures is the debt to equity ratio that is calculated from the balance sheet. Coverage ratios, such as the interest coverage ratio, are calculated from the income statement that relates the returns to assets from the business to the debt obligations.

FINANCING CURRENT ASSETS

The majority of current assets in aquaculture are operating inputs. These inputs of feed, fuel, labor, etc., typically are not carried over into the next year. Operating inputs typically do not constitute collateral because they are not easily sold. Thus, the borrower frequently must use other sources of collateral for operating inputs. The one exception in many forms of agriculture is that of livestock. Inventories of feeder livestock can be used as collateral for operating loans; however, swimming inventory of fish is rarely considered an acceptable source of collateral for aquaculture operating loans.

Financing current assets is a high-priority type of financing for a fish farmer. It is essential to have adequate capital to feed fish, and the capital must be approved and in hand when the feeding season is at hand. The most critical terms of lending for an operating loan are the loan maturity and the repayment schedule. Collateral and interest rates must also be planned carefully.

To be a self-liquidating loan, the operating loan will need to be repaid from sales of the products that were financed from the operating loan. Payments on operating loans typically are made with each shipment of fish delivered to the processing plant.

FINANCING DEPRECIABLE ASSETS

Depreciable assets primarily include equipment such as aerators, vehicles, feed wagons, and other types of equipment used on fish farms. Pieces of equipment are used over a number of years and are replaced either as they wear out or as new technology becomes available. Typically, replacement of equipment can be planned over a period of time to set up the financing in a way that the payments are scheduled when cash is readily available.

For a loan to acquire a depreciable asset to be self-liquidating, the terms of lending (loan maturity and repayment schedule) should match the revenue that is produced by the asset. Some industries schedule equipment loans to match the depreciation of the assets. Most equipment loans have a loan maturity of 5–7 years.

FINANCING FIXED ASSETS

Fixed assets in aquaculture include land, ponds, wells, office buildings, hatcheries, and grading sheds. Of these, land is the only nondepreciable asset. Thus, land is not used up in the production process. Real estate loans can have maturities that range from 15 to 30 years. Payments of farm mortgages and other real estate loans are fixed payments. Fish farms that are highly leveraged will have high levels of fixed loan payments that must be met. The need to make these payments reduces management flexibility, particularly during times when fish prices are low, or feed prices are high. During these times, it may be more profitable to reduce feeding and stocking rates; however, farms that are highly leveraged will need to continue to stock and feed at high rates to produce yields that will enable them to continue to meet the fixed loan repayments.

SOURCES FOR LOAN REPAYMENT

Real estate loans typically are repaid from retained earnings of the aquaculture business. A variety of sources can be drawn upon as sources of capital for loan repayment. Ideally, sources of funds for an operating loan would come from cash that has been set aside to pay operating expenses. Equipment loans would be paid from cash set aside to account for depreciation while retained earnings (profits less income taxes and family withdrawals) frequently are used as the major source of funds to repay real estate loans.

CREDIT PROBLEMS FACED BY SMALL-SCALE AQUACULTURE GROWERS IN DEVELOPING COUNTRIES

Small-scale aquaculture growers in developing countries face special problems related to acquiring sufficient credit to finance their businesses. Investment capital generally is a major obstacle to development of aquaculture businesses in developing countries (Hishamunda and Manning 2002). There are numerous factors that exacerbate the credit problems faced by small-scale aquaculture growers in developing countries.

Lack of collateral typically is one of the first obstacles to gaining access to credit. Small-scale growers frequently lack the level and types of assets considered as acceptable collateral by banks. Land, homes, and other types of assets are preferred sources of collateral that often are not available for use by small-scale growers.

Many developing countries contend with excessively high interest rates on loans. Interest rates of 40–60% are not uncommon. Businesses that can successfully pay off loans at such high rates are very rare. The high interest rates can stem from unstable national economies that result in a risky financial market or restrictive monetary policies and supply of money that

result in very high costs of capital. High rates of default on loans also put upward pressure on interest rates.

Many small-scale growers also lack the knowledge and skill required to prepare a proper business plan and loan proposal. Required information may not be readily available and the expertise to prepare financial statements properly may not be available.

Group lending to small-scale growers has proven to be an innovative approach to overcome some of their obstacles. The Grameen Bank pioneered these efforts in the mid-1970s in Bangladesh (Grameen 2001). It loans to the poorest of the poor, to borrowers without collateral. By 2001, the Grameen Bank had grown to 2.38 million borrowers lending $31 million a week in very small loans, with a repayment rate of 90%. The approach is based on peer pressure, small weekly repayments, and personal contact with borrowers. The concept has spread to 58 countries.

In addition to group lending, village banks and solidarity groups have been used to counter the problems of collateral. The village bank receives one loan that is shared with all members. It must be repaid in full before it receives another.

RECORD-KEEPING

The aquaculture farm business should keep records of each proposal for a loan and the financial statements used to acquire the loan. Once the loan is obtained, records should include the history of debt servicing that includes each payment made, the amount of the payment made on the principal, and the amount paid for interest. These records will assist the farmer when preparing other financial statements such as the balance sheet, income statement, and the cash flow budget.

PRACTICAL APPLICATION

This practical application is based on a 431-acre catfish farm that is a startup farm that borrowed 65% of its real estate and equipment capital to enter the catfish business. In doing so, the pro forma income statement showed that the farm was profitable, but profit levels were relatively low. There was substantial financial risk generated by the low profit levels because any adverse economic conditions such as unexpectedly low prices or disease problems would result in losses. The losses would also create cash flow problems that would make it difficult to meet debt-servicing requirements.

In consultation with extension personnel and reviewing the latest research results, it became apparent that this farm had entered business with too few aerators. Research had shown that by doubling the aeration capacity of the farm from 1 to 2 hp/acre total yield of catfish would increase. This would result in increased revenues and profits. The increased revenues would further improve cash flow of the business and reduce the risk associated with meeting debt-servicing obligations.

Acquiring an additional 10-hp aerator for each of the farm's 43 ponds would require an additional equipment loan of $150,500. To apply for the loan would require an analysis of the financial structure and position of the business.

The 431-acre farm's balance sheet showed adequate, although low, liquidity with a current ratio of 1.13 (Table 5.1). Working capital was $21,275, low for this

Table 5.1. Catfish Farm Balance Sheet, 431-acre Farm, End-of-Year.

Category	Value
Assets	
1. Current assets	
Cash on deposit	$31,000
Fish inventory	$150,000
Total current assets	$181,000
2. Noncurrent assets	
Equipment	$606,035
Ponds	$602,538
Wells	$81,000
Land	354,282
Total noncurrent assets	$1,643,855
3. Total assets	$1,824,855
Liabilities	
4. Current liabilities	
Payments on debt due and payable over next year	
Equipment	$80,914
Real estate	$78,811
Total current liabilities	$159,725
5. Noncurrent liabilities	
Equipment loan	$393,923
Real estate loan	$674,583
Total noncurrent liabilities	$1,068,506
6. Total liabilities	$1,228,231
7. Net worth	$596,624
Current ratio (1/4)	1.13
Working capital (1–4)	$21,275
Equity/asset ratio (7/3)	0.33
Debt/asset ratio (6/3)	0.67
Debt/equity ratio (6/7)	2.06
Debt structure (4/6)	0.13

size of farm. However, the debt/asset ratio was 0.67 and the debt/equity ratio was 2.06. The equity/asset ratio was 0.33. The debt structure was 0.13, indicating that the majority of the debt was in the long-term debt associated with the real estate and equipment loans. In conclusion, the farm's balance sheet does not show adequate equity to support additional borrowing. Long-term borrowing already shows levels that would be deemed unacceptable to the lender. Increasing debt-servicing requirements would decrease working capital and increase financial risk to unacceptable levels.

The owner of the farm has a relative who has been in the catfish business for many years. This relative constructed his own ponds over the previous 20 years and has no outstanding debt related to its real estate. The relative's farm also owns all the equipment on his farm. The equipment was financed with retained earnings from the farm business. Thus, the debt obligations of the relative's business are low. In conversations over the difficulties of developing a fish farming business with high levels of debt financing, the two farmers decide to merge their farming businesses into one business. They believe that they will achieve economies of scale and improve overall production efficiencies. However, both farms, once merged, will need to increase aeration levels to 2 hp/acre. This will require $301,000 in capital that the owners believe will need to come from debt capital.

Table 5.2 presents the revised balance sheet for the now-merged farm business. The new equipment loan of $301,000 has been added to the noncurrent liabilities and the annual payment for the new loan has been added to current liabilities. Noncurrent assets have also been increased by $301,000 to account for the new aerators that have been acquired.

The overall financial position of the merged business, even with the new equipment loan for the aerators, has improved considerably. The current ratio has improved to 1.63 and working capital has increased to $140,448. These liquidity measures show a reduction in financial risk associated with the business' cash flow position. The equity/asset ratio has increased to 0.60, the debt/asset ratio has decreased to 0.40, and the debt/equity ratio has decreased to 0.67. The debt structure shows little change; much of the debt still remains in longer-term debt. However, the liquidity has improved substantially and the longer-term financial indicators are within acceptable ranges. These ranges will likely be viewed as acceptable levels of risk to the lender. Thus, the merger has created a financial position for the business that is likelier to result in

Table 5.2. Catfish Farm Balance Sheet, 862-acre Merged Farm, Including New Equipment Loan to Expand Aeration Rates to 2 hp/acre, End-of-Year.

Category	Value
Assets	
1. Current assets	
Cash on deposit	$62,000
Fish inventory	$300,000
Total current assets	$362,000
2. Noncurrent assets	
Equipment	$1,513,070
Ponds	$1,205,076
Wells	$162,000
Land	708,564
Total noncurrent assets	$3,588,710
3. Total assets	$3,950,710
Liabilities	
4. Current liabilities	
Payments on debt due and payable over next year	
Equipment	$142,741
Real estate	$78,811
Total current liabilities	$221,552
5. Noncurrent liabilities	
Equipment loan	$694,923
Real estate loan	$674,583
Total noncurrent liabilities	$1,369,506
6. Total liabilities	$1,591,058
7. Net worth	$2,359,652
Current ratio (1/4)	1.63
Working capital (1–4)	$140,448
Equity/asset ratio (7/3)	0.60
Debt/asset ratio (6/3)	0.40
Debt/equity ratio (6/7)	0.67
Debt structure (4/6)	0.14

favorable consideration by a lender. The addition of the aerators should result in greater profits and revenue for the business.

OTHER APPLICATIONS IN AQUACULTURE

Analysis of the financing requirements and repayment feasibility is vital to adequate planning for successful aquaculture businesses of all types. Pomeroy et al. (2004) analyzed the financial feasibility of aquaculture cage businesses for two species of grouper. The analysis showed that, overall, cage culture of grouper is financially feasible. However, financing of the businesses is likely to be difficult, particularly on a small scale. This is because the capital required for the various stages of production (broodstock, hatchery, or vertically integrated systems) exceeds the borrowing capacity of small-scale producers. Loans would likely be necessary for growout production, particularly for small-scale farmers. However, the analysis of cash flow demonstrated that the loans could be repaid within one year of production.

SUMMARY

This chapter has described lending primarily from the perspective of the lender. Commercial banks are for-profit companies that seek to provide capital to others for a fee (interest rate). Success in obtaining loans is higher when the prospective borrower understands the perspective of the lender and requirements of the bank. Each bank is limited in the amount of loans that can be made and in the size of individual loans. These limits are related to the reserves maintained by the bank. Loan defaults require the banks to spend reserves to cover the losses, thus reducing the overall amount of reserves and the amount of loans that can be made against those reserves.

Banks use various methods to evaluate the credit worthiness and credit capacity of prospective borrowers. However, whether an individual bank uses a formal or more subjective process of evaluating credit, the potential profitability, the repayment capacity, the amount of assets available for use as collateral, the management expertise of the owner, and the owner's previous history in business will typically influence the decision of whether to grant the loan.

Ideal loans are self-liquidating and asset-generating. Current, depreciable, and fixed assets have different characteristics that require various adaptations to be self-liquidating. Moreover, limits imposed by bank examiners on lenders may result in a lender encouraging borrowers to agree to lending terms that result in a loan that is not self-liquidating. Careful analysis is required of the various alternatives associated not only with terms of lending, but with the type of lender, with definitions of credit capacity, and with the types of assets to be financed.

REVIEW QUESTIONS

1. Contrast the difference in perspective of the lender from that of the borrower.

2. What is a self-liquidating loan? Provide an aquaculture example.

3. What is an asset-generating loan? Provide an example from aquaculture.

4. Describe differences in characteristics of borrowers and savers.

5. What are credit capacity and credit reserves? Describe why these are important and under what conditions they may vary.

6. List several determinants of credit.

7. Contrast several different types of major lenders to aquaculture in terms of their charters, limitations on lending, and how different types of lenders are regulated.

8. Contrast the differences in financing current, depreciable, and fixed assets.

9. Describe several sources of loan repayment funds.

10. Describe how a new, young farmer can establish and develop credit for the business.

REFERENCES

Grameen. 2001. Grameen Fund. Found at http://www.grameen-info.org.

Hishamunda, Nathanael and Manning, Peter. 2002. *Promotion of Sustainable Commercial Aquaculture in Sub-Saharan Africa*. Volume 2. Investment and Economic Feasibility. FAO Fisheries Technical Paper 408/2. FAO, Rome, Italy.

Pomeroy, Robert S., Agbayani, Romeo, Duray Maretta N., Toledo, Joebert D., and Quinitio, Gerald F. 2004. The financial feasibility of small-scale grouper aquaculture in the Philippines. *Aquaculture Economics and Management* 8(1/2): 61–83.

6
Managing Cash Flow

INTRODUCTION

Cash flow problems result in more aquaculture businesses' failure than any other problems. Managing cash flow adequately and properly can make a difference between the success and failure of a business. Liquidity is a financial concept that is linked closely with cash flow. It refers to whether the business will have the cash when needed to make its payments. Cash flow is a strong measure of liquidity.

A balance sheet can provide some limited indication of liquidity in the farm business through calculating the current ratio and level of working capital. However, the balance sheet summarizes the values of all current assets and liabilities over the course of the year. If loan payments are due in the early part of the year but fish are not sold until later in the year, there will be a liquidity problem that will not be evident from the balance sheet. An analysis that accounts explicitly for the timing of receipt of cash and the expected payments is needed. This chapter compares profitability with liquidity, identifies common types of cash flow problems in aquaculture, discusses the use of cash flow statements and budgets to manage cash flow, and outlines the need for record-keeping.

LIQUIDITY

Farms that are profitable and viable businesses can also have occasional cash flow problems. Adequate liquidity is essential to an aquaculture business because it is used as a source of cash to meet demands for cash payment, as a safety margin for adverse financial conditions, and as a source of capital for investment in other alternatives. It measures the capacity of the business to produce enough cash to make financial payments when they are due. Positive cash flow is necessary to meet operating expenses and to meet the cash needs of the farmer and his or her family.

PROFITABILITY VERSUS LIQUIDITY

Many business owners believe that the success of the business is determined solely by its profitability. Successful businesses must be profitable, but inadequate cash flow can result in business failure. In fact, cash flow problems likely cause more severe financial problems than overall issues of profitability.

Cash flow problems are problems of liquidity, not profitability. Because cash flow measures the timing of receipt of revenue and of expenses, it is a prime determinant of liquidity. However, cash flow budgets include only cash items; noncash items such as depreciation are not included. Cash flow budgets do not measure profitability.

COMMON SOURCES OF CASH FLOW PROBLEMS IN AQUACULTURE BUSINESSES

There are a number of different types of cash flow problems that cause difficulty for aquaculture businesses. A common cause of liquidity problems involves the timing of product sales. Aquaculture businesses with seasonal production, that is harvests occur in only one period of the year, can face liquidity problems because revenue is not available throughout the year to cover expenses when due. In a similar manner, the timing of expenditures can result in liquidity problems. Borrowing and loan payments, leverage and use of credit, and business expansion also contribute to liquidity problems. Moreover, farm planning can be hampered if the projections of sales, expenditures, and payment schedules are forecasted incorrectly.

Aquaculture Economics and Financing: Management and Analysis, Carole R. Engle, © 2010 Carole R. Engle.

TIMING OF PRODUCT SALES

Product sales that occur several months later than when production expenses are incurred are a common source of cash flow problems. For example, a salmon crop is stocked at one point in time, typically, with one size of fish. The fish are fed, kept free of disease, and monitored until they reach market size, often 9–15 months after stocking (Bjorndal 1990; Grisdale-Holland et al. 2004). The salmon farm must pay all bills associated with feeding, disease control and treatment, and cost of the smolts, labor, and other expenses, but does not receive revenue until the salmon are sold. This creates a clear cash flow problem that must be accounted for when planning the farm business. Production expenses must be covered for the time period until sales generate revenue.

TIMING OF EXPENDITURES

The timing of expenditures can also create cash flow problems. If a shrimp farm is planning to purchase new aerators across the farm and prepares to make the purchase at the beginning of a production cycle, the investment will result in a substantial outlay of cash at the beginning of the production cycle, before shrimp have been produced and sold. Thus, there is no cash revenue to match the cash outlay at that time of the year. If shrimp prices fall or feed prices increase, the farm may no longer have a surplus from the previous year to cover these adverse conditions because the surplus funds were spent on the new aerators. Cash flow would be improved by waiting to purchase the new aerators until the end of the production cycle, using revenue from shrimp sales.

BORROWING AND LOAN PAYMENTS

The characteristics and terms of loans on the farm can have a strong effect on liquidity. The interest rates charged and the terms of payment (such as the length of the loan) will affect the amount of the periodic payment. Longer-term loans will have lower payments while shorter-term loans will have higher payments. Restructuring debt can be used to improve liquidity by reducing the annual payment by extending the length of the loan. However, restructured loans will result in greater total interest paid over the life of the loan. Serious liquidity problems may require restructuring of loans, but this must be approached carefully.

The *structure of the debt* in the business refers to how the debt is apportioned among short- and long-term loans. A high percentage of debt in short-term current liabilities can be another cause of liquidity problems. Refinancing can be a way to move the debt to longer term by amortizing the repayment over a longer period. The general rule for structuring loans is that the repayment terms should match the useful life of the assets. For example, if a piece of equipment is expected to last for 7 years, then the loan should be structured to be repaid within 7 years.

Scheduled debt repayments can result in cash flow problems. If payments on major loans, such as the real estate or equipment loans on the farm, are scheduled to be paid in the early part of the production season, cash flow problems may result. Cash flow is improved by scheduling payments on loans to coincide with the time of the year that is likeliest to have product sales that generate revenue at the time the payments are due. A repayment schedule that matches the timing of revenue and cash flow in the business can also reduce interest expenses.

LEVERAGE AND USE OF CREDIT

The financial risk of increased borrowing can result in liquidity and cash flow problems. Financial analysts often talk about the trade-off between leverage and liquidity. The term *leverage* refers to the proportion of borrowed to equity capital. The higher the degree of borrowed capital relative to the amount of capital owned by the farm, the more highly leveraged the business. Thus, leverage increases with the debt/asset ratio.

Greater amounts of borrowing can deplete credit reserves. Credit reserves are one of the management alternatives that can be used to prepare and plan for adverse cash deficits. Reductions in the amount of credit reserves reduce the business' ability to withstand downside risks. Increased borrowing also increases the repayment obligations in the future.

BUSINESS EXPANSION

Liquidity is important as a source of funds to take advantage of new investment opportunities. This may include having capital to purchase additional land as it is put on the market, purchasing equipment at good prices at auctions and farm sales, or funds to use as a hedge against adverse commodity prices in the marketplace.

Business expansion can result in cash flow and liquidity problems. For example, the farm may decide to add a hatchery to the farm because careful analysis has shown that vertically integrating the food fish growout operation with a hatchery and fingerling enterprise will result in greater profits. The hatchery will require a new building, new ponds, and development of broodstock for spawning. Selecting broodstock from current stocks of fish may decrease the weight of fish available to be sold. Putting ponds into fingerling production reduces the acreage of water available for food fish growout. Construction of the hatchery building and the ponds also requires cash outlays. The combination of the reduced revenue from diverting fish to broodstock and reduced acreage and the cash outlays to construct the hatchery building may result in a temporary cash deficit. Over time, as the new ponds are built and stocks of fish are built back up to former levels, the cash deficit will disappear and the increased profits will be realized.

INCORRECT ANTICIPATION AND OVERLY OPTIMISTIC ESTIMATION OF COSTS, YIELDS, AND REVENUES

The accuracy of cash flow projections and budgets will depend on the accuracy of the yields, prices, and costs used to make the projections. The farm may base yields on historical farm records. However, in any given year, disease problems, adverse weather conditions, or other unanticipated events may result in lower-than-anticipated yields. If yields achieved on the farm are less than those anticipated and projected, the revenues projected will also be less than anticipated and cash flow problems will likely result.

Fish prices are determined in the market and are outside the control of the majority of aquaculture producers. While close attention on price trends and cycles will generally result in more accurate estimation of prices, there are many forces that can result in unanticipated price changes. Lower prices received, as compared to those projected, will result in lower revenue than projected with cash flow deficits and problems likely resulting.

It can be difficult for many farmers to be disciplined while forecasting prices and expenditures. The temptation is strong to be overly optimistic about prices, yields, and costs. One method to guard against misleading estimates is to have a frank and objective assessment of cash flow projections developed by a third-party individual. An outside analyst can often be relied upon to inject some realism into the projections. Long-term datasets of prices and yields can also be helpful to maintain price and cost projections within realistic ranges.

NONFARM FACTORS

Many farms are managed as a family operation with family expenses and obligations mixed in with the farm's expenses and obligations. If the farm family has a sudden increase in medical expenses or tuition expenses for college, these expenses can create a cash deficit that spills over into the farm business.

STATEMENT OF CASH FLOWS

The statement of cash flows summarizes the actual cash inflow and outflow of the farm business over the course of the production year. It presents an assessment of the cash position for each month of the business. Thus, the statement of cash flows calculates the liquidity for each month. This statement should be prepared over the same fiscal year as the balance sheet and the income statement.

Table 6.1 illustrates the format of a statement of cash flows. Cash farm income and expenses are summarized in the first section (Cash flows from operating activities). The second section (Cash flows from investment activities) includes any investment revenue and expenses. The third section summarizes cash obtained from loans and the cash outflows from debt-servicing and repayment expenses. The fourth section (Cash flows from nonfarm activities) is for cash inflow and outflow from nonfarm items, while the final section balances the cash on hand. The cash flow statement does not include any noncash items. Noncash items are important in the determination of profitability and net worth, but are not included in a statement of cash flows because they are not cash items.

The example of a statement of cash flows in Table 6.1 is for a farm that has no nonfarm revenues or expenses. This farm received $604,800 in revenue from sale of fish and paid $564,433 for farm operating expenses. Net cash from operating activities was $40,367. The farm did not purchase or sell any broodstock, equipment, or land. Cash obtained from financing activities provided $173,650 from an operating loan. Principal payments on longer-term debt were $98,058. When the principal payments of the operating loan were included, the net change in cash was negative, at −$57,692. Thus, this farm does

Table 6.1. End-of-Year Statement of Cash Flows with 25% Market-Sized Fish Off-Flavor.

Item	Total
Cash flows from operating activities	
Cash received from farm operations	$604,800
Cash paid for farm operating expenses	$564,433
Net cash provided by operating activities	$40,367
Cash flows from investment activities	
Cash received from sales of:	
Broodstock	0
Equipment	0
Real estate	0
Cash paid to purchase:	
Broodstock	0
Equipment	0
Real estate	0
Net cash provided by investing activities	0
Cash flows from financing activities	
Operating loan received	$173,650
Term principal payments	$98,058
Operating debt principal payments	$173,651
Net cash from financing activities	−$98,059
Cash flows from nonfarm activities	
Cash received from nonfarm income	0
Cash paid for nonfarm expenses	0
Cash withdrawals for family living	0
Net change in cash	−$57,692

not have adequate liquidity to pay off its operating loan in full over this time period.

The operating line of credit can be determined and monitored from the statement of cash flows. Its purpose is to be certain that the business has adequate liquidity in the months expected to result in cash deficits. An operating line of credit is designed to provide cash in months of cash deficits, and to repay deficits in months of cash surpluses.

The statement of cash flows indicates the amount of the operating line of credit required for each month. The statement of cash flows also shows the amount of outstanding operating capital debt for each month of the year. Increasing amounts of operating capital debt indicate increasing problems of liquidity. A preferable position is for the amount of operating capital outstanding to decrease throughout the year, ideally, to zero. A feasible business is one in which adequate cash surpluses exist during enough months to pay off the operating capital debt accumulated during months of cash deficits. If the outstanding operating capital

does not decrease throughout the year, there is a liquidity problem and there is likely a shortage of working capital.

MANAGEMENT ALTERNATIVES AND PLANNING STRATEGIES TO OVERCOME CASH FLOW PROBLEMS

The first step to managing and planning to avoid or overcome cash flow problems is to monitor and control cash flow from the beginning. This requires planning, organizing, directing, and controlling the cash flow of the business. The last step, control, is to implement the actions that are necessary to be certain that goals and objectives of the overall business plan are met.

This process involves establishing some initial performance standards. The business performance must then be compared against the standards and corrections made as necessary.

MONITORING CASH FLOWS

Table 6.2 presents a sample of a form that could be used to monitor cash flow for the aquaculture business. The total amount of cash flow projected is entered into the first column. The amount budgeted up to the date the report is prepared is entered in the next column, while the actual cash flow to date is entered in the third column. Ideally, this report is prepared on a monthly basis for frequent monitoring of the cash flow situation. If the owner cannot schedule adequate time for monthly monitoring, quarterly monitoring would still provide some degree of control throughout the year. Excessive outflows of cash will become readily apparent with frequent monitoring, and a course of corrective action can be identified to avoid the development of more serious deficits. Monitoring cash flow only once every 6 months, or once a year, is not adequate. The earlier the diagnosis of a potential problem, the sooner the corrective action and steps can be taken. Not monitoring cash flow can result in business failure.

Year-end estimates can be revised as necessary. At the end of the year, the past year's cash flow performance can be used to develop a more realistic projection for the upcoming year. Deviations in borrowings and loan payments should be evaluated primarily based on their effects on the ending cash balance.

Measures of Repayment Capacity

There are two measures of repayment capacity that are recommended by the Farm Financial Standards Council (FFSC): (1) term debt and capital lease coverage ratio, and (2) capital replacement and term debt repayment margin. The term debt and capital lease coverage ratio is calculated with the following equation:

Term debt and capital lease coverage ratio

= cash available for term debt payments for

the past year/total of the term debt payments

due in the next year.

Term debt covers scheduled, amortized payments. The cash available for term debt payments is calculated by adding the total nonfarm income, depreciation expense, and interest paid on term debt and capital leases to net farm income from operations and subtracting out withdrawals for family living and personal income taxes. A value greater than 1.0 indicates adequate cash flow to meet the scheduled term debt payments. It is

Table 6.2. A Sample Form for Monitoring Cash Flows on Aquaculture Businesses in March.

Item	Projected annual total	Projected total to date	Actual cash flow to date
Beginning cash			
Pounds of catfish sold, lb	1,152,000	138,240	103,680
Receipts from catfish sold, $	806,400	96,768	72,576
Operating cash expenses			
Feed	278,264	13,913	13,913
Fingerlings	72,832	72,832	72,832
Labor			
Year-round, full-time	40,560	9,735	9,735
Seasonal, full-time	10,140	0	0
Plankton control	3,687	0	0
Gas, diesel fuel, and oil	18,944	2,273	2,273
Electricity	56,832	3,979	3,979
Repairs and maintenance	24,832	4,470	4,470
Bird depredation	1,600	720	720
Seining and hauling	57,600	6,912	5,184
Telephone	2,688	591	591
Office supplies	2,816	394	394
Farm insurance	6,476	1,554	1,554
Legal/accounting	1,562	406	406
Total operating expenses	578,833	117,779	116,051

similar to the current ratio calculated from the balance sheet. However, the current ratio does not include all positive and negative cash flows.

The capital replacement and term debt repayment margin is the difference between the cash available and the total term debt payments due. It is similar to the working capital measured from the balance sheet, but differs in that all cash inflows and outflows are considered.

MANAGEMENT ALTERNATIVES

Some farm management strategies result in more positive cash flows than do other types of management strategies. In catfish production, for example, many farmers use a multiple-batch system of production in which submarketable fish from one year are carried over to the second year of production with a second crop of smaller fingerlings "understocked" in the same pond. This system results in the presence of several year classes and sizes of fish, from small fingerlings to market-sized fish, in each pond at most times. Thus, whichever pond is "on-flavor" can be harvested and sold to the processing plant during any month of the year.

Very large farms can also stock different sizes of fish in different ponds and stagger production schedules to have a number of ponds with market-sized fish available at all times. However, it is difficult to manage farms to maintain only single batches at all times. A problem in any one pond, whether a power failure, a disease problem, or some other adverse condition, can have an adverse effect on the schedule across the entire farm.

Cash flow deficits can be overcome with several alternative financial management techniques. Firstly, the farm's expenses can be subsidized with off-farm income. Many farmers operate their farms as a lifestyle choice, not necessarily because of the farm's business or investment properties. Some fish farms have been in the family for generations and there are inheritance and family traditions involved in maintaining the farm business. When these reasons are strong, a fish farm family may choose to work off farm to generate the cash needed to meet the farm's cash flow problems.

A second means to meet cash deficits is to sell capital assets. This can be particularly successful if the farm is overcapitalized or if a careful analysis has indicated that downsizing may improve efficiencies and profitability. Surplus equipment sales will bring cash into the business. Timing the sales so as to generate the cash needed to overcome deficits can assist with overall cash flows. However, selling off equipment needed to operate the farm at an efficient and profitable level will have serious adverse effects on the future of the business.

Cash deficits can be overcome by borrowing additional operating capital or by restructuring the debt of the business. The decision to incur additional debt should be made after careful analysis of the balance sheet, the overall financial and debt structure of the business, and the farm's ability to absorb additional debt. Analysis of the effects on cash flow of additional interest payments on the higher operating capital debt is essential before approaching a lender to request new loans.

Crop and livestock inventories can also be sold to cover cash deficits. During times of low fish prices or high feed prices, selling off part of the fish inventory will reduce stocking densities on the farm. Lower stocking densities allow the remaining fish to grow faster. The faster growth may result in greater turnover of the fish crop. In such a case, then, the combination of revenue from selling off a portion of the crop as well as the higher turnover of the fish crop may result in improved cash flow and lower cash deficits. Older broodstock can also be sold to generate revenue and can be replaced with smaller, less expensive broodstock.

Savings accounts can be drawn upon to cover cash deficits. This is a temporary measure that can relieve the pressure of cash deficits. Clearly, if the fundamental problems that generate the cash flow problems are not corrected, the savings account will become depleted over time if used continuously.

Other sources of outside cash such as gifts or inheritances can be used to cover cash flow problems. However, use of these sources of cash is not a sustainable solution. Over time, the business must generate adequate cash flow to be feasible and viable.

FINANCIAL CONTINGENCY MEASURES

Every business should have plans in place to handle unexpected adverse conditions that may result in severe cash flow difficulties. Contingency plans can take many forms. Every family or business should have some degree of savings that can be used to meet short-term cash shortfalls. Maintaining some type of liquid reserves is one of the most important financial contingency measures because these types of reserves do not affect the overall scale or scope of the business.

The business should also maintain a reserve of credit that would allow for additional borrowing in the event of a difficult cash flow crisis. Maintaining a credit reserve requires the farm not to borrow up to its maximum borrowing limit.

Part of a contingency plan is to use cash surplus obtained in years of favorable earnings and cash inflow to prepare for future downturns. This can be done by paying off some of the debt at a faster rate when the business has a cash surplus. Paying off debt increases the credit reserve as well as reduces interest rates and payment amounts.

In times of liquidity problems, nonfarm expenditures may need to be decreased or, alternatively, off-farm income may need to be increased. Maintaining adequate insurance for crop losses, medical problems, and casualties can help to prevent the cash flow problems that can result from these types of nonfarm activities.

During times of cash shortfall, additional cash revenue can be obtained by selling assets that are less productive and less profitable than others. In some cases, assets that are sold can be leased back to maintain the size of the operation.

Relatives or other personal friends can be sources of future financing. One other option is to declare bankruptcy and to then work out a plan to repay creditors over time.

PROJECTED AND ACTUAL CASH FLOWS

Pro forma (projected) cash flow statements provide an idea of the expected cash flow position of the business expected over the course of the year. The most effective use of the pro forma cash flow budget is to use it to assess potential alternative management options to select those options that will work best for the business. Specific decisions that should be examined with a pro forma cash flow budget are decisions related to purchasing or replacing equipment, renovating ponds, and to evaluate repayment capacity of the business.

For example, projecting a cash flow budget provides a means for the farmer to change the timing of receipts from product sales or of business expenses or loan payments and scheduling. Making these changes before production starts can help avoid potential problems. However, comparing the projected and actual cash flows during the production year provides a means

to alter plans in the middle of the production cycle to head off cash flow problems.

The format for a pro forma cash flow budget is presented in Chapter 12. Development of the pro forma cash flow budget requires that a full whole-farm plan be in place. Chapter 17 presents details on developing whole-farm plans. The whole-farm plan includes a crop production plan. The crop production levels and feed requirements must be estimated. Cash receipts from the aquaculture crops must be estimated along with other income anticipated, such as interest, government payments, or insurance payments. The cash operating expenses are then estimated. Personal and nonfarm cash expenses required for family living must be estimated. Plans for any purchase or sale of capital assets such as ponds or equipment should be estimated. The amortization schedule that specifies the principal and interest payments on each loan needs to be available. Finally, the pro forma cash flow budget requires an estimate of the beginning balances of all cash accounts, including checking, savings, and other accounts. All this information is then entered into the cash flow budget template.

Table 6.2 illustrates the statement of cash flows that is used to record and monitor the actual cash flows throughout the production season. The values assigned to the projected annual total come directly from the total column for each line item on the pro forma cash flow budget. The projected total to date column adds up the projected amounts by month up to the date of the statement being prepared. Problems that exist will begin to be visible as the projected total to date column is compared side by side with the actual expenditures to date column.

The example illustrated in Table 6.2 shows that most cash expenditures matched with those projected for this particular business. However, fewer fish were sold than what had been projected. Thus, revenue and cash inflow were lower than what had been projected. Seining and hauling costs were also lower because seining and hauling costs are charged per pound of fish sold.

The timing of receipt of revenues and expenditures will frequently vary from that projected. Maintaining a record of how actual cash flow is different from what was planned for the year will make it easier to identify emerging problems. Problems will be detected more quickly and actions needed to correct the problem can be implemented more quickly. It is often easier to make more frequent, smaller corrections than to wait until the cash flow problem has developed into a major financial concern.

CASH FLOW DEVIATION REPORT

A cash flow deviation report can be prepared by comparing the actual cash flow monitored in the business with the projected cash flow budget. The report shows the amount of the deviation as well as the source (specific change in receipts or expenses that differ from what was originally projected). It also shows the magnitude of the difference. The cash flow deviation report can serve to alert the manager to substantial changes in cash flow and trigger a more complete analysis of its effects on the business.

Table 6.3 presents a cash flow deviation report prepared from the statement of cash flows in Table 6.2. The report is developed for the first quarter of the year; hence, the first quarter and year-to-date values are the same. Fish sales in the first quarter were 25% lower than that projected. Thus, cash inflow was down by 25%, −$24,192. Seining and hauling costs were also down by 25% (because these are charged on a per-pound basis). Thus, total cash outflow for this quarter was $1,728 less, providing a cash flow benefit. Overall, the net change in cash flow for the first quarter was negative, −$22,464.

The first challenge is to identify the cause of the deviation in cash flow. The second is to determine whether the problem is likely to occur again. In the example illustrated in Table 6.3, the deviation report showed that the problem was lower-than-expected fish sales during that quarter. It is then important to carefully assess whether this situation is an unusual, one-time occurrence or if it is an early warning of a larger problem. If the latter is suspected, immediate corrective action is required.

RECORD-KEEPING

Records necessary for managing cash flow in the aquaculture business include a cash flow budget that projects cash flows into the production year as well as a statement of cash flows that records the actual cash flows as they occur.

PRACTICAL APPLICATION (DIFFERENT TIMING OF EQUIPMENT PAYMENTS)

Table 6.4 demonstrates a cash flow budget for a 256-acre catfish farm. This farm has borrowed 30% of the total amount of capital needed for the business (operating, real estate, and equipment). Payments for the real estate and equipment loans are made in March (real estate loans) and April (equipment loans). The cash receipts for the business have been affected by off-flavor on the farm that has restricted sales to 75% of the total amount produced.

Table 6.4 illustrates several cash flow problems created by the reduced sales due to off-flavor of fish on the farm. Cash deficits occur in April, June, July, August, September, November, and December. These deficits require new borrowing to meet all cash expenses. The new borrowing increases the outstanding amount of operating capital across the year. The outstanding operating capital loan increased from $166,705 in January to $322,241 at the end of December.

Catfish sales from farms typically increase through the production season and more fish are available to be sold at the end of the year. However, this farm has loan payments scheduled for the earlier part of the year (March and April). Table 6.5 examines the effect of moving these loan payments to the end of the year (November for the equipment loan and December for the real estate loan). With this change, the number of months with cash deficits decreases from seven months of the year to two (November and December), the months in which the loan payments are due for the real estate and equipment loans. Moreover, across the year,

Table 6.3. Cash Flow Deviation Report.

	First quarter				Year-to-date			
			Deviation				Deviation	
	Budget	Actual	$	%	Budget	Actual	$	%
Fish sales	$96,768	$72,576	−$24,192	−25%	$96,768	$72,576	−$24,192	−25%
Seining and hauling	$6,912	$5,184	+$1,728	+25%	$6,912	$5,184	+$1,728	+25%

Net change in cash flow = −$22,464.

Table 6.4. Cash Flow Budget with 25% of Market-Sized Catfish Off-Flavor and Loan Payments in March and April.

Item	January	February	March	April	May	June	July	August	September	October	November	December	Total
Beginning cash	173,650	175,837	142,106	41,873	10,000	20,871	10,000	10,000	10,000	10,000	37,993	10,000	
Pounds of catfish sold, lb	34,560	34,560	34,560	155,520	86,400	86,400	51,840	86,400	86,400	146,880	60,480	0	864,000
Receipts from catfish sold, $	24,192	24,192	24,192	108,864	60,480	60,480	36,288	60,480	60,480	102,816	42,336	0	604,800
Total cash inflow	197,842	200,029	166,298	150,737	70,480	81,351	46,288	70,480	70,480	112,816	80,329	10,000	1,257,130
Operating cash expenses													
Feed	2,783	5,565	5,565	19,479	13,913	44,522	44,522	55,653	55,653	16,696	8,348	5,565	278,264
Fingerlings	0	36,416	36,416	0	0	0	0	0	0	0	0	0	72,832
Labor													0
Year-round, full-time	3,245	3,245	3,245	3,245	3,245	3,245	3,650	3,245	3,650	3,245	3,650	3,650	40,560
Seasonal, full-time	0	0	0	0	1,521	2,535	2,535	2,535	1,014	0	0	0	10,140
Plankton control	0	0	0	0	922	0	922	0	1,843	0	0	0	3,687
Gas, diesel fuel, and oil	1,137	568	568	568	947	1,516	2,463	2,463	3,031	3,031	1,326	1,326	18,944
Electricity	1,137	1,137	1,705	2,273	2,842	5,683	8,525	9,661	9,661	9,661	2,842	1,705	56,832
Repairs and maintenance	2,980	497	993	993	1,242	2,483	3,228	3,228	2,483	1,242	1,242	4,221	24,832
Bird depredation	240	240	240	160	80	0	0	0	0	160	240	240	1,600
Seining and hauling	1,728	1,728	1,728	7,776	4,320	4,320	2,592	4,320	4,320	7,344	3,024	0	43,200
Telephone	215	215	161	269	269	215	215	215	215	215	215	269	2,688
Office supplies	225	28	141	56	563	563	282	141	141	113	141	422	2,816
Farm insurance	518	518	518	518	518	518	583	518	583	518	583	583	6,476
Legal/accounting	156	125	125	125	125	125	125	125	125	125	125	156	1,562
Total operating expenses	14,364	50,282	51,405	35,462	30,507	65,725	69,642	82,104	82,719	42,350	21,736	18,137	564,433

(*Continued*)

Table 6.4. (*Continued*)

Item	January	February	March	April	May	June	July	August	September	October	November	December	Total
Debt servicing													
Real estate													
Interest	0	0	44,835	0	0	0	0	0	0	0	0	0	44,835
Principal	0	0	20,544	0	0	0	0	0	0	0	0	0	20,544
Subtotal	0	0	65,379	0	0	0	0	0	0	0	0	0	65,379
Equipment													
Interest	0	0	0	24,726	0	0	0	0	0	0	0	0	24,726
Principal	0	0	0	77,514	0	0	0	0	0	0	0	0	77,514
Subtotal	0	0	0	102,240	0	0	0	0	0	0	0	0	102,240
Operating													
Interest	695	695	695	3,126	1,737	1,737	1,042	1,737	1,737	2,952	1,216	0	17,369
Principal	6,946	6,946	6,946	31,257	17,365	17,365	10,419	17,365	17,365	29,521	12,156	0	173,651
Subtotal	7,641	7,641	7,641	34,383	19,102	19,102	11,461	19,102	19,102	32,473	13,372	0	191,020
Total debt servicing	7,641	7,641	73,020	136,623	19,102	19,102	11,461	19,102	19,102	32,473	115,612	65,379	526,258
Total cash outflow	22,005	57,923	124,425	172,085	49,609	84,827	81,103	101,206	101,821	74,823	137,348	83,516	
Cash available	175,837	142,106	41,873	−21,348	20,871	−3,476	−34,815	−30,726	−31,341	37,993	−57,019	−73,516	
New borrowing	0	0	0	31,348	0	13,476	44,815	40,726	41,341	0	67,019	83,516	
Ending cash balance	175,837	142,106	41,873	10,000	20,871	10,000	10,000	10,000	10,000	37,993	10,000	10,000	
Summary of debt outstanding													
Real estate													
Equipment													
Operating	166,705	159,759	152,813	152,904	135,539	131,650	166,046	189,407	213,383	183,862	238,725	322,241	

Table 6.5. Cash Flow Budget with 25% of Market-Sized Catfish Off-Flavor and Loan Payments Moved to November and December.

Item	January	February	March	April	May	June	July	August	September	October	November	December	Total
Beginning cash	173,650	175,837	142,106	107,252	146,271	157,142	132,795	87,980	47,254	5,913	33,906	10,000	
Pounds of catfish sold, lb	34,560	34,560	34,560	155,520	86,400	86,400	51,840	86,400	86,400	146,880	60,480	0	864,000
Receipts from catfish sold, $	24,192	24,192	24,192	108,864	60,480	60,480	36,288	60,480	60,480	102,816	42,336	0	604,800
Total cash inflow	197,842	200,029	166,298	216,116	206,751	217,622	169,083	148,460	107,734	108,729	76,242	10,000	1,824,906
Operating cash expenses													
Feed	2,783	5,565	5,565	19,479	13,913	44,522	44,522	55,653	55,653	16,696	8,348	5,565	278,264
Fingerlings	0	36,416	36,416	0	0	0	0	0	0	0	0	0	72,832
Labor													0
Year-round, full-time	3,245	3,245	3,245	3,245	3,245	3,245	3,650	3,245	3,650	3,245	3,650	3,650	40,560
Seasonal, full-time	0	0	0	0	1,521	2,535	2,535	2,535	1,014	0	0	0	10,140
Plankton control	0	0	0	0	922	922	922	0	1,843	0	0	0	3,687
Gas, diesel fuel, and oil	1,137	568	568	568	947	1,516	2,463	2,463	3,031	3,031	1,326	1,326	18,944
Electricity	1,137	1,137	1,705	2,273	2,842	5,683	8,525	9,661	9,661	9,661	2,842	1,705	56,832
Repairs and maintenance	2,980	497	993	993	1,242	2,483	3,228	3,228	2,483	1,242	1,242	4,221	24,832
Bird depredation	240	240	240	160	80	0	0	0	0	160	240	240	1,600
Seining and hauling	1,728	1,728	1,728	7,776	4,320	4,320	2,592	4,320	4,320	7,344	3,024	0	43,200
Telephone	215	215	161	269	269	215	215	215	215	215	215	269	2,688
Office supplies	225	28	141	56	563	563	282	141	141	113	141	422	2,816
Farm insurance	518	518	518	518	518	518	583	518	583	518	583	583	6,476
Legal/accounting	156	125	125	125	125	125	125	125	125	125	125	156	1,562
Total operating expenses	14,364	50,282	51,405	35,462	30,507	65,725	69,642	82,104	82,719	42,350	21,736	18,137	564,433

(*Continued*)

Table 6.5. (Continued)

Item	January	February	March	April	May	June	July	August	September	October	November	December	Total
Debt servicing													
Real estate													
Interest	0	0	0	0	0	0	0	0	0	0	0	44,835	44,835
Principal	0	0	0	0	0	0	0	0	0	0	0	20,544	20,544
Subtotal	0	0	0	0	0	0	0	0	0	0	0	65,379	65,379
Equipment													
Interest	0	0	0	0	0	0	0	0	0	0	24,726	0	24,726
Principal	0	0	0	0	0	0	0	0	0	0	77,514	0	77,514
Subtotal	0	0	0	0	0	0	0	0	0	0	102,240	0	102,240
Operating													
Interest	695	695	695	3,126	1,737	1,737	1,042	1,737	1,737	2,952	1,216	0	17,369
Principal	6,946	6,946	6,946	31,257	17,365	17,365	10,419	17,365	17,365	29,521	12,156	0	173,651
Subtotal	7,641	7,641	7,641	34,383	19,102	19,102	11,461	19,102	19,102	32,473	13,372	0	191,020
Total debt servicing	7,641	7,641	7,641	34,383	19,102	19,102	11,461	19,102	19,102	32,473	115,612	65,379	358,639
Total cash outflow	22,005	57,923	59,046	69,845	49,609	84,827	81,103	101,206	101,821	74,823	137,348	83,516	
Cash available	175,837	142,106	107,252	146,271	157,142	132,795	87,980	47,254	5,913	33,906	−61,106	−73,516	
New borrowing	0	0	0	0	0	0	0	0	0	0	71,106	83,516	
Ending cash balance	175,837	142,106	107,252	146,271	157,142	132,795	87,980	47,254	5,913	33,906	10,000	10,000	
Summary of debt outstanding													
Real estate													
Equipment													
Operating	166,705	159,759	152,813	121,556	104,191	86,826	76,407	59,042	41,677	12,156	71,106	154,622	

the outstanding balance on the operating capital loan goes down from \$166,705 to \$154,622. Thus, moving the payments on the real estate and equipment loans eliminates five months of cash deficits and allows the farm business to pay down a portion of its outstanding operating capital debt.

OTHER APPLICATIONS IN AQUACULTURE

Adams et al. (2001) used an annual cash flow budget as the basis for an investment analysis to assess the economic feasibility of small-scale, commercial culture of the southern bay scallop (*Argopecten irradians concentricus*). A capital replacement schedule was used to compare annual capital replacement costs of racks for cages and for longline systems to culture bay scallops. Growout cages were identified as the largest capital replacement cash cost that occurs in Year 6. Sensitivity analyses showed that the economic life of the growout cages had a substantial effect on overall net returns. Extending the economic life through proper care and maintenance increased returns while decreased economic life of growout cages due to poor maintenance substantially decreased net returns.

SUMMARY

This chapter discussed how to use cash flow budgeting techniques to manage cash flow and avoid serious cash flow problems. Cash flow problems can result from a number of sources. These can include problems associated with the timing of receipt of revenues, payment of farm expenses, and debt payments. The price, yield, and income risk in farming result in price and cost fluctuations that can cause cash flow problems. Prompt actions can often head off more serious cash flow problems.

Pro forma cash flow budgets are used to project out monthly cash flow for the upcoming year while the statement of cash flows records the actual cash inflows and expenditures. Comparing the projected and actual cash flow to date by month allows the farm manager to spot deviations early and devise actions to effectively correct and avoid further cash flow problems.

Corrective actions for cash flow problems can take many forms. Changing pond management strategies can help when fish can be sold that can change the timing of receipt of revenue. Effective planning can identify ways to change the timing of payment of expenses. Debt payments can be rescheduled.

One of the key steps is to maintain proper records that include projected and actual cash flow as well as deviations from planned and anticipated cash flow.

REVIEW QUESTIONS

1. What is the difference between a cash flow budget and a statement of cash flow?

2. Define and explain liquidity and its importance to the farm business.

3. Describe the relationship between leverage, use of credit, and cash flow, using aquaculture examples.

4. Explain how rescheduling loan payments can be used to improve cash flow for an aquaculture farm business.

5. Describe two aquaculture production management options that can result in improved cash flow.

6. Describe two measures of repayment.

7. Describe how to develop a cash flow deviation report.

8. Describe the types of records needed to compare projected and actual cash flows.

9. Compare profitability and liquidity.

10. Describe some of the problems involved in accurately forecasting prices and costs of aquaculture products and inputs.

REFERENCES

Adams, Charles M., Leslie Sturmer, Don Sweat, Norman Blake, and Robert Degner. 2001. The economic feasibility of small-scale, commercial culture of the southern bay scallop (*Argopecten irradians concentricus*). *Aquaculture Economics & Management* 5 (1/2): 81–97.

Bjorndal, Trond. 1990. *The Economics of Salmon Aquaculture*. Oxford: Blackwell Scientific Publications.

Grisdale-Helland, Barbara, Ståle J. Helland, and Kurt Kolstad. 2004. Salmon. In W.G. Pond and A.W. Bell (eds). *Encyclopedia of Animal Science*. Philadelphia: Taylor & Francis, pp. 777–780.

7
Managing Capital Assets in Aquaculture Businesses

INTRODUCTION

Aquaculture operations tend to have large amounts of capital assets. The magnitude of the capital assets in aquaculture businesses require that these assets be managed carefully. Key questions that need to be asked with regard to the management of capital assets are: (1) how much total capital should be used; (2) how to decide when to replace capital assets; (3) whether to purchase or lease capital assets like ponds and equipment; and (4) how to monitor the use of capital assets. This chapter discusses issues related to decisions on the capital requirements for various scales of aquaculture production. A format for making decisions related to replacing capital equipment and costs of capital is presented. Trade-offs among purchasing, leasing, or custom hiring are compared. The chapter concludes with a discussion of monitoring the use of capital and record-keeping requirements.

HOW MUCH TOTAL CAPITAL SHOULD BE USED?

Determining the amount of total capital to be used is primarily a process of long-run planning. In the long run, many different changes can be made. For example, the farm can be sold, it can double in size, or it can be divided into half. It is because of this flexibility that economists say that all resources become variable in the long run.

ECONOMIES OF SCALE

The question of the total amount of capital to be used in a business is also a question of how large the business should be. This is because larger businesses require greater amounts of capital. A primary determinant of the economically optimal (best) size of a business is the extent of economies or diseconomies of scale.

Economies of scale are represented by a decrease in the per-unit cost of production as output (total production) increases (Figure 7.1). Economies of scale are used to identify the optimal or best size of farm, the farm size that results in the lowest cost per pound of production across the farm. The first step in an analysis of economies of scale is to develop an enterprise budget for each possible farm size. The farm size that results in the lowest cost per pound is the most economical size. If the costs of production per pound decrease as the farm size increases, then economies of scale exist. If, however, the cost of production increases as the farm size increases, then diseconomies of scale exist. When economies of scale exist, the output increases by a greater percentage than the increase in each of the inputs as farm size increases. Economies of scale exist when there are increasing returns to scale such that production is more efficient (lower cost) due to larger farm sizes.

Economies of scale are found in industries with high capital costs. The high capital costs result in higher annual fixed costs; greater levels of production spread these fixed costs over the increased units of production.

Economies of scale in aquaculture businesses can result from the adoption of larger pieces of equipment or more advanced technologies. For example, a larger feed wagon may allow the manager to feed more ponds before returning to the feed bin and may result in lower costs per pound of fish produced than feeding with a smaller feed wagon. Similarly, a larger farm may be able to hire enough workers to seine their own ponds as needed rather than waiting to schedule a custom

Aquaculture Economics and Financing: Management and Analysis, Carole R. Engle, © 2010 Carole R. Engle.

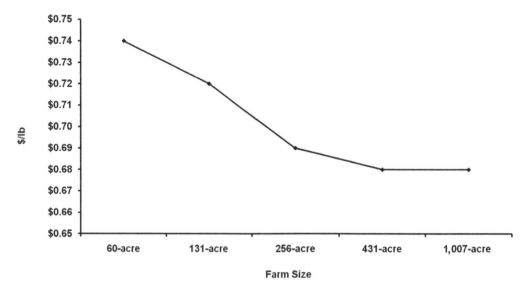

Figure 7.1.   Economies of scale in catfish foodfish production.

harvester who may not be available as often as would be best.

Economies of scale can also result from division and specialization of labor. Larger sizes of businesses can allow the labor to become specialized in certain tasks and thus become more efficient at that task. For example, a large farm with a dedicated seining crew can devote time to properly train that crew to seine efficiently whereas a smaller farm with a generalized work force may not develop the same level of efficiency or may have to contract with a custom harvesting company that may increase costs.

Other factors can contribute to economies of scale. These can include purchasing supplies in bulk at a discounted price, or spreading a manager's salary over greater acreage and greater total pounds sold. For example, Figure 7.2 demonstrates that building

fish ponds in a "windowpane" fashion can reduce the cost/acre of pond construction. The major investment cost in a pond-based aquaculture farm is building the levees; the more levees, the higher the fixed costs of constructing the pond.

Constructing a 1-acre pond requires four levees, or four levees per acre. However, if two 1-acre ponds are constructed side by side, as in Figure 7.3, with a shared levee, the total number of levees required is seven, or 3.5 levees per acre. Thus, the cost of levee construction will be less per acre by constructing ponds in a way that minimizes the total number of levees to be constructed.

Diseconomies of scale can also occur when businesses become too large to manage efficiently. The evidence of diseconomies of scale is an increase in per-unit costs of production as the business gets larger

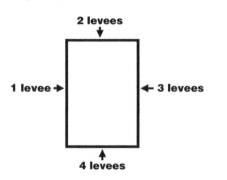

Figure 7.2.   Levees for a single pond.

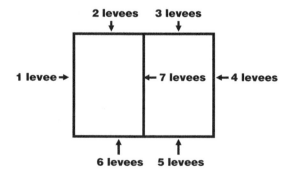

Figure 7.3.   Levees for two ponds with a shared levee.

and generates greater output. Coordinating and controlling the business becomes more difficult as the size of the overall business gets larger. The top manager becomes more separated from the direct operations, and efficient coordination becomes more difficult with larger scales of business.

FIRM GROWTH

In businesses with economies of scale, larger firms have a cost advantage in that the per-unit costs are lower for larger than for smaller farms. The realities of lower costs that result from economies of scale must be considered carefully in long-run planning for the growth of the farm.

The decision whether to increase the size of the business depends on what the optimal scale of production is for that particular farm business and its market. Once the optimal size of the business has been determined, the extent of expansion that would be ideal will be the difference between the optimal size of the business and the current size of the business. This section describes how to identify the optimal size of a business; however, credit capacity and other factors may limit both the amount and rate of growth that is possible for the business.

Farms in industries with economies of scale have incentives to grow to a larger size if the business has resources available, its resources are allocated in a suboptimal manner, or it has savings from disposable income. One major incentive for business growth is to operate more efficiently by achieving economies of size to enhance income-generating capacity.

The optimal scale or size of the business is determined by developing a long-run average cost curve (LRAC). Figure 7.4 illustrates the development of an LRAC for a shrimp hatchery. The LRAC consists of a series of short-run situations that can each be analyzed and combined into a comprehensive analysis of the effects of a variety of long-run options. Each short-run average cost curve (SRAC) corresponds to a particular size of the business. The LRAC line is drawn by connecting the lowest points of each SRAC. The lowest point of the LRAC indicates which scale of business is optimal.

In Figure 7.4, the two SRACs correspond to two different sizes of shrimp hatcheries. In this example, the larger hatchery results in lower costs of production and the firm's overall business plan should include plans to expand to achieve that size if the current business is smaller or plans to contract to that size if the current business is larger.

Increased financial leverage can result in increased firm growth, but the additional borrowing used to finance new investment also increases financial risk. Risk comes from increased potential losses of equity capital and increased variation in expected returns. High financial risk can affect the rate of growth in equity.

HOW TO DECIDE WHEN TO REPLACE CAPITAL ASSETS?

The decision to replace or trade a piece of equipment should involve consideration of a number of factors. Should equipment be replaced only when fully worn out or should it be replaced earlier? Considerations should include ownership and operating costs, reliability of the equipment, and the effects on cash flow and taxes. As equipment wears out, its repair costs increase and can reach a level where it is more economical to replace that piece of equipment. New, more efficient, equipment may be more economical as long as acquiring the capital necessary to upgrade the

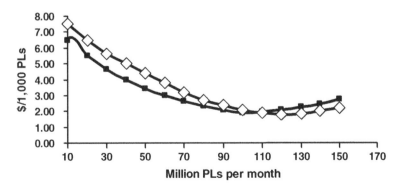

Figure 7.4.   Developing a long-run average cost curve for a marine shrimp hatchery.

technology does not result in adverse financial effects on the farm. If the farm has expanded in size, new equipment may be needed to operate the larger farm size efficiently.

Capital budgeting techniques can be used to make decisions as to when to replace equipment. Using capital budgeting, the net present value (NPV) for each number of years in the replacement period can be calculated. The period that yields the greatest NPV should be selected. Another approach is to conduct a marginal analysis that compares the gains from keeping the asset for another time period with the gains from replacing it.

Equipment on aquaculture farms is a major portion that requires capital investment. In the enterprise budgets described in Chapter 10, the total equipment investment ranged from $138,100 to $1,471,121 and from $1,461/acre to $2,302/acre on the 60-acre to the 1,007-acre catfish farms. Capital spent on equipment cannot be used for other investments or expenses. The value of what would be earned from using the capital for some other use is referred to by economists as an *opportunity cost*. The opportunity cost of investment capital used to purchase equipment is part of its total cost.

The costs associated with owning equipment include the initial purchase cost and interest rates if funds are borrowed to acquire the equipment while operating costs include fuel and lubricants, repairs, and maintenance of the equipment. Many of the fixed ownership costs consist of a one-time large outlay of cash for the initial purchase that may be quite large. Another large part of the cost of owning equipment is its depreciation, a noncash cost that is easily overlooked or ignored in financial analysis and planning. *Depreciation* is a charge that reflects the loss in value of a capital good due to wear and tear, obsolescence, and age. Accounting for the costs of depreciation is a way to spread the purchase price of a piece of equipment across its useful life to recover the original investment amount.

The annual ownership costs of equipment can be about 10–15% of its original purchase price. However, the annual ownership costs will vary with the type of equipment, how it is maintained, its age, and current interest rates.

Calculating the ownership costs of a particular piece of equipment can be complicated. The American Society of Agricultural and Biological Engineers (ASABE) provides guidance to estimate these costs (www.asabe.org). The calculation involves the annual capital recovery cost which is a measure of the annual ownership costs of equipment.

$$ACRS = (\text{purchase price} - \text{salvage value}) \times \text{capital recovery factor} + (\text{salvage value} \times \text{interest rate})$$

The salvage value (the value of the equipment after its useful life) is calculated by multiplying the original purchase price by the remaining value percentage found on the ASABE's web site. For example, a 10-year-old tractor may have a remaining value of 37%. If the tractor originally cost $60,000, its salvage value after 10 years of use would be $60,000 times 0.37, or $22,200. The capital recovery factor (CRF) is an annuity factor that can be found in Table 14.6. Thus, the annual capital recovery factor (ACRF) for the $60,000 tractor at an interest rate of 8% is:

$$ACRF = (\$60,000 - \$22,200) \times 0.14903$$
$$+ (\$22,200 \times 0.08) = \$7,409$$

It is important to note that the interest expense on the value of the equipment is calculated to account for the use of investment capital in that particular piece of equipment.

Taxes (property and sales), insurance, and equipment storage costs are additional ownership costs of equipment and pond assets. Some states do not charge personal property taxes on equipment, and sales taxes are variable across states. Insurance on equipment typically costs about 0.8% of its average value with a range of 0.5–1.0%. Facilities to store equipment typically cost about 2% (ranging from 0.5 to 3%) of the average value of the equipment.

The annual ownership costs for each piece of equipment can be compared by dividing the annual cost by the estimated annual use (in hours used for each piece of equipment). The result is a measure of the total ownership cost per hour of use. Fuel efficiency use of different pieces of equipment can be compared by dividing the gallons of fuel consumed by each piece of equipment by the hours of consumption to obtain a measure of fuel cost per hour. These can be compared with the performance guidelines published by the ASABE for different types of equipment and engine sizes.

The repair and maintenance costs of equipment can be based on the accumulated repair costs over the life of the piece of equipment. The engineering cost is calculated by multiplying the list price by the repair and maintenance cost percentage, which is calculated by dividing the accumulated repair and maintenance cost by the total hours of use over its useful life.

Operating costs of equipment include fuel and lubrication, repairs, and maintenance costs. Farm records provide the best information on fuel and lubrication costs. Tractor tests show that fuel consumption can be estimated from the maximum power takeoff horsepower (hp) of a tractor. For gasoline tractors, the gallons of gasoline used per hour are 0.06 times the hp of the power takeoff unit, while for diesel units, it is 0.044 times the hp of the power takeoff unit. Lubricants and filters add an additional 10–15% of fuel costs.

Repair costs vary for various types of equipment and for various levels of usage. However, agricultural engineers estimate that repair costs can range from 0.5 to 9% of the original purchase price of the equipment for each 100 hours of use. Repair costs typically increase over time as equipment ages and increasing amounts of repairs are required.

Repairs and Maintenance versus Depreciation

One of the difficult decisions for a farm business is how much capital to spend on repairs and maintenance. Over time, adequate maintenance and repair of equipment can allow that equipment to be used for a greater number of years than if it is not maintained and wears out more quickly. Some farmers attempt to survive by minimizing as many costs as possible, and repairs and maintenance can be some of the first expenses to be foregone in difficult economic times.

Monitoring the efficiency of equipment can provide some indication of when it may need to be replaced. One indicator of machinery efficiency is machinery investment cost per crop acre. This is calculated by dividing the current value of all crop machinery by the number of crop acres on the farm. Another measure, total machinery cost per crop acre, compares the operating cost of the equipment. It is calculated by dividing the total annual machinery costs, both ownership and operating, by the number of crop acres.

The costs of repairs and maintenance will begin to increase as equipment and ponds wear out. The wear and tear on equipment and ponds must be replaced and accounted for at some time or the farm will not be able to continue to survive. Eventually, the depreciated value of the asset is very low and the costs to maintain and repair the tractor or truck become excessive. A careful comparison of the annual costs of replacing the capital item with the average annual costs

to maintain and repair that piece of equipment will provide guidance on the best time to replace it. The timing of the replacement of equipment should take into consideration the pattern of cash flow and time the purchase to coincide with a time period in which there is greater cash inflow. Debt payments should be scheduled for ease of payment. If fish prices are expected to be low, deferring equipment replacement for an additional year may be preferable.

The Cost of Acquiring Capital

The costs a business pays for its financial capital are referred to as the *costs of capital*. The liabilities listed on the business's balance sheet comprise the business's financial capital. Debt and equity together constitute the degree of a business's financial leverage and the relationship of debt and equity levels in the business reflect the business's capital structure. The cost of debt capital should be lower than the cost of the equity capital.

For family-owned farms, equity capital is provided by retained earnings (reinvesting farm profits) and unrealized capital gains from the farm. However, in contrast to national financial markets that provide capital for large corporate businesses, retained earnings are highly uncertain with a rate of growth not fast enough for large investments. Agricultural farm businesses have thus relied heavily on borrowing and leasing.

Use of retained earnings also incurs an opportunity cost in that the funds are invested back into the business instead of spent by the family for some other activity. Costs of retained earnings can be thought of as the rates of return needed to compensate for the family consumption that is foregone.

Acquiring capital results in interest charges typically paid to the lender from whom the loans are obtained. Various credit plans can be compared by calculating the cost of the various plans.

To compare the cost of different loans designed to acquire capital assets, the actual percentage rate (APR) of each can be calculated. The APR is calculated by manipulating the following equation:

Periodic loan payment = net amount borrowed
× amortization factor.

This equation is solved for the amortization factor. The net amount borrowed is calculated by subtracting any non-interest charges and loan points or signing fees from the original loan amount. The periodic loan payment is then divided by the net amount borrowed to identify the amortization factor. The value of the amortization factor calculated is then used to identify the interest rate relative to the number of periods for the loan by using Table 14.6. If there is more than one payment a year, the interest rate is multiplied by the number of payments per year to obtain the annual rate.

The net present value (NPV) and the internal rate of return (IRR) can be used to compare the returns to the possible alternatives to purchasing capital assets. The true cost of borrowing can be compared with the use of these indicators. Chapter 13 presents details of the calculation and interpretation of NPV and IRR.

How Much Debt is Too Much?

New loans are frequently required to finance new capital assets needed to expand the business by building or acquiring new ponds or purchasing new equipment. Each farmer or farm business has a certain level of credit capacity. It will be difficult for farmers to borrow capital beyond their credit capacity because the increased leverage would increase financial risk to an unacceptable level.

The business owner can use the financial statements described in this book to calculate important financial ratios that shed light on this type of decision. Additional debt results in two different types of concern: (1) short-term cash flow requirements and short-term debt; and (2) solvency of the business and longer-term effects of increased debt.

The short-term cash flow requirements and debt are evaluated with a cash flow budget (see Chapter 12 for details on preparation and evaluation of cash flow budgets). The debt-servicing ratio and the cash flow coverage ratio, calculated from the cash flow budget, provide quick indicators of the effects on cash flow. The debt-servicing ratio divides the interest and principal payments by the cash inflow. Thus, smaller values are preferred and increasing values indicate increasing risk of inadequate cash flow to meet financial payments when due. The cash flow coverage ratio divides the cash inflow by the sum of the interest and principal payments. Higher values are clearly preferred and decreasing values over time indicate increasing financial risk to the business.

In addition to these formal indicators, a quick look at several lines on the cash flow budget can be useful. In particular, the line items that indicate the cash available, new borrowing, and outstanding operating capital debt on the cash flow budget provide insight into the effects of additional borrowing. Declining cash available, increased new borrowing, and increasing amounts of outstanding operating capital debt all

indicate that cash flow risk is increasing. These are a cause for concern unless there is some realistic expectation of a major change in the next year.

The effects of additional debt to acquire new capital assets on long-term debt, financial structure, and solvency of the business must be evaluated from the balance sheet (see Chapter 11 for details on preparation of and interpretation of balance sheets). The net worth on the balance sheet should generally increase over time. Acquisition of new capital assets that require new borrowing will increase both the liabilities and the assets on the balance sheet. The result often is a short-term decrease in net worth. However, wise investments in new capital assets are those that increase profits over time. Increasing profits will result in increased net worth. However, if a new capital acquisition results in a large negative net worth, the financial risk may be unacceptable.

The debt/asset ratio provides a quick view of the solvency of the business following additional capital acquisitions. Lower debt/asset ratios are preferable and debt/asset ratios greater than one indicate that the business is no longer solvent. Thus, if a new capital acquisition increases the debt/asset ratio, but the value remains below one, the additional debt level is not excessive. A debt/asset ratio that exceeds one indicates potential solvency problems. In such a case, a full evaluation is needed to assess whether the increased debt/asset ratio is a temporary condition or whether it would likely be lasting longer. Long-term exposure to debt/asset ratios above one is an unacceptable level of financial risk that few businesses can sustain.

PURCHASING, LEASING, OR CUSTOM HIRING CAPITAL ASSETS SUCH AS PONDS AND EQUIPMENT

There are alternatives to purchasing equipment. If the farm's financial structure cannot withstand additional debt, machinery tasks can be either hired out to custom operators or the equipment can be leased.

There is no one rule of thumb for deciding whether leasing equipment and ponds is better than purchase of capital goods. The alternatives must be analyzed carefully before making the decision. The decision as to whether to purchase or lease a piece of equipment such as a tractor should be based on the estimated total overall costs associated with the possible alternatives. A tractor needed for the fish farm will provide the same benefits in terms of production and revenue regardless of whether it is purchased or leased. However, the costs will be different. Careful analysis should compare the cash outlay for purchase and leasing, adjusted for any tax savings that would be obtained from being able to deduct expenses or depreciation. The lowest cost option would be the preferred option.

Leasing or renting equipment or ponds can be an alternative to purchasing capital items. It provides an opportunity for a farmer to increase the size of the business or to adopt more efficient technology without having the capital to acquire it outright. Leasing can be a feasible way for young farmers to start out in the business. One principal benefit of leasing or renting is that lower initial payments may provide a better cash flow pattern to cover other expenses in the year. Some lease rates may result in lower costs. Moreover, leasing may result in greater liquidity for the farm business, particularly in the earlier years. Leasing can also conserve working capital because the only liabilities incurred are those of the rental rates.

On the other hand, a high percentage of its capital assets will have less collateral to use as a basis for increasing its credit capacity. Ownership of ponds and equipment also increases equity for the farm business. Increasing equity enhances the balance sheet and contributes to owner's wealth.

The costs of leasing are handled differently from costs associated with capital acquisitions in the analysis of the farm's finances. Rent and other expenses from leases are variable or operating costs rather than fixed or ownership costs and are cash, not noncash expenses. A lease will result in higher variable or operating costs. If fish prices are expected to decrease in the near term, revenue would decrease, and a lease may not be the best choice. Moreover, high lease rates may not be profitable. Engle (2007) showed that catfish farming was not profitable at certain pond lease rates but was profitable at other lower levels. The farmer must analyze his/her options carefully.

The annual amount of the rent and other lease expenses is entered into the enterprise budget and the income statements as a variable cost. Leased assets, however, are not included on the balance sheet either as assets or liabilities. Thus, leasing does not affect the financial ratios of the business that are calculated from the balance sheet.

Leasing may increase a farmer's credit over time if working capital is conserved. However, if the cash flow demands of a lease are excessive, the lease may be detrimental to the farm's cash position. Such a situation could negatively affect a farmer's credit.

A farm that is highly leveraged (operating with a high debt load) and with low equity may not have the financial strength to assume additional debt. Leasing ponds, tractors, or aerators may be the best option for a few years. During this time, some of the debt load can be paid down to a level that would allow for new loans to pay for purchasing new equipment.

Pond leases should include a legal description of the ponds and land to be leased, the length of time over which the lease is valid, the amount of rent to be paid, how and when payments are due, the names of the owner (known as the lessor) and the tenant (known as the lessee), and the signatures of both. Pond leases should specify whose responsibility it is to maintain the levees, mow the grass, and who pays for adding gravel to maintain the roads on top of the levees. Most farm leases are written for 1 year but contain a clause that provides for automatic renewal from year to year. Pond lease rates in the southern portion of the United States can vary from about $100 to $350 per acre.

The lease agreement should stipulate when rent is to be paid. Lease payments should be entered into the cash flow budget to be certain that the lease payments can be made in a timely fashion. Cash flow problems can occur if lease payments are required before the crop will be marketed. If cash is not available, then the lease payments may need to be financed through short-term operating loans that increase total costs from the additional interest charges. The timing of rent payments should be negotiated along with the rental rate. If the rent is due earlier in the production season, the risk is shifted to the grower from the owner and there should be an accompanying decrease in the amount of the rent. Some leases require a prepayment and/or a security deposit that must be accounted for in financial planning.

MONITORING CAPITAL USE

Part of managing capital efficiently involves continuous analysis of its use. The first general approach to such an analysis is to compare the performance of the farm's capital assets from one year to the next. Chapter 4 presents details of methods of doing this. This type of analysis shows the business' progress from one year to the next and how economic changes affect the business. However, it can be difficult to sort out whether the business's performance is due to its management or due to external economic effects. Another disadvantage of this approach is that it relies entirely on the past.

Another approach to monitoring capital use is to compare the business's performance to that of other farms. This method allows the farm to compare their performance indicators to those of other, similar farms. There are private firms and some university extension services that operate programs for a fee to do this. Farms then receive periodic statements that allow them to compare their farm's performance on a number of different indicators with other farms that raise the same crops.

RECORD-KEEPING SUGGESTIONS

Record-keeping is essential for managing capital assets. The fundamental records necessary to evaluate the use of capital in the business are balance sheets (Chapter 11) and cash flow budgets (Chapter 12). From these, the relevant ratios and indicators, described in Chapter 4, should be calculated.

Managing capital assets also requires planning for the use of capital on the farm. Hours used for each piece of equipment allow the manager to calculate the total ownership cost per hour of use and compare this across different types of equipment. Records of hours of machinery use can be used to calculate machinery hours per acre or per pound of fish or shrimp produced. These records can form the basis for comparing the potential value of newer or larger pieces of equipment.

Machinery cost records that include both ownership and operating costs should be maintained. A depreciation schedule should be set up for each piece of equipment and adjusted annually to show the loss in value over the year as well as the current value. With no repairs and maintenance, equipment and ponds will wear out more quickly and will then need to be replaced in their entirety. It can be more expensive to renovate or repair ponds and equipment than it is to purchase them.

Tax and insurance costs, lease payments for any equipment rental, and the cost of capital should be recorded, preferably by piece of equipment. Operating costs that should be monitored for equipment on the farm include fuel and oil expenses, costs of spare parts in inventory, repairs, and any custom hires. Maintaining records of fuel use by piece of equipment provides a means to calculate and compare fuel efficiency across machinery on the farm. Increasing values over time indicate when it would become profitable to replace that piece of equipment.

Complete records for a farm include repair and maintenance costs for each type of capital asset in the business. This list should match the depreciation schedule on capital assets that is used for income tax purposes.

The profitability of capital is typically measured by the return on capital. Chapter 13 presents a more detailed discussion on its calculation. The return on capital and the internal rate of return on capital use can be calculated and maintained in the farm's records each year.

PRACTICAL APPLICATION

The frequency of rebuilding or renovating ponds is a question relevant to pond-based farms and their strategy for managing capital assets. When pond levees become too narrow for a feed wagon, hauling truck, or harvesting equipment to pass, production and output of fish will suffer. Moreover, ponds silt in over time through erosion of the main levees. This results in the ponds becoming too shallow. Shallow, silted-in ponds result in weed problems and reduced capacity to store oxygen. The result can be a reduction in fish yields and possibly an increase in the feed conversion ratio.

An analysis of when it is appropriate to incur the investment required to renovate ponds would begin with a tabulation of the costs and benefits that would be incurred through the renovation. Costs would include the costs of contracting the equipment and labor for the actual renovation, any lost production from having the ponds out of production for a time, and pumping costs required to fill the ponds again. Benefits would consist of improved feeding rates, increased yields, and possibly lower feed conversion ratios.

Equipment used to renovate ponds can include a bulldozer, bulldozer and box blade, bulldozer and dirt buggy, or a trackhoe to pull up the levees. Renovating a pond can cost on average $575/acre, but costs can range from $350 to $1,000/acre. Taking the pond out of production will result in lost production for that period of time. For example, it may take 5 months to dry out the pond, but this time can range from 3 to 12 months. The actual renovation may take only 2 weeks with a range of from 0.5 to 3 weeks.

To estimate the costs associated with the time that the pond is out of production requires several assumptions. In the best case scenario, the pond could start drying out in February and could be dried by April. Thus, 1 month of production would be lost for drying, an additional 2 weeks for renovation, and another 2 weeks for pumping up the pond again. In the worst case, the entire growing season could be lost. This would entail a 6-month period that extends from mid-April through mid-October. Net daily yield in ponds can be 19 lb/acre/day for a 7,000 fish/acre stocking density, but can be as much as 24 lb/acre/day across all densities. In the worst case scenario, net daily yield from single-batch production with 0.6 lb stockers was 31–40 lb/acre/day.

Table 7.1. Renovation Costs for a Variety of Scenarios.

Category	Best case scenario costs	Worst case scenario costs	Large fish scenarios (31 lb/acre/day)	(40 lb/acre/day)
Renovation	$7,475	$7,475	$7,475	$7,475
Lost production	$815	$3,011	$3,906	$5,040
Pumping cost	$1,040	$1,040	$1,040	$1,040
Total	$9,330	$11,526	$12,421	$13,555
lb catfish to cover costs	13,328	16,466	17,744	19,364
lb/acre	1,025	1,267	1,365	1,490

Putting these together for a 13-acre pond with 2 months of lost production (best case scenario) would have renovation costs of $7,475, lost production of $815, and pumping costs of $1,040, for total costs of $9,330 (Table 7.1). It would take an additional weight of 13,328 lb of catfish to cover these costs, or 1,025 lb/acre (calculated at a price of $0.70/lb). In the worst case scenario, the pond is out of production for a longer time and the lost production costs increase to $3,011 for a larger total cost of $11,526. Thus, for the worst case scenario, 16,466 lb of catfish are required to cover costs, or 1,267 lb/acre. With yields of 31 lb/acre/day, the cost of the production lost increases further to $3,906 and would require a total of 17,744 lb of catfish, or 1,365 lb/acre. If net daily yield was 40 lb/acre/day, the lost production cost would increase to $5,040 and would require 19,364 lb of catfish to cover costs, or 1,490 lb/acre/day.

In summary, the longer the pond is out of production, the greater the renovation cost. The more productive the pond, the greater the overall renovation cost.

In terms of the benefits, there was no significant difference in the feed conversion ratio between 1 and 6 years following renovation (Figure 7.5). Figure 7.5 shows the additional feed cost if the feed conversion ratio increases, but the yield stays the same.

The decision as to when to renovate ponds requires an investment analysis because renovation is a long-term capital decision. For the investment analysis, returns and costs, including renovation costs, must be estimated for each year for 10–20 years. The NPV is calculated to measure the profitability of the proposed investment.

Figure 7.6 shows the net returns by year if the pond is not renovated, with hypothetical feed conversion ratios. Figure 7.7 shows the net returns for 7 years following renovation. In contrast, Figure 7.8 shows the pattern of net returns if ponds are renovated every 10 years.

How do we determine which scenario is the most profitable? Table 7.2 compares the NPV without renovation, renovating every 7 years, 10 years, or 13 years. Figures 7.7 and 7.8 demonstrate negative net returns in those years that ponds are renovated, due to the cost of the renovation. The highest NPV (most profitable) occurs when ponds are renovated once every

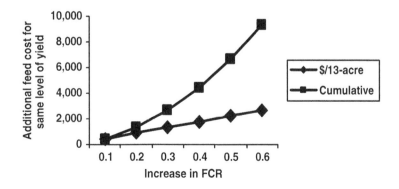

Figure 7.5. Additional feed cost if feed conversion ratio increases but yield remains the same as pond ages.

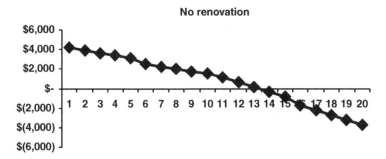

Figure 7.6. Net returns by year with hypothetical feed conversion ratios if ponds not renovated over time.

10 years. This analysis will show different results if the estimates of costs and benefits are changed.

Other factors to consider other than a strict cost–benefit analysis include: (1) can you still operate the pond? (2) income taxes; (3) cash flow; (4) production contracts; (5) overall debt load and credit capacity; and (6) risk of emergencies. The final decision should be made after considering all relevant factors, including the economic analysis.

Renovation is expensive. It is critical to establish vegetation quickly for longer-lasting levees. Data are not currently available as to whether feed conversion ratios increase or yields decrease following renovation and how quickly they change. Careful assessment of pond records following renovation might reveal these. Annual cash flow budgets with NPV analysis will show the most profitable pattern of renovation.

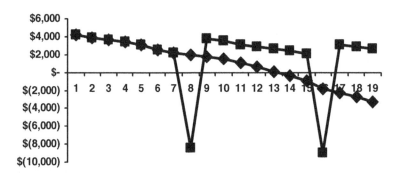

Figure 7.7. Annual net returns for 7 years following pond renovation.

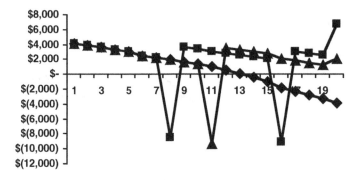

Figure 7.8. Annual net returns for 10 years following pond renovation.

Table 7.2. Net Present Value with Different Frequencies of Pond Renovation.

Option	Net present value
No renovation	$16,754
After 7 years	$20,150
After 10 years	$20,931
After 13 years	$20,531

OTHER APPLICATIONS IN AQUACULTURE

Cyrus and Pelot (1998) developed a database planning software system for management of scallop farms. Simulations demonstrated the required allocation of capital resources among various types of nets. It traced the use of optimal capital through different stages in the production process until the operation reached a steady state.

SUMMARY

This chapter reviewed decisions related to managing the capital assets in the farm business. These assets are those that are used by the farm business for more than a year. The fundamental concept of economies of scale is reviewed because many aquaculture businesses exhibit strong economies of scale due to the relatively high proportion of capital costs in aquaculture businesses. Economies of scale exist due to spreading pond construction costs across greater levels of output, sharing equipment and management across greater levels of output, and the opportunity to purchase inputs such as feed at a lower cost. The long-run average cost curve (LRAC) is used to identify the optimal size of a business by finding the size of business associated with the short-run average cost curve that has the lowest cost.

Managing capital assets efficiently includes making decisions to be able to use the proper size of equipment, maintain it in good operating condition, and to make appropriate decisions related to its timely replacement. The decision when to replace equipment is complicated and the timing of replacement can affect the financial structure of the business as well as its cash flow position.

Capital acquisition can occur through custom hire arrangements or through leasing ponds and other facilities as well as equipment. The decision to purchase or lease should be made after careful analysis of the effect on the business's financial position and its cash flow. Assessment of how much debt is too much is needed to avoid incurring debt loads that entail excessive financial risk.

REVIEW QUESTIONS

1. What are economies of scale? Give several examples of aquaculture industries that exhibit economies of scale.

2. What creates economies and diseconomies of scale? Give aquaculture examples of each.

3. How do economists identify the optimal size of a business? Draw the appropriate curve and describe how it is derived.

4. What are the trade-offs between purchasing and leasing ponds and equipment? How are equipment costs treated differently in financial analyses if equipment are purchased as compared to leasing?

5. How does a farm owner assess whether incurring additional debt will incur excessive financial risk?

6. Describe two indicators of machinery efficiency and explain how each is calculated.

7. Compare the types of costs associated with ownership of equipment and those associated with operating the equipment.

8. What types of records should be maintained on capital assets of the business, both long-term facilities and equipment?

9. What are the costs associated with borrowing capital?

10. Which financial indicators provide the best indication of whether additional debt is feasible or not?

REFERENCES

American Society of Agricultural Engineers. www.asabe.org. Accessed March, 2009.

Cyrus, Pemberton and Ronald Pelot. 1998. A site management system for shellfish aquaculture. *Aquaculture Economics and Management* 2(3):101–118.

Engle, Carole R. 2007. *Arkansas Catfish Production Budgets*. Arkansas Cooperative Extension Program MP466. Pine Bluff, AR: University of Arkansas.

8
Managing Risk in Aquaculture Businesses

INTRODUCTION

The level of risk in an aquaculture business requires that managers plan to make adjustments when negative outcomes occur so that the business will be successful. The goal of managing the risk in a business is to improve the probability of success. However, even with the very best management, profits are not guaranteed. Well-thought-out and carefully planned decisions can still result in poor outcomes. Moreover, higher profits are frequently associated with greater risk of losses. The key is to evaluate decisions based on the objectives of the business, not in reaction to previous decisions. Good decisions and strategies over time will result in positive outcomes in spite of the occasional negative outcomes.

This chapter focuses specifically on strategies to manage risk. Chapter 16 presents details of what constitutes risk, various sources of risk, and how to measure risk. For those unfamiliar with how economists define and measure risk, reading Chapter 16 prior to this chapter would be worthwhile.

PRIORITIZING RISK

A manager of an aquaculture business must account for risk in each phase of the business. Planning begins with careful identification of the types of risk that the business is likely to face in the upcoming planning term. The second step is to identify the possible outcomes from each type of risk.

The process of prioritizing risk requires assessment of the potential impact on the business as a result of the various sources of risk and an estimate of the probabil-

ity of that type of effect occurring. Careful evaluation of the combined effects of both these criteria allows the manager to identify those likeliest to cause the greatest problems.

Consider Figure 8.1 related to assessment of the risk associated with the discovery of viral hemorrhagic septicemia (VHS) in one's state. This disease broke out in the United States in the Great Lakes area and has resulted in development of regulations that prohibit the transport of live fish from VHS-positive states. A fish farm whose market requires interstate shipment of live fish in the United States should carefully weigh management options related to the risk of their state becoming VHS-positive. According to Figure 8.1, a farm located in a state adjacent to a VHS-positive state would have a higher probability of VHS spreading. The potential impact would be very high because the result likely would be a loss of all sales to other states because regulations would prohibit interstate shipment. Thus, steps to develop biosecurity protocols or to engage in an inspection and certification program would likely be recommended to minimize these risks. However, food fish growers who sell to a processing plant in the same state may still have a high probability of experiencing some effect, but the potential negative impact on the business would be lower. Thus, the need for action would be less.

METHODS TO ORGANIZE RISKS AND STRATEGIES

A decision tree is a useful way to organize the various types of risks and the alternative strategies for managing them. Figure 8.2 illustrates a decision tree for an example of alternative strategies for coping with risk on a shrimp farm. The decision tree consists of

Aquaculture Economics and Financing: Management and Analysis, Carole R. Engle, © 2010 Carole R. Engle.

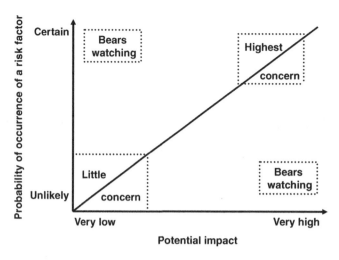

Figure 8.1. Prioritizing risk.

branches and nodes that denote decisions and is drawn from left to right. In Figure 8.2, there are three possible outcomes (good, average, poor) for each alternative option.

The options considered in Figure 8.2 consist of three stocking options for the shrimp ponds: (1) 1 postlarvae/m²; (2) 1.5 postlarvae/m²; or (3) 2 postlarvae/m².

The probability of disease incidence varies from year to year and with stocking density. The price received for shrimp is based on the size of the shrimp produced and the size of shrimp produced varies with the stocking density. Figure 8.2 shows the probabilities of good, average, or poor outcomes from each stocking density and the net returns associated with each outcome for

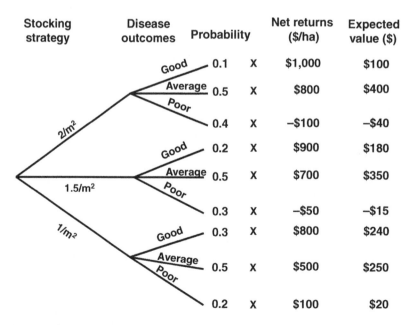

Figure 8.2. Decision tree for shrimp stocked at different densities.

Table 8.1. Payoff Matrix for Shrimp Stocking Rate.

Disease Outcomes	Regrets			Stocking density					
				$1/m^2$		$1.5/m^2$		$2/m^2$	
	$1/m^2$	$1.5/m^2$	$2/m^2$	P	$	P	$	P	$
Good	$200	$100	$0	0.3	800	0.2	900	0.1	1,000
Average	$300	$100	$0	0.5	500	0.5	700	0.5	800
Poor	$0	$150	$200	0.2	100	0.3	−50	0.4	−100
Expected value	n.a.	n.a.	n.a.	n.a.	460	n.a.	515	n.a.	510
Minimum return	$0	$100	$0	n.a.	100	n.a.	−50	n.a.	−100
Maximum return	n.a.	n.a.	n.a.	n.a.	800	n.a.	900	n.a.	1,000
Range	n.a.	n.a.	n.a.	n.a.	700	n.a.	950	n.a.	1,100
Maximum regret	$300	$150	$200	n.a.	n.a.	n.a.	n.a.	n.a.	n.a.
Expected regret	$210	$115	$80	n.a.	n.a.	n.a.	n.a.	n.a.	n.a.

each strategy. The expected net returns are determined by multiplying the probability of occurrence of each by its expected net returns and adding up the results. Thus, the expected net returns for a stocking rate of 1 postlarvae/m^2 are $510/ha (sum of $240/ha, $250/ha, and $20/ha), $515/ha for 1.5 postlarvae/m^2 (sum of $180/ha, $350/ha, and −$15/ha, and $460/ha (sum of $100/ha, $400/ha, and −$40/ha) for the 2 postlarvae/m^2 stocking density.

Another way to organize information on the relative values of alternative strategies is to develop a payoff matrix. A payoff matrix is developed in the form of a table instead of a decision tree (Table 8.1). The payoff matrix indicates the returns possible from a particular action. The resulting values are identical to those in Figure 8.2. Typically, a payoff matrix includes just one decision and one set of events. However, Table 8.1 shows a payoff matrix for the three shrimp stocking densities from Figure 8.2. The possible events are listed on the left side and the choices are indicated across the top. The returns expected from each choice are calculated for each combination.

A regret matrix indicates the potential returns lost by not choosing the best option. For example, Table 8.1 also shows the regrets the manager would have by not choosing the other shrimp stocking densities. For example, if outcomes in that year are good, and the manager chose to stock at a low density, he/she would have earned $200/ha less than if he/she had chosen the highest density. However, if outcomes would be poor that year, the manager would have no regrets (0) if he/she had stocked at the low rate. If he/she had stocked at the high rate, the regrets would be $200/ha

because of the differences in net returns between the $1/m^2$ and $2/m^2$ stocking densities when outcomes are poor. The regret matrix also includes a summary of the maximum regrets for each stocking density and the expected regrets. The expected regrets are calculated by multiplying the regret value for each outcome by its probability and then summing them. Decision makers seek decisions with higher payoffs and lower regrets. A regret matrix can be combined with a payoff matrix in the same table with net returns demonstrated for both payoffs and regrets, as in Table 8.1.

DECISION CRITERIA FOR DECISION MAKING

Economists have developed several types of decision rules as a framework for making decisions related to risk. It is important to recognize that there are differences in the results obtained from use of the different types of decision rules. Factors that must be considered in order to select the appropriate decision rule include: (1) the decision maker's attitude to risk; (2) the financial condition of the business; and (3) the cash flow requirements, among others.

The various decision rules include: (1) maximin (maximization of the minimum possible returns); (2) minimax (minimization of maximum possible regrets); (3) most likely outcome; (4) maximum expected value; (5) risk and returns comparison; (6) safety first; and (7) break-even probability. With the maximin rule for returns or payoffs, the option chosen is the one with the largest minimum returns. Thus, this decision rule focuses on downside risks and selects those options that will generate higher returns during the worst-case outcomes. The minimax criterion also focuses on downside risk but is based on regrets rather than payoffs. It is used to select the option with the smallest value of maximum regrets. The most likely outcome is based on the outcome with the highest probability of occurrence and then selects the strategy with the highest net returns for that particular strategy. This is one of the easiest strategies to apply.

The maximum expected value decision rule will select the alternative that has the greatest expected value (weighted average of the net returns multiplied times its expected net returns). This decision rule does not account for the variability that may occur with the different outcomes. Thus, the maximum expected value decision rule would be used primarily by risk-loving managers.

The risk and returns comparison takes into consideration both the level of risk and the level of returns expected from each alternative strategy. It results in rejecting strategies with lower expected returns and higher risks than other strategies. This decision rule is preferred by managers who are more risk averse.

The safety-first decision rule results in selection of the best possible results of the worst-case scenario. In this rule, the emphasis is on managing the most negative outcomes because outcomes that are more positive than expected do not result in financial problems for the business. Financial problems arise from the negative, not the most positive outcomes. This rule is particularly useful for businesses in difficult financial positions, those that may not survive another bad year.

The break even probability decision rule weighs the risk of losses against the expected average return. Cumulative density functions are calculated for each alternative that allows for comparison of the various alternatives.

TECHNIQUES TO HANDLE RISK

Techniques to reduce risk attempt to reduce the variability of outcomes. This can be achieved by: (1) setting a minimum income or price level; (2) maintaining as much flexibility for decision making as possible; or (3) improving the ability of the business to bear risk. There are a variety of alternative strategies that have potential to mitigate each type of risk. Each business should identify those that are feasible and practical for its situation.

The risk associated with each needs to be estimated and the expected returns and trade-offs evaluated for each alternative. The alternative strategies for coping with the risks identified are known as decision strategies.

TOOLS FOR MANAGING PRODUCTION RISK

Skilled Management

The riskiness of aquaculture requires skilled management. Skilled managers can reduce risk by maintaining facilities and equipment in good repair. For example, aerators that function properly will minimize the risk of losses due to oxygen depletion. Skilled managers who subscribe to trade journals, market news services, and those who attend educational meetings will be in a better position to minimize risk by making better-informed decisions than managers who do not stay abreast of current conditions and the latest research. Contracting experts to monitor specific aspects of the business can help to be prepared to manage adverse conditions. Lawyers, accountants, and fish health experts may be valuable resources.

Flexibility

Flexibility, in and of itself, does not affect risk, but a firm that is more flexible has the ability to make changes that can help it to find ways to adjust and cope with risk. Aquaculture businesses tend to have relatively high fixed costs. The high fixed costs reflect the level of long-term assets that are more difficult to adapt and change as conditions change. Short-term assets can be sold more quickly and frequently in response to adverse conditions. However, there can be a trade-off between economies of scale (lower per-unit costs of production) that are typically achieved with greater capital investment and higher fixed costs. The greater efficiencies obtained from larger-sized operations are often achieved at the expense of flexibility. On the other hand, greater flexibility may reduce risk but will likely result in a less-efficient business.

Farm investments that can be used for multiple purposes can increase flexibility of the farm. Buildings and equipment that can be used for multiple enterprises will provide greater flexibility than buildings and equipment that are highly specialized.

Crops with shorter production cycles reduce risk by providing greater flexibility and opportunities to recapture costs sooner. This is because revenues are created more quickly and more frequently. Greater cash inflow to the business provides greater liquidity and provides cash reserves to adjust to changing conditions. Raising fingerlings, for example, can result in a faster cash turnaround time than raising food fish.

Share Leases

Crop and livestock share leases apportion risks between the tenant and the landowner. This is because both expenses and revenue from the sale of crops are typically split between the tenant and the landowner. Share leases are not as common in aquaculture as in other forms of agriculture, but can offer a way to spread production cost risks.

LONG-TERM CONTRACTS WITH SUPPLIERS AND BUYERS

Long-term contracts for input supplies will reduce risk by stabilizing prices of inputs or products. For example, farmers who book feed in advance enter into a long-term contract. The farmer then is protected against rising prices of feed for the quantity of feed booked.

Diversification

Identifying stable enterprises is another technique that can be used to reduce production risk. Some types of crops are more stable in terms of generating income than others are. Diversifying production on the farm can reduce risk by producing more than one crop if the price of that crop varies in an opposite direction from the other. If the second crop produces high revenues in years when revenues from the first crop are likely to be lower, producing both will minimize overall price risk across the business. The key is for the prices and yields of the different crops produced to have highs and lows at different times.

Diversification can reduce the total variability in net returns by pooling several types of income-generating activities. Investments that entail greater risk usually also have greater potential for higher net returns. Thus, strategies designed to reduce risk are usually accompanied by lower returns.

Correlations can be calculated to measure how prices and yields are related. If prices and yields of the two crops are positively correlated, it means that they will both be high at the same time or that they will both be low at the same time. With positively correlated prices and yields, diversification will have little effect on reducing risk. Risk reductions will oc-

cur when the prices and/or yields of the two crops are negatively correlated. Crops with strong yield correlations frequently will also have positive correlations for price because the quantity sold of a commodity will affect its price.

However, diversification across the farm may increase costs by complicating management. A farm that specializes in only one type of product may have higher returns resulting from improved efficiencies obtained through specialization.

Larger aquaculture companies may be able to diversify across geographic areas. Geographic diversification can result in greater reductions in risk because production is spread over a wide area with different seasons and different price variations.

Diversification strategies may also involve nonfarm sources of revenue to counteract the effects of price and production risk. These activities can include investing in stocks and bonds, working off the farm to generate off-farm revenue, or hedging in futures contracts. Renting versus owning capital assets required for aquaculture production may have different effects on risk, particularly if the farm business has considerable financial risk due to being highly leveraged.

TOOLS FOR MARKET RISK

Market risk results primarily from fluctuations in market prices. Techniques that can be used to manage market risk include: (1) inventory management; (2) contract sales and hedging; (3) forward contracting; and (4) marketing cooperatives.

Inventory Management

Holding fish to wait for higher prices can be a strategy to manage market risk because prices of many commodities are seasonal. This can be particularly advantageous for farmers raising new aquaculture species that are competing with wild-caught product. If the supply of the wild species is seasonal, prices likely will be seasonal. If the aquaculture grower can manage to sell inventories during the times of high prices, there could be an advantage, at least until aquaculture production increases to high levels. For example, Chen et al. (2008) demonstrated that monthly prices of summer flounder range from a low of $0.88/lb in January to $1.26/lb in April. Some entrepreneurs have taken advantage of this price seasonality by transferring flounder caught in the wild to tanks on shore in Virginia, on the east coast of the United States (Kauffman 2004). The flounder were held and fed for several months and then sold in the live fish market in the northeast of the United States. Gross margins of $4/lb were obtained as compared to $0.25/lb or lower in the iced fish market.

Managing inventory can provide opportunities to spread sales over time. Such a strategy sometimes can spread the risk of varying market prices by selling both during times of low and high prices. Continuous production systems, as practiced in aquaculture segments such as channel catfish in the United States, have the potential to sell throughout the year. Spreading sales is more effective with crops that are not seasonal.

This strategy provides no protection from downside risk because prices could fall as inventories are held. Moreover, holding inventories of live product incurs additional risk of losses due to disease. Diseases could reduce the quantity of marketable yields. Holding crop inventories also subjects the farmer to the risk that prices will fall further in the future.

Forward Contracting

Contract sales can stabilize prices, especially if contracts signed are long-term. Contracts may also specify management practices to be associated with the production of that commodity.

Forward price contracts exist for many grain and livestock buyers. Forward contracting protects against declines in prices. The disadvantage is that forward contracting also prevents the individual from gaining during times of rising prices.

Contracting with a buyer to sell product at a certain price in the future has the benefit of stabilizing market prices for the farm; more stable prices reduce market risk. Moreover, a forward sales contract may be viewed favorably by lenders who are concerned with downward price risk.

However, forward contracts restrict the farm to specific production performance standards, harvest conditions, and terms of sale. If the contract calls for delivery of a specified amount and yields fall short, the farmer may need to purchase additional product to meet the terms of the contract.

Futures Contracts and Hedging

Futures contracts are common for grain commodities. For example, the Minneapolis Grain Exchange has traded wheat for over 110 years. Futures markets tend to enhance cash markets, but no one is required to physically deliver or receive the product.

The prices of futures contracts tend to move in the same manner as cash prices. The difference between the futures contract price and the cash market price is referred to as the *basis*.

A *futures contract* is an agreement to either deliver or receive a certain quantity and grade of a specific commodity during a designated delivery period (the contract month). The contract includes information on where the commodity would be delivered and any adjustments in price from substituting a different species or size.

Selling a commodity futures contract instead of the commodity itself allows a buyer or seller to set a market price for that quantity of product. The process of buying and selling futures contracts is known as *hedging*. Hedgers enter the market to protect themselves from market risks. Hedgers participate to reduce market risk associated with either rising or falling shrimp prices.

To protect against price declines, a farmer would buy the right to sell a futures contract (known as a "put" option) at an established price. A shrimp grower can buy a "put" option to insure against decreases in cash prices. The "strike" price is the level of price protection selected. A put option with a strike price of $4.50 and a premium of $0.20 enables the farmer to protect a net sales price of $4.30/lb (strike price—premium). For example, a shrimp producer who intends to sell shrimp during the next month could enter the futures market to sell futures contracts. This is known as a "short" position. A shrimp grower, in a short position, is locking in a price for future sales.

Call options (the right to buy a futures contract at an established price) can be used to establish a maximum purchase price for a commodity. Buying a call option provides insurance against price increases. A call option with a strike price of $4.50 and a premium of $0.20 enables the individual to protect a net purchase price of $4.70/lb. These can be useful for inputs, such as soybean meal for fish feeds.

For example, a buyer for a major restaurant chain who will need to buy shrimp to serve in the restaurants a month from now could enter the futures market to buy futures contracts. This is known as a "long" position. A shrimp buyer in a long position (the buyer of futures contracts) is locking in a price for future purchases of shrimp. A buyer has the right, but not the obligation, to buy or sell futures contracts at a predetermined price at any time until the option's expiration date.

Speculators provide the capital, risk capital, and liquidity needed for those who wish to use the futures market as a way to hedge prices. Speculators are betting that the market will rise or fall.

Hedging for aquaculture growers is difficult because there are no futures markets for the majority of aquaculture products. Brokers require margins that decrease the benefits to farmers. Moreover, sometimes the product specifications may not match well with the capabilities of the farm and preclude participation in futures markets.

There have been some experiments with futures markets for aquaculture commodities such as shrimp, with varying amounts of success. The Minneapolis Grain Exchange began to trade futures contracts for farm-raised and wild white shrimp in 1993 and then added a contract for farm-raised giant tiger shrimp in 1994.

There were two main contracts for shrimp: (1) 5,000 lb of raw, frozen, headless, shell-on 41–50 count white shrimp (*Penaeus vannamei, P. occidentalis, P. schmitti, P. merguiensis,* and *P. setiferus)*; (2) 5,000 lb of raw, frozen, shell-on 21/25 count farm-raised giant tiger shrimp (*P. monodon).* The contracts were standardized so that they could be exchanged. However, the shrimp futures market was ultimately closed. A major problem of the shrimp futures market was the lack of homogeneity of the product (Martinez-Gamendia and Anderson 2001). Shrimp come in many different sizes and the fixed premium/discounts used to account for the price differentials turned out to be inadequate.

Marketing Cooperatives

Marketing cooperatives have had some success in stabilizing farmers' returns. Forming a cooperative provides an opportunity to pool product from across different farms. Cooperative members can develop markets jointly. By doing so, storage, sales, and pricing functions can be transferred to a larger company. The larger company can supply larger buyers, develop longer-term contracts, and take advantage of economies of scale with regard to marketing functions. Marketing cooperatives require knowledgeable management. With skilled management, the marketing cooperative can increase profit through its size and specialization.

TOOLS FOR MANAGING FINANCIAL RISK

Management plans that result in adequate liquidity and solvency can help to reduce financial risk. Maintaining adequate levels of liquid reserves will reduce financial risk. Cash accounts that are maintained to provide cash during emergency situations are a way of providing insurance for the farm from the farm's own resources.

Using fixed interest rates stabilizes the amount of interest payments. Fixed interest rates may be higher than variable rates, but they protect against rising interest rates. Knowing what the interest payments will be over time allows the manager to plan for timely payments.

With careful planning, capital expenditures can be postponed if financial conditions are not favorable and then made when cash is available. Flexibility in scheduling the purchase of assets will help to reduce risk in the business.

Use of self-liquidating loans (those repaid from sale of the loan collateral) will reduce financial risk. Establishing operating loans based on the inventory of fish raised would be an example of a self-liquidating loan. However, many banks are hesitant to allow use of swimming inventory as loan collateral, primarily due to difficulties in accurately estimating the quantity of fish inventory on the farm.

Credit Reserves

Maintaining a credit reserve will reduce risk because it provides for a ready source of additional borrowing capacity. Maintaining a sound relationship with one's banker will help to maintain a reserve of credit (additional borrowing capacity that can be drawn upon during adverse financial times). A sound relationship with the lender may help to obtain approvals to carry over loans, defer payments during adverse conditions, or to refinance high debt loads. Thus, working with one's lender is an important risk management strategy for many farmers.

Sensitivity analyses of liquidity and financial risk can be done to assess the effects of variations in farm revenue, farm expenses, and interest rates on the financial risk in the business. The percent that farm revenue can decline and still meet cash flow requirements is calculated by dividing the excess cash available by the total cash available. These values are obtained from the cash flow budget. Greater detail on the mechanics of cash flow budgets can be found in Chapter 12. The percent that farm expenses can increase and still meet cash flows is calculated by dividing the excess cash available by the cash operating expenses. The percent that interest rates can increase and still meet cash flows is measured by dividing the excess cash available by the total liabilities.

Long-Term Debt to Equity

Increasing equity in the business also reduces risk because it provides the solvency and much of the liquidity of the business. Dividing the long-term debt by the equity (and multiplying by 100) provides a measure of the relative importance of long-term as opposed to short-term debt. These values are obtained from the balance sheet (Chapter 11). Long-term debt can impose a constant drain on the finances of the business in terms of interest costs and debt servicing, thereby increasing financial risk.

AQUACULTURE INSURANCE

Crop insurance can provide some assistance to reduce risk on aquaculture farms because it constitutes a form of liquidity. Purchase of insurance can provide a source of funding to protect against an adverse event such as a loss of property due to disaster events. Insurance reduces downside risk or the risk of a negative occurrence.

Most insurance programs function by establishing a minimum income or price level and charging a fee (premium) to do so. Thus, purchasing insurance is a strategy that shifts the risk to the insurance company in exchange for the price of the premium. Premiums typically increase with the amount of risk.

There are several types of insurance that can be used for farming businesses. For example, property insurance can be purchased to protect the business from losses from fires or storms. Liability insurance can protect the farm against lawsuits and liability claims.

Multiple peril crop insurance is available for a number of agricultural crops in the United States and is backed by the U.S. Department of Agriculture (USDA). The USDA program provides guarantees to purchase up to 85% of the proven yield for the farm. Losses that are covered include those from disaster events such as floods, hail, and others. Various private companies provide this type of insurance.

Secretan (2003) identified 24 countries where insurance for aquaculture businesses either exists currently or once existed.[1]

However, overall, the opportunities to insure aquaculture crops are limited. One of the difficulties with aquaculture insurance is that many aquaculture farms are small-scale businesses that do not maintain the types of stock control and production records necessary for insurance. Providing proof of loss that clearly documents the numbers, sizes, and value of fish, shrimp, or clams lost is difficult. Even in large aquaculture sectors such as the shrimp industry, insurers view stock control methods as insufficient (Secretan 2003). Thus, insurance for aquaculture is a small class of insurance for which few standards have been developed.

The salmon industry is one of the few sectors of aquaculture with insurance coverage programs (Secretan 2003). Norwegian banks, for example, require insurance protection of salmon stocks. English, Norwegian, French, Dutch, Spanish, Greek, and Italian companies offer insurance coverage to salmon farms. While a few companies exist in Chile, Australasia, Japan, and India, most of the insurance coverage is dependent on European reinsurance.

In the United States, international salmon producers and some small salmon growers have purchased insurance through the European providers. Attempts have been made by U.S. insurance agents to offer insurance through Lloyd's of London to other types of aquaculture businesses in the United States, but most of these attempts have failed.

The first federal crop insurance program for aquaculture in the United States was the Cultivated Clam Pilot Insurance Program that began in 2000. The program covers the growout phase of clam production and covers specific perils such as: oxygen depletion, disease, freeze, hurricane, and decrease in salinity, tidal wave, storm surge, and ice floe (Beach and Viatar 2008). Claims are based on inventory records. However, the program suffered high loss ratios in its initial years that prompted program changes. The result was higher premium rates that in turn reduced participation levels.

Some insurance programs for agriculture have been developed to protect against decreases in revenue. In such programs, the amount of payment from the insurance company is based on the yield multiplied by the market price.

The decision to purchase insurance should involve an analysis of the potential losses on the farm and the probability of those losses. This must be compared with the cost of the insurance, the ability of the business to recover from that loss, and its overall impact. Insurance can be provided from an independent insurance company or by the business itself, through investment of retained earnings in liquid financial reserves.

TOOLS FOR MANAGING CASH FLOW RISK

Cash flow budgets can be used to evaluate scenarios that represent varying levels of risk from a cash flow perspective. The expected net returns estimated for various outcome categories of different scenarios are entered into cash flow budgets. Details on preparing a complete cash flow budget are presented in Chapter 12. The expected net returns form the base scenario of the pro forma budget for the farm business. The net return values for the worst-case and other scenarios can be substituted sequentially into the cash flow budgets. The effects on various cash flow and financial indicators are then recorded, compared, and evaluated.

Financial ratios that can be calculated to help evaluate the effects of risk-rated outcomes include: (1) operating cash flow divided by current liabilities; (2) operating cash flow divided by total liabilities; (3) operating cash flow divided by the largest short-term credit balance; (4) current ratio; (5) rate of return on assets; (6) rate of return on equity; (7) ending cash balance; (8) ending operating loan balance; and (9) ending net worth. Details on calculating these ratios can be found in Chapter 4.

A system referred to as the risk-rated management strategies system was developed to evaluate the effects of risks that result from price and production risk in agriculture (Ikerd and Anderson 1983). The system involves identifying the possible types of outcomes, that is, best, optimistic, expected, pessimistic, and worst. The outcome that is the likeliest to occur is the expected outcome. For the other categories, the manager estimates the probability of its occurrence. This can take the form of, for example, a percent chance of occurrence of the worst possible outcome. For example, the worst possible outcome for a fish farm could be the loss of 50% of the fish on the farm. Such a loss would be represented by the level of negative net returns that would result from the 50% loss of fish. That level of negative net returns is multiplied times a 1% chance of its occurrence to obtain the expected net returns from that outcome. Developing these same expected returns for each of the categories of outcomes and then summing them develops an estimate of the expected net returns that accounts for the riskiness of the total enterprise.

MANAGING RISK THROUGH LEGAL TOOLS

The form of business organization can provide differing levels of protection from damages due to risk. Corporations and limited liability companies and cooperatives can provide greater protection than can a sole proprietorship. Splitting the farm business into separate legal entities can result in increased eligibility for various programs of assistance. Careful estate planning will reduce losses at the time of transfer of a farm business.

RISK ANALYSIS THROUGH THE USE OF RISK SOFTWARE

The concept of risk in agriculture enterprises refers to the level of uncertainty under which the different farm operations are carried out. In aquaculture farms, risk can be introduced by many factors. For instance, a failure in the aeration systems may lead to lethal dissolved oxygen (DO) concentrations during the evening hours, resulting in a massive fish kill. Incidents of this type bring about a reduction in the expected production levels. Risk is then introduced as the farm manager sees gross receipts drop to a level that may not be sufficient to cover operating expenses. Repeated episodes of low DO concentrations could cause substantial financial losses, eventually leading to the closure of the operation.

Uncertainty regarding the technical aspects of aquaculture production is not the exclusive source of risk. Fluctuations in input and output prices can make the difference between profit and losses. Of course, a higher degree of variability in prices will translate into a higher degree of risk in the operation. Unfortunately, risk cannot be measured from the enterprise budgets because these budgets are based on average values and reflect the expected level of net income generated by the farm. Sensitivity analyses represent an approach to measure risk. These analyses evaluate at what degree enterprise profitability is affected by changes in specific budget items. However, this type of analysis demands considerable time and effort, particularly if one wants to evaluate the effect of two or more changes.

Fortunately, there are a number of commercially available software products that allow the user to incorporate a variability component into spreadsheet-based enterprise budgets, and to measure the resulting risk level. Programs such as Crystal Ball™ (an add-in program to Microsoft Excel) or @RISK (an add-in program to Lotus) are risk analysis tools that can be used with standard spreadsheet software programs.

These programs expand the capabilities of spreadsheets by allowing the user to define probability distributions in the spreadsheet cells instead of just single values. Fish prices may fluctuate widely within relatively short periods of time, for example, and this variability is difficult to capture in a single enterprise budget analysis. If the long-run mean and standard deviation of prices, yields, or other parameters can be calculated, means and standard deviations can be entered into the spreadsheet. If the only data available are a likeliest, minimum, and maximum value, these can be used as a triangular distribution. The various

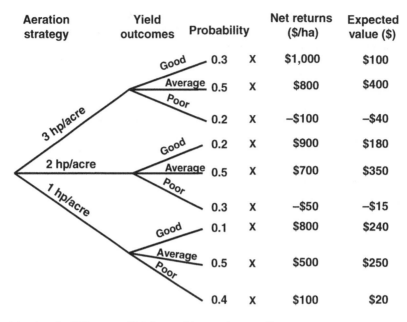

Aeration strategy	Yield outcomes	Probability		Net returns ($/ha)	Expected value ($)
	Good	0.3	X	$1,000	$100
3 hp/acre	Average	0.5	X	$800	$400
	Poor	0.2	X	−$100	−$40
	Good	0.2	X	$900	$180
2 hp/acre	Average	0.5	X	$700	$350
	Poor	0.3	X	−$50	−$15
	Good	0.1	X	$800	$240
1 hp/acre	Average	0.5	X	$500	$250
	Poor	0.4	X	$100	$20

Figure 8.3. Decision tree for 256-acre catfish farm, of increasing aeration.

risk analysis programs allow for a wide variety of distributions to be used.

The programs then run the enterprise budgets for a specified number of iterations. Each iteration uses different values for the parameters for which distributions have been entered. The result is a distribution of possible outcomes of profitability. The probability that net returns are positive can then be calculated.

RECORD-KEEPING

The record-keeping section of Chapter 16 presents information related to records to measure risk. To manage risk, additional records will need to be maintained for the various alternative production and management options that have potential to reduce risk in the business. Detailed information on the seasonality of prices of products with potential to diversify the operation along with details on the variation in yields of alternative crops will provide the basis for identifying risk-reducing alternatives. Files on insurance options and nonfarm alternatives to reduce risk will assist in making good decisions over time to manage risk.

PRACTICAL APPLICATION

Figure 8.3 illustrates a decision tree for the example of 256-acre catfish farm facing a decision to increase

aeration capacity. Increased aeration would allow the manager to maintain higher levels of DO in the pond. Especially during hot summers, yields would likely be higher with greater aeration. Thus, in Figure 8.3, yields increase by 500 lb/acre as aeration increases from 1 to 2 hp/acre. Increasing aeration rates to 3 hp/acre does not necessarily increase yields, but would likely increase the probability of achieving good yields. However, increasing the rates of aeration increases financial risk by increasing both fixed (capital costs associated with additional aerators) and variable costs (increased use of electricity). Overall, the expected value in this example of catfish production with 1 hp/acre of aeration is −$95/acre (summing the expected values of good, average, and poor yield outcomes) and $60/acre for both the 2 and 3 hp/acre options. The values for the probabilities and the net returns will vary with the overall management strategies used on the farm as well as the electricity rates faced by individual farmers.

OTHER APPLICATIONS IN AQUACULTURE

The onset of Taura syndrome virus (TSV) and white spot syndrome virus (WSSV) in the Honduran shrimp industry resulted in decreased net farm income by more than 80%. Valderrama and Engle (2004) developed a linear programming model to evaluate the effects of strategies to manage viral diseases in shrimp farms. The management strategies evaluated included reducing water-exchange rates, reducing stocking rates, and chemical treatment of ponds before stocking (Jory 1999). The model simulations of the alternative management strategies indicated that implementing the prevention programs increased net farm income by 47% above the levels of those of disease-affected scenarios.

SUMMARY

Aquaculture businesses must successfully manage a variety of risks. This chapter discussed a variety of techniques that can be used to manage these risks. Skilled managers, who use flexible production and marketing strategies, are better positioned to adjust and cope with risk. Share leases, strategic inventory management, contract sales, hedging, and marketing cooperation can be used to manage market risk.

Diversifying the business can help to offset production risks. The key is to produce different crops with prices and yields that vary in opposite ways. Establishing farms in different geographic areas and taking advantage of nonfarm revenues and investments can also result in reduced production risk.

Management can reduce financial risk by using fixed as compared to variable interest rates, self-liquidating loans, and maintaining liquid and credit reserves. While insurance can reduce risk for many crops, it is not readily available for aquaculture. This is due primarily to problems associated with stock control.

Decision trees, payoff matrices, and regret matrices can be used to help make decisions that incorporate the riskiness of alternatives. Cash flow budgets can also be used to provide insights into the effects of risk.

REVIEW QUESTIONS

1. Describe how flexible management can help to reduce risk in aquaculture.

2. Describe how contract sales and hedging can help to reduce risk.

3. Describe efforts to develop futures markets for aquaculture products, including salmon and shrimp, and indicate the outcomes of these efforts.

4. Explain how diversification can reduce risk on an aquaculture farm. Use several aquaculture examples.

5. Describe several types of financial responses to risk, with aquaculture examples.

6. Describe a decision tree. Draw one for an aquaculture decision to be made.

7. Develop a payoff and regret matrix for an aquaculture business decision.

8. List five decision criteria used to make decisions under risk.

9. Describe several insurance programs for aquaculture businesses.

10. Describe how to use cash flow budgets to plan to reduce cash flow.

ENDNOTE
1. Australia, Canada, Chile, Denmark, Finland, France, Germany, Greece, Holland, Honduras, Iceland, Ireland, Italy, Japan, South Korea, Mexico, New Zealand, Norway, Portugal, Spain, Sweden, Turkey, UK, and USA.

REFERENCES
Beach, Robert H. and C.L. Viatar. 2008. The economics of aquaculture insurance: an overview of the U.S. pilot insurance program for cultivated clams. *Aquaculture Economics and Management* 12:25–38.

Chen, Wei, Dan Kauffman, Daniel B. Taylor, and Everett B. Peterson. 2008. *Managing Flounder Openings for Maximum Revenue.* Paper presented at the annual meeting of the American Agricultural Economics Association, Orlando, FL.

Ikerd, John E. and Kim B. Anderson. 1983. *Risk-rated Management Strategies for Farmers and Ranchers.* OSU Extension Facts No. 159. Stillwater, OK: Cooperative Extension Service, Oklahoma State University.

Jory, Daryl E. 1999. Proper pond management for prevention of whitespot, Part 1. *Aquaculture Magazine* 25(5):92–95.

Kauffman, Dan. 2004. *Finding a Good Niche Market: What Producers Have Done to Create an American Market for Live Fish.* Roanoke, VA: 5th International Conference on Recirculating Aquaculture.

Martinez-Gamendia, Josue and James L. Anderson. 2001. Premiums/discounts and predictive ability of the shrimp futures market. *Agricultural and Resource Economics Review* 30(2):160–167.

Secretan, Paddy. 2003. *The Availability of Aquaculture Crop (Stock Mortality) Insurance*. Report to USDA-Risk Management Agency. Starkville, MS: Mississippi State University.

Valderrama, Diego and Carole R. Engle. 2004. Farm-level economic effects of viral diseases on Honduran shrimp farms. *Journal of Applied Aquaculture* 16(1/2):1–26.

9
Managing Labor

INTRODUCTION

Labor in aquaculture, while not the single greatest cost, does constitute one of the larger costs of aquaculture production. For example, it is the fifth largest cost in catfish production (Engle 2007), second largest cost in trout production in the United States (Engle et al. 2005), fourth largest cost in baitfish production (Stone et al. 2008), third largest cost in shrimp farming (Valderrama and Engle 2001), and fourth largest cost in salmon production (Bjorndal 1990). Perhaps even more important is that mistakes and errors committed by farm workers can have substantial effects on the overall farm business. Careful attention to management of labor can result in a more efficient and profitable business.

What is it that employees need to be motivated to be efficient and productive employees? According to Olson (2004), attention should be paid to: (1) responsibility; (2) authority; (3) accountability; and (4) compensation when managing labor on the farm. To have a sense of belonging, each employee needs to be responsible for specified tasks. At the same time, each needs to have the authority to do the jobs assigned to them. With responsibility also comes accountability. Setting clear standards for employees and measuring the employee's performance against those standards is the key to effective employee management. The level of compensation should match the levels of responsibility, authority, and accountability assigned to each employee.

This chapter discusses some of the changing requirements of labor for aquaculture businesses, estimating the quantity of labor needed, planning for the seasonality of labor demand and supply, risks associated with managing labor, accounting for the costs and efficiency

of labor, and substituting capital for labor. It concludes with a discussion of issues related to hiring, supervising, and evaluating labor as well as applicable state and federal laws.

CHARACTERISTICS OF FISH FARMING LABOR

Aquaculture production varies worldwide in its cost structure. In some countries, aquaculture continues to provide food for household consumption and some sales to generate cash for the family. In these situations, the labor force for fish production tends to be primarily the family itself with little or no hired labor. At the other end of the extreme are corporate farms in which all management and labor are hired. However, proper planning is required for adequate quantities of labor for all aquaculture production, whether the farm is a large commercial farm or a subsistence farming operation. Even if fish farming is primarily for subsistence or household consumption, labor shortages can result in a failed aquaculture enterprise.

Labor, unlike capital and some other types of resources used in businesses, cannot be stored. It is a flow input that must be used when available or it will be lost. Thus, it is important to schedule labor carefully to have adequate labor on hand when needed and to fully use labor when it is available.

While there are some exceptions, many fish farms continue to be family owned and operated. The exceptions are found in some concentrated segments of aquaculture, such as the salmon industry. However, in many sectors of aquaculture, the owner is actively involved in the farm. Spouses and children often provide key management and labor services as well. What is different about this type of resource is that family labor is frequently not compensated directly with wages and benefits. The benefits accrue from use of farm revenue

Aquaculture Economics and Financing: Management and Analysis, Carole R. Engle, © 2010 Carole R. Engle.

to cover living expenses. Thus, the level of compensation varies as net farm income varies from year to year.

Full-time employees and especially managers are not divisible types of resources. Economists refer to these types of resources as "lumpy" because they can be acquired only in whole units. Adding or losing a full-time manager can constitute a major change in the organization of the farm business. This is particularly true of smaller farms that have previously been run with just the owner-operator.

QUANTITY OF LABOR

Experienced managers have a good idea about how many workers of various skill levels are needed to run their business at its particular size and scope.

Those starting a new aquaculture enterprise will need to rely on other sources of information to identify the amount of labor that will be required. Enterprise budgets, extension personnel, other managers, owners experienced in the business, and consultants can provide valuable advice. Information on both the quantity of labor needed as well as the level of skill and training necessary for each position will be important.

The information obtained must then be developed into an objective estimate of the amount of labor required for the particular business under consideration. Each task required in the business must be itemized. The number of hours or days required to complete each task should be estimated.

Table 9.1 provides an example of a form that can be used to organize a schedule or plan for the use

Table 9.1. Form that could be Used to Develop a Labor Schedule and Assess Whether Adequate Labor is Available.

Labor category/task	Total hours available per year	Catfish	Hybrid striped bass	Total hours of labor available per year	Surplus or deficit of labor hours
Owner					
Family					
Manager					
Full-time employee					
Part-time employee					
Management					
Planning					
Marketing/flavor checks					
Feeding					
Analysis of farm performance					
Full-time labor					
Dissolved oxygen monitoring					
Mowing					
Equipment repair and maintenance					
Seining/harvesting					
Total					

of the labor on the farm. Specific tasks required are listed along with the estimated numbers of days and weeks required. The current workforce is itemized and the difference compared to identify whether there is a surplus or deficit of labor in the business.

The listing of the tasks required in the aquaculture business should be subdivided into tasks that are best performed by the owner, manager, full-time employees, and part-time help. In the U.S. catfish industry, for example, much of the feeding is done by the owner of the farm. This is because feeding is the only time that the owner can view the fish and assess their condition. Moreover, feed efficiency is one of the most important farm-level efficiencies. Insufficient feeding will result in suboptimal growth and yield while excessive feeding will result in wasted feed, deterioration of water quality, increased demand for aeration, and increased feed costs per pound of catfish produced.

The work tasks should also be subdivided by the types of enterprises in the overall farm business. For example, if a fish farm raises two species of fish as well as two types of row crops, then the labor required for each fish crop and each type of row crop should be divided out separately.

Once the tasks required on the farm and the amount of time required for each is completed, the total quantity of hours of labor required are summed and divided by the average workweek or the average number of work hours a year to determine the number of workers required. This is done for each type of labor category (i.e., management, full-time employees, and part-time employees). Estimates for the typical number of standard man-days or full-time equivalents (FTEs) per acre of land have been published for many types of row crops. However, this type of information is not typically available for aquaculture enterprises.

The FTE is a useful measure because it allows for summing the total labor used across both full-time and part-time employees. For example, one full-time worker employed year-round represents one FTE of labor. One seasonal worker who works full time for 6 months of the year represents 0.5 FTE. Similarly, a part-time worker employed year-round but who works only 25% of the time represents 0.25 FTE. The total number of FTEs of labor across the farm business can be summed.

The estimated number of employees for each category of employment should be compared with the number of employees available in each category. Deficits in a category will indicate the need to hire

additional workers and should provide an indication of the types of skills needed for additional employees.

LABOR SCHEDULE

There often is some degree of seasonality in both the labor available for the farm and the need for labor on the farm. Greater quantities of labor are frequently required during spawning season than during other times of the year. Similarly, harvesting activities typically are more labor-intensive than some other types of activities. These seasonal demands for labor may be difficult to meet for farms that are located in areas without a readily available skilled labor force. In some areas, the temporary workforce is composed primarily of high school students. Students often are not available until school is over. If the school year extends beyond the beginning spawning season, for example, it may not be possible to meet spawning season labor requirements with temporary student workers.

In some cases, seasonal labor demands can be met by hiring custom seiners and harvesters, or increasing the length of the workday. If hiring temporary workers is not an option, it may become necessary to purchase additional equipment to reduce the demand for labor during those periods.

The labor schedule that lists the jobs and tasks required on the farm should also schedule these tasks out by month or season of the year. A farm that also operates a hatchery will need to apportion its workers between the pond operation and the hatchery during spawning season. Insufficient labor during spawning season may result in a hatchery operation that is less efficient than it could be. Thus, for a farm with labor needs, as illustrated in Table 9.1, the table should be extended to divide the tasks and the amount of time required into estimates by month.

Thus, the goal of the labor schedule is to calculate the amount of labor required for each enterprise for each month of the year. The unit used typically is a man-hour that provides a common denominator to allow the manager to add up the labor requirements for each enterprise for each month.

Developing the labor schedule by month also demonstrates the months in which there may be a labor shortfall. A monthly labor schedule can be thought of as a cash flow budget in that periods of labor availability and shortages can be identified. A carefully thought out monthly labor schedule provides a sound basis from which to make plans in advance to hire additional workers to avoid labor shortages and costly downtime.

The labor schedule also indicates which types of crops and enterprises are the most labor-intensive and which contribute the most to labor deficits. Alternative actions that can be taken to manage labor deficits may include hiring additional seasonal or part-time help during those periods, changing from labor-intensive to less labor-intensive crops and enterprises, improving output and productivity per worker, or substituting equipment for labor.

QUALITY REQUIRED OF LABOR

Aquaculture technologies tend to be more complex and sophisticated than those in some other types of farming businesses. As fish farms have intensified, the complexity of the business and the skill level required of workers has also increased. Increasing regulations for aquaculture businesses have also increased the requirement for businesses to select employees with the capability of complying with the sampling, monitoring, and record-keeping practices mandated by regulatory authorities. Other requirements may include training for pesticide applications, Hazard Analysis of Critical Control Points (HACCP) plans, and other employee certifications.

Fish farms require a variety of types of skills. Managers must be familiar with fish culture practices, water quality, repairs and maintenance, purchasing, marketing, accounting, and labor management. Individual employees with each of these types of skills will be needed. Many farms require a dedicated mechanic to service and maintain the various types of equipment on the farm. Accountant and legal support will be necessary either on a full-time or contract basis.

ASSOCIATED RISKS

Hiring labor is a management decision that entails a certain degree of risk, as do other decisions. The immediate risk associated with labor involves not having sufficient labor available on the farm when needed to complete essential tasks in a timely manner. However, poor quality labor is another source of risk. Poor quality labor results in lower levels of efficiency and possibly expensive mistakes with equipment, fish mortalities due to poor handling, or poor morale due to conflicts among employees.

If the farm experiences high rates of labor turnover, the retraining costs and periods of insufficient labor can be costly to the business. Employee conflicts among each other can also result in reduced efficiencies and

higher costs. Risks associated with hiring labor can be reduced by planning carefully, maintaining open lines of communication, and maintaining the business current with the laws associated with labor.

LABOR COSTS

Labor costs can be treated as either fixed or variable costs, depending on the particular situation of the farm, the type of labor required, and the type of compensation structure and plan used on the farm. For example, the cost associated with a manager whose salary is supported by a long-term contract is a fixed cost. This is so because if the contract provides for that salary to be paid regardless of whether there is a good or poor crop that year, then the cost of the salary is a fixed cost. If part-time or seasonal workers are hired only if a good crop is expected and are laid off if there are problems with the crop, then those costs are variable costs because they vary with the amount of product produced.

LABOR-CAPITAL SUBSTITUTION

Purchasing additional equipment may reduce the need for labor in the business. Partial budgets can be used to compare the net change in the farm business from hiring labor or substituting some of the labor with new, labor-saving equipment.

Capital to purchase equipment has been used to replace labor over time in many developed countries. Wage rates have increased over time and the relative costs of labor have increased as a consequence. Thus, in some cases, equipment can be less expensive than labor to accomplish the same tasks. In other cases, equipment is substituted for labor to increase the scale and total volume of production. Use of equipment can ease the more burdensome tasks associated with aquaculture production and make work on a fish farm easier and more attractive. Appropriately sized equipment can increase the efficiency of labor, resulting in shorter times required to complete tasks.

LABOR EFFICIENCY

The efficiency of labor on the fish farm is affected by the type of business and the required skill level, the number and types of enterprises on the farm, the amount of equipment on the farm, and the level of training and skill of the workforce. The efficiency of labor can be measured in a number of different ways. FTEs can be calculated as a metric of the use of labor. Other indicators that can be used include: (1) value of farm production per person; (2) labor cost per crop acre; and (3) crop acres per person.

The value of farm production per person divides the total value of all the production on the farm by the FTEs on the farm. This indicator will be affected by the size of the business, the level of equipment available, and the specific types of fish crops raised on the farm.

The labor cost per crop acre divides the total labor cost on the farm by the number of crop acres on the farm. Reducing this value over time indicates that the efficiency of use of labor is increasing. It is important to include the opportunity cost of all family labor used on the farm.

The number of crop acres per person is calculated by dividing the total number of acres in production by the FTEs of labor. Increasing efficiency of labor will result in higher numbers of crop acres managed for each FTE.

In some cases, increasing the use of equipment on the farm can increase the efficiency of labor. However, this will only be the case if the substitution increases revenues or if total costs are reduced.

At times, labor efficiency can be increased by making policies and procedures simple. Well-maintained equipment, having all necessary tools and equipment available on the farm, and clearly defined instructions can all improve the efficiency of labor. Efficient communications are facilitated with adequate cell phones or other communication mechanisms for all relevant workers. Safe and comfortable working conditions can also enhance labor efficiency by improving employee morale and maintaining conditions more favorable to preventing injuries and illness.

MANAGING LABOR

Satisfied employees will generally be more productive employees who stay with the business longer. Maintaining low employee turnover will reduce the costs associated with training new employees. Job satisfaction stems from factors that include the ability to do a task well and be recognized for it. It also requires that the individuals find the job interesting and useful. Most individuals also need to feel that they have some control over some part of their tasks. Lastly, the opportunity for advancement is important to many aspects of job security.

On the other hand, job dissatisfaction can result from disagreement with rules and procedures, high stress, and dislike of and disagreement with one's boss. Low compensation frequently is a source of dissatisfaction as well as poor or unsafe working conditions.

Duties and responsibilities
1. Supervise other employees on farm
2. Stock, sample, and harvest fish
3. Feed fish daily
4. Have fish flavor checked
5. Monitor water quality
6. Purchase feed and other inputs
7. Maintain records and annual analyses of farm performance
8. Keep owner informed of progress and status of crop

Authorities
1. Purchase feed, fingerlings, and other inputs
2. Decide on stocking densities, feeding rates, aeration rates
3. Ensure that equipment is maintained properly

Authority
1. Manager will report to owner
2. Expenditures made must be appropriate and within budget
3. Machinery has to be maintained
4. Ponds and fish must be cultured and maintained properly

Evaluation standards
1. Equipment and tractors must be maintained properly
2. Fish must be fed properly
3. Ponds must be mowed properly
4. Purchases remain within budget
5. Tasks must be completed on time

Minimum qualifications
1. Experience in fish culture
2. Able to order and buy parts and supplies
3. Experience in supervising other employees

Figure 9.1. Job description for a fish farm manager for a 500-acre fish farm.

Figure 9.1 provides an example of a job description for a fish farm manager. The job description lays out the expectations and responsibilities of the position. This will help in recruitment, interviewing, and ultimately in selecting the individual for the position.

Moreover, the description of the various types of tasks required for the position provide an outline for the types of training that will be needed. Even if the worker has been employed in a similar position previously, training will assist the employee to adjust to the specific farm and its individual conditions and equipment. Maintaining an attitude of continuous improvement through training is an important characteristic of successful businesses. Training on safety procedures, for pesticide applicator certification, and attending extension programs and workshops are all important parts of managing labor in a way that will keep farm practices current and up-to-date.

All workers must also be evaluated at least annually and often more frequently. More frequent corrections and affirmation of proper actions are important for overall, long-run performance. More frequent evaluation sessions are needed early on to ensure that employees are trained properly and often will reduce the need for supervision over the longer term.

The basis for evaluation should be the specific tasks and performance standards specified in the original job description. Standards that are very specific and lack vague terms are more effective so that the employee understands exactly what is expected. Evaluations should reflect both the quantity and the quality of the work performed.

The evaluation conference is the time to discuss whether the employee met the standards or not. Constructive suggestions for how to improve performance are often more effective than strictly negative comments. If the overall evaluation score is negative, action must be taken to retrain the employee, reassign him/her, reevaluate the employee as to whether he or she may have just had a bad day, or to terminate his or her employment.

COMPENSATING EMPLOYEES

While most employees view compensation primarily in terms of the dollar amount of the wage, the total compensation may include other incentives such as

housing, a vehicle, and other benefits. Wage rates are affected by the cost of living in the area, prevailing wage rates, minimum wage rates, and other regulations that affect labor. If wage rates in the area are more than the farm can afford, alternatives to labor such as acquiring additional equipment should be considered.

Compensation should be based on the work that is needed from the employee. Many fish farms pay a bonus for production above a certain base level. This provides an incentive for the manager or worker to attempt to maximize yields. In industries with substantial economies of scale, this may be important to achieve a low-cost scale of production. However, if bonuses for increasing production lead the farm to produce at a level above the profit-maximizing level, a profit-sharing plan may be more appropriate. Farm units that operate under a corporate type of structure may have production quotas that may perform better with a bonus incentive structure based on the cost per pound of production.

Partial budgets can be used to evaluate various types of incentives. For example, Table 9.2 illustrates a partial budget to compare a profit-sharing bonus incentive program for the farm manager to a production bonus incentive program. The example is for a 500-acre catfish farm that currently averages 6,000 lb/acre/year, with the 6,000 lb/acre/year yield corresponding to the profit-maximizing stocking and feeding rates. However, the farm is capable of producing a maximum yield of 7,000 lb/acre/year. If an additional $0.10/lb/year is paid to the manager who then achieves the maximum possible yield of 7,000 lb/acre/year, the additional production would be 1,000 additional pounds per acre or 500,000 additional pounds across the farm. At an average price of $0.70/lb, additional revenue would be

Table 9.2. Partial Budget to Compare a Bonus Incentive Program Based on Sharing Profit as Compared to One Based on Yields above a Base Level, 500-acre Catfish Farm.

Additional costs	Additional revenue
Increase in costs of production of $0.09/lb for 7,000 lb/acre = $315,000	1,000 additional lb/acre/year @ $0.70/lb = $350,000
Salary increase of manager by $0.10/lb = $50,000	
Total additional costs = $365,000	Total additional revenue = $350,000
Reduced revenue = $0	Reduced costs = $0
Total additional costs = $365,000	Total benefits = $350,000
Net change = −$15,000	

Yields increase from 6,000 to 7,000 lb/acre/year.

$350,000. The manager would then receive the additional $0.10/lb, for a total salary increase of $50,000. However, because the 7,000 lb/acre/year yield is higher than the profit-maximizing level of yield, costs have risen faster than revenue. Additional aeration equipment are required and feed conversion ratios decrease with the result that the costs of production increase by $0.09/lb. Total additional costs are $365,000. There is no reduced revenue; thus, the total cost of switching to a production bonus system is $365,000, $15,000 greater than the additional revenue. The proposed change is not warranted, based strictly on the relative costs and benefits that would result from this change.

LABOR RELATIONS

There are a number of federal laws that regulate labor and wage rates. In the United States, minimum wage laws apply to employers who use more than 500 person-days of labor a quarter. Agricultural workers are not entitled to overtime pay, but state laws can enact more stringent levels. Social security (FICA, Federal Insurance Contributions Act) must be withheld from the wages, and the employer must match the amount charged to each employee. Farm owners are also obligated to withhold federal income taxes.

Some states require that employers maintain workers' compensation insurance for workers. Workers' compensation pays a fixed compensation for illnesses or injuries. Employers also typically pay both state and federal unemployment tax. These taxes are used to pay part of an employee's income if the employee becomes unemployed.

Other federal regulations (Fair Labor Standards Act in the United States) require that employees be at least 16 years old to be employed to do farm work after school hours. Moreover, additional restrictions apply to operating farm equipment or working with hazardous substances such as pesticides and other chemicals. The Occupational Safety and Health Act (OSHA) in the United States requires that workplaces be safe and healthy for employees.

In the United States, there has been an increasing proportion of fish farm workers hired from Mexico and other Latin American countries. This change in the labor force requires that farm owners become familiar with and understand the Immigration Reform and Control Act. While some Hispanic families have settled permanently in the United States, others prefer to return home for several months over the winter.

There are attorneys who specialize in providing immigration assistance to farm owners who hire Hispanic workers.

RECORD-KEEPING

Records to be kept on labor include information on each employee that includes the basic types of personnel information on contact and personal identification information. Hours worked by week and month should be recorded along with periodic performance evaluations.

PRACTICAL APPLICATION

A partial budget was developed to examine whether it was more or less profitable for a catfish farm to hire a crew for on-farm seining than to use custom harvesters for the various farm sizes. To do their own seining, farmers would need to purchase the equipment needed (seine, seine reel, boat, trailer, motor, and fish loader) and have an adequate amount of labor available. The seining equipment would cost about $13,100, with an annual depreciation of about $1,588. The smaller farm sizes would need only one set of equipment. On the 431-acre farm, one full-time seining crew could handle seining each pond twice a year. However, to seine three times a year would require a second seining crew. A 1,007-acre farm would need three sets of seining equipment and crew under these assumptions.

Table 9.3 itemizes the additional labor that would be required to do all the seining on farm. Ponds were assumed to be 10 acres each. Seining was assumed to require six people all together. The number of additional workers required was obtained by subtracting the number of individuals available (including unpaid family labor) from the six required. Ponds were assumed to be seined either two or three times a year. It was also assumed that a working day was 10 hours/day, and that the additional labor was valued at $6.50/hour.

Hiring enough full-time, permanent employees to provide for on-farm seining appears to be possible only on the two largest farm sizes, 431 and 1,007 acres (Table 9.3). The 60-acre farm, for example, would need to hire 4.5 additional workers, but only needs 720 hours of labor for seining. Hiring 4.5 full-time individuals will provide 2,600 hours of labor a year from each worker, or 11,700 hours a year, but only 720 hours are needed for seining, if ponds are seined twice a year (1,080 hours if seined three times a year). There would be enough seining work to justify hiring

Table 9.3. Labor Requirements and Value to add a Seining Crew to Varying Sizes of Catfish Farms.

Farm size (acres)	Workers available on farm	Number of ponds[*]	Additional workers required for seining and available hours[†]		Labor required for seining and value				
						Two times/year[‡]		Three times/year[§]	
	Number		Number	Hours	Hours[¶]	$	Hours	$
60	1.5	6	4.5	11,700	720	4,680	1,080	7,020
131	2	13	4	10,400	1,560	10,140	2,340	15,210
256	3	25	3	7,800	3,000	19,500	4,500	29,250
431	6	43	0	0	5,160	33,540	7,740	50,310
	6	100	6	15,600	5,160	33,540	7,740	50,310
1,007	14	100	4	10,400	12,000	78,000	18,000	117,000

[*] Assuming most ponds are 10 acres each in size.

[†] Seining crews are assumed to require 6 men each; 1 crew for 60-, 131-, 256-, and 431-acre farms. A second crew would be needed to seine each pond three times per year on the 431-acre farm and three crews would be needed for the 1,007-acre farm.

[‡] 10 hours/day (2,080/8 hours/day = 260 work days in a year); = 2,600 hours/year per worker. 6 men × 10 hours to seine, stake, and load a pond once = 60 hours × 2 times/year = 120 hours/pond = 180 for 3 times/year.

[§] Seining three times a year would require a second seining crew.

[¶] Labor is valued at $6.50/hour. Management is assumed to be present on farm.

the additional people needed for seining only on the 431- and 1,007-acre farms. Smaller farm sizes may be able to hire hourly labor as needed if there is an adequate local pool of labor. Similarly, for the 131-acre farm, four additional workers would be needed, but only 1,560 hours of labor are needed. In areas of labor shortages, the smaller-scale farms may need to depend on custom harvesters.

Table 9.4 presents the partial budgets developed to estimate the total net benefit of switching to on-farm seining from custom harvesting. There was no additional revenue. There would be a reduced cost of $0.02/lb of fish harvested that would no longer be paid to the custom harvester. The hauling cost of $0.03/lb would continue to be charged. The additional costs consist of the annual depreciation of the seining equipment (seine, seine reel, boat, trailer, motor, and fish loader), interest on the investment in the additional equipment (to account for either an additional equipment loan or the value of using that capital for something else), and the additional labor. There was no reduced revenue. An on-farm seining crew was assumed to be as proficient at seining as a custom harvester. The

total additional costs increased from $7,578 to $86,694 as farm size increased.

Total net benefits (total additional revenue − total additional costs) were negative for the two smallest farm sizes (60 and 131 acres). It was more profitable for the larger farm sizes to hire seining crews. Thus, based strictly on costs, it was more profitable for the smaller farms to use custom harvesters. However, there may be other reasons for on-farm seining, depending on markets and management strategies.

These results will vary with yields on the farm. The more fish seined on the farm, the lower the cost/lb of hiring a seining crew. Table 9.4 also indicates how high yields would have to be on the various farm sizes to justify the additional costs associated with on-farm seining. These ranged from 6,322 lb/acre for the 60-acre farm to 4,375 lb/acre for the 256-acre farm. Farms with yields higher than these breakeven yields may find it more profitable to switch to on-farm seining, if they can find appropriate amounts of labor. On the 60-acre farm, for example, this would mean being able to hire hourly labor as needed for seining events.

Table 9.4. Partial Budgets of Switching from Custom Seining to On-farm Seining.

Partial budget category	Farm size (acres)				
	60	131	256	431	1,007
Additional revenue	0	0	0	0	0
Reduced costs	$5,400	$11,790	$23,040	$38,790	$90,630
Total additional revenue	$5,400	$11,790	$23,040	$38,790	$90,630
Additional costs					
Annual equip. depreciation[†]	1,588	1,588	1,588	1,588	4,764
Interest, additional capital	1,310	1,310	1,310	1,310	3,930
Labor	4,680	10,140	19,500	33,540	78,000
Reduced revenue[‡]	0	0	0	0	0
Total additional costs	$7,578	$13,038	$22,398	$36,438	$86,694
TOTAL NET BENEFITS	−$2,178	−$1,248	$642	$2,352	$3,936
Breakeven production (lb)[§]	379,350	651,900	1,119,900	1,966,800	4,479,600
Breakeven yield (lb/acre)	6,322	4,976	4,375	4,563	4,448

[*]$0.02/lb harvested charged by custom harvesters.
[†]Total cost of one set of equipment needed for seining is $13,100. This includes a seine, seine reel, boat, trailer, and motor, and a fish loader. The 431-acre farm will need one set if seining twice a year and two if seining three times a year. The 1,007-acre farm will need three sets of seining equipment.
[‡]An on-farm seining crew is assumed to be equally proficient as a custom harvester.
[§]Divided additional costs by the $0.02/lb saved by on-farm seining. Assuming hiring the hours required to do all the seining and seining two times/year.

OTHER APPLICATIONS IN AQUACULTURE

Agriculture and aquaculture in Rwanda are primarily subsistence farming activities (Jones and Egli 1984). Engle et al. (1993) examined the trade-offs between the benefits and labor resources required for fish production in Rwanda.

A survey was conducted of 55 fish farmers in Rwanda. The most striking result from the survey was that the principal resource used to produce fish in Rwanda was labor. Ponds were constructed by hand by family members and thus represent an investment of labor, not capital as in cash economies. The primary variable resource used to produce fish was compost made from vegetative matter that was cut or obtained from weeding other crops. Thus, its availability was related to the labor used to acquire it. The greatest quantities used of labor were to build the ponds and to cut grass and pull weeds to make compost. Measures of labor efficiency were estimated to compare the attractiveness of fish production to that of producing other types of crops and to produce fish across different climatic zones. Such an indicator provides a means for evaluating management changes related to the allocation of labor to improve farm-level efficiencies.

SUMMARY

This chapter reviewed some basic principles related to managing labor resources in an aquaculture business. In some types of aquaculture businesses, labor costs can be substantial. Much of the work in aquaculture requires high degrees of skill and a certain degree of specialized skill. It is critical to plan carefully to be certain to have the quantity and quality of labor required for the farm business to be efficient. Carefully prepared lists of job tasks and a detailed labor schedule by month will provide a basis for the manager to assess the likelihood of labor shortages at certain times of the year. Options to manage labor shortages can include hiring custom operators, temporary workers, or even purchasing additional equipment to substitute for labor. Various indicators of labor efficiency are presented and discussed. The chapter concludes with a discussion of regulations that affect labor.

REVIEW QUESTIONS

1. What are some of the unique characteristics of agricultural labor?

2. Describe the best ways to plan out the quantity and quality of labor needed for the farming business.

3. What is an FTE and how is it used?

4. Describe the effects of the seasonality of labor on the overall efficiency and operation of the farm business.

5. Are labor costs fixed or variable costs? Use examples from aquaculture to justify your answer.

6. Can capital be used to substitute for labor? Give examples from aquaculture farms.

7. List five different indicators of the efficiency of use of labor and what each describes.

8. List the major federal regulations that affect farm labor.

9. Describe various types of risk associated with managing labor.

10. Describe the management process involved in recruiting, selecting, and retaining farm workers.

REFERENCES

Bjorndal, Trond. 1990. *The Economics of Salmon Aquaculture*. Oxford: Blackwell Scientific Publications.

Engle, Carole R., Marcie Brewster, and Felix Hitayezu. 1993. An economic analysis of fish production in a subsistence agricultural economy: the case of Rwanda. *Journal of Aquaculture in the Tropics* 8:151–165.

Engle, Carole R. 2007. *Arkansas Catfish Production Budgets*. Cooperative Extension Program MP466. Pine Bluff, AR: University of Arkansas.

Engle, Carole R., Steeve Pomerleau, Gary Fornshell, Jeffrey M. Hinshaw, Debra Sloan, and Skip Thompson. 2005. The economic impact of proposed effluent treatment options for production of trout *Oncorhynchus mykiss* in flow-through systems. *Aquacultural Engineering* 32(2):303–323.

Jones, William I. and Roberto Egli. 1984. *Farming Systems in Africa; the Great Lakes Highlands of Zaire, Rwanda and Burundi*. Technical Paper No. 27. Washington, DC: The World Bank.

Olson, Kent D. 2004. *Farm Management: Principles and Strategies*. Ames, IA: Iowa State University Press.

Stone, Nathan, Carole R. Engle, and Eric Park. 2008. *Production Enterprise Budget for Golden Shiners*. Southern Regional Aquaculture Center Publication No. 122. Stoneville, MS: Southern Regional Aquaculture Center.

Valderrama, Diego and Carole R. Engle. 2001. Risk analysis of shrimp farming in Honduras. *Aquaculture Economics and Management* 5(1/2):49–68.

Section II
Economic and Financial Analysis of Aquaculture Businesses

INTRODUCTION TO SECTION II: ECONOMIC AND FINANCIAL ANALYSIS OF AQUACULTURE BUSINESSES

The chapters in the following section will present detailed information on how to develop the key financial statements needed in the economic and financial analysis of an aquaculture business. This section is written for students and those who wish to learn how to develop and complete each type of analysis, not just interpret it.

The specific analyses included in the following chapters include: enterprise budgets (Chapter 10), par-
tial budgets (Chapter 10), balance sheets (Chapter 11), income statements (Chapter 11), cash flow budgets (Chapter 12), investment analyses (Chapter 13), and loan amortization schedules (Chapter 14). Each chapter begins with an introduction to the analytical methods presented. The bottom line of each analysis is described along with its interpretation. The unit of analysis and structure of each analysis are outlined in each chapter. Valuation approaches for each are discussed and the mechanics of each developed in detail. Challenges, limitations, and pitfalls associated with each are discussed.

10
The Enterprise Budget and Partial Budgeting in Aquaculture

INTRODUCTION

The enterprise budget has been the cornerstone of most production economic analysis. Indeed, some aquaculturists consider the enterprise budget to be the only type of economic analysis in aquaculture. This is because aquaculture, at least in the Western hemisphere, is still a relatively new type of production activity or enterprise. For new activities, the first economic question that arises is: is it profitable?

For example, in the U.S. catfish industry, early catfish budgets were developed to answer a simple question of whether or not catfish farming was profitable. There are numerous examples of these estimates in the literature in the late 1960s and early 1970s. The majority of these showed that it was possible, even with the technologies available at that time, to make profits from catfish production. Later efforts developed more comprehensive enterprise budgets that enabled researchers to explore fundamental economic relationships.

Enterprise budget analysis continues to be an important first step in analysis of the economics of production, particularly as new technologies continue to be developed.

To a farmer, any published enterprise budget analysis should be viewed strictly as a guide to finding the answer for a particular farm. The budget itself will not have the answer for any particular farm. This is because all enterprise budgets are based on a host of assumptions for each individual budget. Budget analysis is a necessary step in most production economics research, but it is a static analysis that cannot take into account factors such as: (1) fluctuations in prices, yields, and

costs; (2) farming systems interactions among labor, marketing, and resource constraints; (3) issues related to the production or financial risk of a particular technology; (4) social, economic, or welfare effects of the technology; or (5) market factors.

Enterprise budgets include all the relevant costs and returns from a particular enterprise. However, to analyze the effect of making a small change in the production process, a partial budget is often more appropriate. The partial budget is used to calculate the expected change in profit for a proposed change in the farm business. A partial budget analysis includes only those income and expense items that will change if the modification to the farm plan is made.

This chapter describes the structure of an enterprise and partial budgets and illustrates how to calculate the various costs and revenues required. It will also provide examples of how to interpret and use the information provided by enterprise and partial budgets.

THE ENTERPRISE BUDGET

Enterprise budgets provide a fairly simple and readily understandable measure of profitability of a specific production activity (enterprise). The enterprise budget indicates whether the activity proposed is profitable or not. It also indicates how profitable the enterprise is likely to be. Enterprise budgets typically also include an analysis of breakeven prices and yields. These provide valuable information related to the price that must be obtained for the enterprise to be profitable (breakeven price) and the yield that must be produced for the enterprise to be profitable (breakeven yield).

An enterprise budget is an estimate of all income and expenses associated with a specific enterprise and an estimate of its profitability (Kay et al. 2008). A budget is a snapshot that freezes all assumptions and values at

Aquaculture Economics and Financing: Management and Analysis, Carole R. Engle, © 2010 Carole R. Engle.

one point in time; thus, it is a static analysis. It provides an estimate of the potential revenue and expenses for a single enterprise on the farm, and also an estimate of profitability for the set of assumptions and values selected for the budget. Different levels of production and different technologies each require a separate enterprise budget. Thus, a shrimp enterprise may have a series of budgets, each for a different stocking rate.

The bottom line of the enterprise is what indicates both whether the enterprise is profitable and the degree of its likely profitability. This measure is referred to as net returns. If the estimated value of the net returns is positive, then the enterprise is profitable. If the net returns value is a negative number, then the enterprise is not profitable.

The magnitude of the value of the net returns indicates how profitable the enterprise is. For example, if a tilapia farm shows net returns of $1.00, it is a profitable enterprise. However, at $1.00 of profit for the entire enterprise, it clearly is not a very profitable enterprise. These net returns are not likely to be high enough for an individual to be attracted to this enterprise. Nevertheless, it must still be judged as profitable as long as the value is positive.

An enterprise budget must be developed for a specific type and size of production unit. Writing out a clear description of the specific production unit to be evaluated will aid in identifying the types of costs to be incurred and in developing estimates of the quantities of inputs and resources to be used. A budget can be developed for a variety of sizes and types of enterprises. Thus, a budget can be developed for an individual pond, or a whole-farm budget can be developed for the entire farm. The decision as to the best unit to use for the budget depends on the specific use of the analysis. If the intent is to compare pond-based production methods, a budget for an individual pond may be best, but if the question to be answered is a farm-wide technology, then a whole-farm budget may be most appropriate.

Different production technologies will require separate enterprise budgets. Ponds, raceways, cages, net pens, and tanks will each require separate budgets. The relative importance of different types of costs will vary across different production technologies and systems. Different types of inputs may be used for different production systems. Thus, the first step in constructing an enterprise budget is to identify the specific production system, species to be raised, and critical production and input levels. These include the size of fish to be stocked, the stocking density, yield (lb/acre) of marketable-sized product, and feed conversion ratio.

Most enterprise budgets are developed for a time period of 1 year, but in cases of a multiyear crop, it may be more useful to develop a 2-year budget.

STRUCTURE OF AN ENTERPRISE BUDGET

A *budget* is generally defined as a structured system for estimating values of the revenue generated and the costs incurred. Thus, the enterprise budget itemizes the types, quantities, and prices of products to be sold by the enterprise. All costs associated with that particular production activity are also itemized by types, quantities, and prices. The revenues are then compared to the costs to determine whether there is adequate revenue to cover all costs. If so, the enterprise is profitable.

The most useful enterprise budgets are those that include sufficient detail to fully understand each specific line item. Column headings typically include the name of the item, its description, the unit of measure used, its cost for the unit specified, the quantity to be used of that unit for the time period specified, and the total cost or value for that particular item for the specified time period, production system, production technology selected, and level of production selected. The first column lists the various items to be included. The second column lists a description of each item. It is important that each item be thoroughly described. The next column indicates the unit used. This must be specified carefully and is usually selected based on the most common unit of purchase or sale.

Estimates of revenue typically are the first category in the budget. A separate line is used for each type of product to be sold. For example, if an enterprise budget is developed for a pond that is in production of both shrimp and tilapia, there would be two separate revenue lines, one for shrimp and one for tilapia. If prices differ by size for shrimp and tilapia, then additional revenue lines are required. If, for example, the price per pound of large tilapia (greater than a pound) is higher than the price of smaller tilapia, then separate revenue lines are needed for each size of tilapia. The name of the item might be tilapia for each row, but the description would specify the size range within which price remains the same. The description might simply be "average weight greater than a pound" or "average weight less than a pound." The unit of sale is specified for each type of product sold. For most types of fish, the unit of sale is often a weight measure, such as a pound or a kilogram. However, other units are sometimes used. Fingerlings may be sold as a price per 1,000 fingerlings or shrimp postlarvae may be sold as a price per million postlarvae. The unit should reflect the most commonly used unit of sale for ease of understanding. Prices used should be based on historical price levels, trends, and price outlooks.

The next column indicates the price of the unit for sale. If salmon is purchased by the pound, then the price should be the price per pound. If the product to be sold is a unit of 1,000 fingerlings, then the price should be the price per 1,000 fingerlings. Similarly, for shrimp postlarvae sold in units of a million postlarvae, the price should be the price per million postlarvae. The next column indicates the quantity to be sold. This quantity should represent the quantity of product projected to be sold. It must be expressed in appropriate units. If the unit of analysis of the enterprise budget is one 10-acre pond, then the pounds of fish to be sold should be the total weight of pounds to be sold from that one 10-acre pond. It is incorrect to specify the quantity as a yield in lb/acre, or to include the total weight from the entire pond if the budgetary unit of analysis is one 10-acre pond. The quantity should be based on historical yields and trends and the type and amount of inputs to be used. The price expected to be received for the products is multiplied by the quantity projected to be sold to obtain the total revenue for the final column.

Once the revenues are estimated, the next broad category for the budget is to estimate the costs. There are two broad categories of costs, fixed and variable costs. These are also referred to at times as ownership (fixed) and operating (variable) costs. The difference between the two types of costs is in their respective relationship to the volume of production. Variable costs increase as production levels increase. For example, if a farmer stocks ponds at a higher rate and produces more pounds of fish, feed costs would go up because the farmer would feed more feed at the higher stocking and production levels. Because the quantity of feed fed increased with the higher production levels, feed would be a variable cost then. Similarly, the farmer would likely have to aerate more at higher stocking and feeding rates. Thus, the total quantity of electricity and fuel used on the farm would increase as the production levels increased. Thus, electricity and fuel costs would also be variable costs.

In the enterprise budget, the variable costs frequently are listed first. Typical types of variable costs include fingerlings, feed, electricity, fuel, repairs and maintenance, any chemical use, and labor, particularly hired labor. These are all examples of the types of costs for which the total amount of the cost will increase with expanded production levels and will decrease if production levels are decreased. Expanding production quantities by building more ponds will require more fingerlings, more feed, more energy, and more labor. Conversely, selling off a portion of the farm will result in a decrease in the total quantity of fingerlings, feed, and energy used. Thus, these types of inputs are variable inputs and their costs are variable costs.

The total cost of each variable cost item is calculated in a manner similar to that described earlier for revenue. For instance, for feed, each type of feed that has a different price and quantity is listed on a separate row. For example, a farm may purchase one type of feed for its growout ponds and a different type of feed for fingerlings. If the budget is for the entire farm, then both types of feed must be listed in the enterprise budget. The unit of sale is included. Typically, the unit of sale for feed is ton. In countries where metric units of measure are used, this would be a metric ton. The price, then, is the price per ton for the growout feed used. The price per ton of fingerling would be specified in the proper row. The price of each feed is multiplied by its respective quantity to obtain the total cost for each type of feed. Input prices can be determined by contacting input suppliers. Fuel use and machinery repairs are frequently calculated by dividing the total expenses for the farm by the number of acres and multiplying by the number of acres per pond. Interest on the operating capital is included in variable costs. It is charged for the period of time it has been borrowed. For example, if operating capital is borrowed for only 10 months at an interest rate of 6%, annual percentage rate, the proportion of the year (10/12) is multiplied times the annual percentage rate (6%). The same process is repeated throughout the budget. When all the variable cost items have been estimated, these are summed to obtain total variable costs.

In the next section of the enterprise budget, fixed costs are specified. Fixed costs include depreciation, interest on investment, taxes and insurance, and any other costs that are not related to the level of production of the business. Inset 10.1 describes the calculations to determine annual depreciation. Fixed costs, on the other hand, are costs that are incurred irregardless of the level of production. Fixed costs are incurred whether there is a bumper crop produced or no crops are produced. Examples of fixed costs are taxes, contractual salaries, and insurance. Land and property taxes would have to be paid even if the entire crop of fish was lost due to disease. The same amount of land and property taxes would be paid if there were no fish produced that year or if there was a bumper crop of fish produced. If the manager or foreman of a farm is contracted in such a way that he/she would need to

Calculating Depreciation

Depreciation must be calculated for all capital goods (goods with a useful life greater than a year). Capital goods such as buildings and equipment are necessary for aquaculture production. However, their use in aquaculture production results in their aging, obsolescence, and become worn out, losing value as a consequence. This loss in value is a business expense because it is related directly to the asset's use to produce revenue and profit. To be considered a depreciable asset, the capital good must have a useful life of more than 1 year and a useful life that can be quantified. From an accounting perspective, the cost of the asset must be matched with the periods when it produces revenue. Calculating depreciation can be thought of as putting aside money within the business to replace equipment and buildings at some point in the future. It is also a way to spread the purchase cost across the years that it will be used; it would not be correct to charge it out all in 1 year. For example, if a piece of equipment will be used for 5 years, it would not be correct to charge the entire cost to purchase that piece of equipment in just 1 year. Thus, calculating annual depreciation is a way to determine an annual replacement cost for each piece of equipment and for each capital item. Annual depreciation costs represent the loss in value of a capital good due to use, wear and tear, age, and obsolescence. Land, on the other hand, is not depreciable because it has an unlimited useful life.

There are different ways to calculate annual depreciation, depending upon the use. For income tax purposes, the double declining balance method (also known as diminishing-balance or reducing-balance method) is frequently used because it allows for greater deductions in the early years. Its calculation is based on rapid decline in value during the early years.

An enterprise budget is developed for only one period of time, however, and represents a "typical" or "average" year. Depreciation methods that calculate different amounts of depreciation for different years pose problems for enterprise budgets. Thus, for purposes of an enterprise budget, the straightline method to calculate annual depreciation is the best. Straightline depreciation results in the same amount of annual depreciation for each year. To calculate depreciation, the initial cost of the asset is needed. Its useful life in the business must be estimated. Finally, its resale value at the end of its useful life must be known. The initial cost is the purchase price. The useful life of equipment can be found by consulting with other fish farmers using that same type of equipment or from the farm's own records.

It is calculated as follows:

$$\text{Annual depreciation} = \frac{\text{initial purchase cost} - \text{salvage value}}{\text{years of useful life}} \tag{10.1}$$

This calculation represents the difference between the initial purchase price and its resale value, divided by the number of years it is likely to be used in the business. In reality, there is little salvage value for most types of equipment on a fish farm. For practical purposes, therefore, the following equation can be used to calculate annual depreciation:

$$\text{Annual depreciation} = \frac{\text{initial purchase costs}}{\text{years of useful life}} \tag{10.2}$$

The value calculated is also referred to as "book value." The term "book value" is used to refer to the value of each asset as recorded in the depreciation schedule. Book value is calculated by subtracting the accumulated depreciation from the asset's cost. However, the actual value of a particularly asset is the market value at the time of sale, regardless of the book value. With the straightline method, the amount calculated for annual depreciation is constant across the years that the piece of equipment will be used. This makes it a preferred method when compiling an enterprise budget. This is because the enterprise budget is based on an average or typical year.

be paid regardless of the level of production, then that salary is also a fixed cost. Thus, the costs are "fixed" with respect to the quantity of product produced.

The individual fixed costs are summed to obtain total fixed costs. These are added to total variable costs to obtain total costs. Net returns are obtained by subtracting total costs from gross returns. This is the measure of profit for this business.

Breakeven prices and yields offer additional insights into the overall feasibility of the operation. Breakeven price above variable cost is calculated by dividing total variable costs by the quantity produced. Breakeven price measures the cost of production of a single unit of the product. If the product can be sold for a price that is more than its cost of production, then a profit is generated.

Breakeven yield is calculated in a similar manner. Breakeven yield above variable cost is calculated by dividing total variable costs by the price. It measures the yield necessary to cover all costs.

Value Estimation of Parameters Used in the Budget

The relevancy and accuracy of budget estimates depend on the choice of values assigned to each item included in the budgets. It is critical to spend adequate time to identify values that are both appropriate and realistic. In general, the price and quantity variables selected should be conservative values. Any errors made should be on the side of caution by selecting prices of input costs that tend toward the high side and prices of products to be sold that tend to be on the lower side. The best approach is to use long-term datasets to estimate 10-year average prices of products to be sold and inputs to be used. Price information for major aquaculture species can be found in the reports online from the National Agricultural Statistics Service, U.S. Department of Agriculture (http://usda .mannlib.cornell.edu/MannUsda/viewDocumentInfo. do?documentID=1375). Long-term price datasets for electricity, gasoline, and diesel can be found on the Department of Energy's web site (www.eia.doe.gov) and wage rates on the Department of Labor's web site (www.bls.gov).

There are two main approaches to estimating the quantities of product sold and quantities of inputs used: economic engineering approaches and cost of production surveys. The majority of aquaculture budgets have been developed using economic engineering approaches. In this approach, a hypothetical farm sit-

uation is defined. Types of input items are itemized based on experience with the production system and species selected and from the aquaculture research literature. Quantities used of various input items typically are based on recommended management practices from Extension specialists and from the aquaculture research literature.

In many cases, the economic engineering approach is the only feasible approach. This is particularly true for newer production systems and species because there are no long-term datasets of prices. Cost of production surveys are not typically done for new systems and often there are no commercial producers to visit.

Budgets developed using economic engineering techniques should be used with caution. Values selected for use in economic engineering budgets frequently are "best case" scenarios in which optimistic parameter values are used. Sometimes, this is done intentionally to demonstrate the profitability under well-managed and optimum conditions. In this manner, economically engineered budgets can serve as a target for others to aspire to achieve. Problems occur, however, when a lender, investor, or farmer believes a budget to be based on average, not "best case" values. Misguided expectations for profit can lead to serious financial losses. These lead to negative perceptions on the part of lenders that may be reflected in an unwillingness to grant additional loans for aquaculture. Chapter 15 on use and misuse of budgets discusses these problems in greater detail.

A preferable approach for the development of enterprise budgets for aquaculture is to use values obtained from cost of production surveys. Surveys of commercial growers reflect the costs, production levels, and farm prices that farmers face in reality. These values reflect the risks, problems, and setbacks that occur in the course of operating a farm and tend to be the most realistic values available.

Unfortunately, while cost of production surveys are done routinely for a number of agricultural and livestock crops in the United States, there has never been a publicly available national cost of production survey done of aquaculture.

Cost of production surveys allow for sorting cost data by farm size, production intensity, labor structure, and other factors. This allows for development of cost estimates that are more accurate and relevant to the appropriate farm situations and circumstances. It should be noted that as with any data collection methodologies, improperly designed questions and questionnaires, poor sampling design, and lack of care in implementation of surveys can result in inaccurate and unreliable data. However, properly designed and implemented surveys often result in reliable data.

There are different accounting practices that can be used to assign values to assets. These include: (1) market value; (2) cost; (3) lower of cost or market value; (4) farm production cost; and (5) cost less accumulated depreciation. Market values are assigned according to current market prices of the product to be sold. Market values work best for assets that will likely be sold in a relatively short time and for which current market prices are available.

The cost method is used for assets that have been purchased recently, particularly if cost records are available. Supplies such as feed and fertilizer are commonly valued with the cost method. Noncurrent assets like buildings and equipment lose value over time and should not be valued with the cost method. The lower of cost or market valuation method compares values from the market and cost methods, and selects whichever method compensates for short-term increases or decreases that may distort values.

The farm production cost method can be used to value assets grown or produced on the farm. Submarketable stocker fish, for example, can be valued this way. Profit and opportunity costs are not included as valuation methods, but rather actual production expenses incurred are used as the basis for the calculation. Depreciable assets can be valued using cost less accumulated depreciation. Book values are used as the current value for this method.

Profitability Measures that can be Obtained from the Enterprise Budget

The most important measure of profitability from the enterprise budget is net returns. Net returns are calculated by subtracting total costs from total revenue. An intermediate measure is to calculate income above variable costs by subtracting total variable costs from gross revenue. Also referred to as gross margins, income above variable costs provide a measure of whether the business can continue to operate in the short run. For long-term profitability, net returns that account for fixed costs must be calculated.

Net returns, if not calculated properly, may not be comparable across enterprise budgets. Net returns represent the earnings of the resources used in the production process. However, there are many different types of farm businesses and business structures, and net re-

turns can be used to indicate which costs have and have not been subtracted out.

The reporting convention used is to report "net returns to" whatever resources have not been subtracted out. For example, if the owner's labor and management costs have not been subtracted out, net returns should then be reported as "net returns to owner's labor and management." The value of the owner's management (opportunity cost of management) is difficult to estimate and is often omitted from the enterprise budget. If the budget has no cost for land, then net returns should be identified as "net returns to land." Few budgets account for risk and, thus, net returns should be expressed as "net returns to risk."

To calculate economic profit, the costs of all resources used in the production process must be accounted for. To do so, opportunity costs, or the value of what that resource would earn if used for something else, should be charged for all resources used. This measure of economic profit is different from accounting profit, as determined by the income statement. Opportunity costs of land can be charged at the value of cash rent or the value of an acre times the opportunity cost of the owner's capital.

Some individuals wish to include a measure of the return on long-term investment in enterprise budgets. Chapter 13 provides details on developing investment analyses and various comprehensive measures of the returns on investment. An enterprise budget is not the best means to estimate a return on investment, as Chapter 15 points out. However, a simple indicator is the rate of return on the average investment. To calculate the rate of return on the average investment, the estimated net returns are divided by the sum of the nondepreciable investment items and half of the depreciable investment items, and multiplied by 100. The value of the depreciable items is cut in half because it loses value (depreciates) over time through use, age, and obsolescence. Because the enterprise budget reflects a single point in time, it is not possible to calculate the long-term return on investment and, hence, an "average" return on investment is used, if necessary, to provide a generalized view of the return. It is preferable to use the methods described in Chapter 13 whenever possible.

Mechanics of the Budget

In addition to the basic calculations in the budget, there are other relationships that can be checked to ensure internal consistency of the budget. The feed conversion ratio (FCR), for example, is a commonly used measure of the efficiency of feed use. It is calculated by dividing the weight of the feed fed by the weight gain of the fish. Thus, for every pound of weight gain of fish, the FCR indicates how many pounds of feed were required to produce that weight of fish. Calculating the FCR from the budget values of quantity of feed fed and quantity of fish sold provides a check on the validity of these values.

Another check for validity is to compare the breakeven prices and quantities with the corresponding budgeted values as a check on the overall conclusion relative to profitability. For example, if net returns are positive, the enterprise is deemed to be profitable. This also means that the breakeven price calculated in the budget must be less than the market price used in the budget. Because the breakeven price is that price that just covers costs, profits are incurred only when the market price is higher than the breakeven price. In a similar fashion, for an enterprise to be profitable, the breakeven yield must be lower than the yield used as the projected sales volume for the enterprise.

PRACTICAL APPLICATION OF AN ENTERPRISE BUDGET

As an example of the application of enterprise budgeting to aquaculture, an example is provided of a comprehensive enterprise budget for a 256-acre catfish farm. This budget was developed from a cost of production survey conducted in Arkansas (Engle 2007), and reflects cost structures on commercial operations at the time. To develop a comprehensive enterprise budget, the first step is to estimate the capital investment items needed. Capital investment items are those that will be used for a time period greater than a year; thus, it is incorrect to charge out the total amount of the investment in a 1-year enterprise budget. Table 10.1

Table 10.1. Capital Investment in Land and Buildings Required for a 256-acre Catfish Farm.

Item	Unit cost $	Number of units $	Total cost $
Land	822	256	210,432
Wells	8,000	6	48,000
Ponds	1,398	256	357,888
Total			616,320

includes the capital investment items for the 256-acre catfish farm. Major investment items on this type of farm include the land, construction of ponds with water supply, drainage, electrical service, and construction of the farm headquarters. The unit of measure is specified (e.g., "acre" for land, m³ for cages) along with the cost per unit and the number of units. Multiplying the cost per unit times the number of units provides the total investment cost for each item.

For depreciable items (those that lose value over time), annual depreciation must be calculated. Inset 10.1 provides additional detail on annual depreciation. Because the purpose of an enterprise budget is to estimate an average annual amount of profitability for a "typical" year, straightline depreciation is the best method for an enterprise budget. Some other methods result in different amounts of annual depreciation for different years; thus, it is difficult to select which amount is appropriate. To calculate annual depreciation, the number of years of useful life must first be estimated. This value will vary from farm to farm depending on the care and maintenance provided by the farmer. If available, farm records can be used as a basis for determining the number of years of useful life for each item. If not, examination of enterprise budgets for other similar types of crops and conversations with experienced farmers can provide a good basis for this value.

Table 10.2 presents the equipment investment required for the 256-acre catfish farm. When developing an equipment list, it is important to include all types of equipment needed for the entire business. Annual depreciation values are developed for each item and summed for the entire farm, as described earlier for the other investment items.

Table 10.3 presents the annual costs and returns and constitutes the enterprise budget itself. For this farm, average annual yield is 4,500 lb/acre/year. Fish are grown in multiple batches with an average annual stocking rate of 5,690 fingerlings/acre/year. The average feed fed is 4.78 tons/acre/year. This amount of feed equates to a feed conversion ratio of 2.12.

Total pounds of catfish sold per year from the farm are 1,152,000 lb. In the long-run, average cost of production, $0.70/lb, total revenue is $806,400/year. Variable cost includes feed, fingerlings, labor, electricity, fuel, repairs and maintenance, and harvesting costs. Of these, feed represents 45% of the total variable costs of production and 36% of the total costs of production. Interest on operating capital is another variable cost of production. This line of item represents the interest

Table 10.2. Equipment Cost and Annual Depreciation for 256-acre Catfish Farm.

Item	Unit cost $	Number of units	Total cost $	Years of useful life Years	Annual depreciation $
Tractors	20,000	7	140,000	20	7,000
Trucks	17,000	2	34,000	8	4,250
Mower	8,365	1	8,365	5	1,673
Electrical aerators	3,460	25	86,500	5	17,300
PTO aerators	2,065	8	16,520	10	1,652
Feeder	5,000	1	5,000	5	1,000
Feed bin	5,500	3	16,500	8	2,062
Pump	1,600	1	1,600	10	160
Office, shop, tools	15,000	n.a.	15,000	5	3,000
Utility trailer	800	1	800	10	80
Storage container	1,000	1	1,000	10	100
DO meter	500	1	500	5	100
Computer	1,650	1	1,650	5	330
Generator	30,000	2	60,000	15	4,000
Total			387,435		42,707

that would be paid on an operating loan or an operating line of credit. This must be included in the budget even if the farmer did not borrow any operating capital because the farmer could have earned interest on this money from investing it elsewhere. This is what economists refer to as an "opportunity cost," or the cost of revenue that could have been earned from its use in some other activity.

The annual variable costs are summed in the budget to $652,473, the total variable costs. Subtracting the total variable costs from total revenue results in $153,927 of income above variable costs. This is not a measure of profit because fixed costs have not yet been considered. However, this value provides an indication of whether the farm can cover its variable costs for the year (if income above variable costs is a positive number). If income above variable costs is negative, the farm loses more money by continuing to produce and loses less money by shutting down. Since this value is positive, the farm can continue to operate in the short run.

The next section of the budget is annual fixed costs. In catfish production, fixed costs include telephone expenses, taxes, insurance, legal/accounting fees, annual depreciation, and interest on investment. Total annual fixed costs are $151,135 for the 256-acre catfish farm. The annual depreciation values for equipment are taken directly from Table 10.2. Interest on investment is charged for the same reason as interest on operating capital is charged. If the farmer is paying

interest on real estate and equipment loans, this charge covers it. Otherwise, interest on investment is an opportunity cost that represents the revenue that would have been earned from some other use of that investment capital. Total costs (sum of total variable and total fixed costs) for this catfish farm are $803,608.

Net returns to operator's labor, management, and risk (total revenue minus total costs) for the farm are $2,792 or $11 per water acre. This budget includes several forms of net returns. Net returns to risk, −$18,502 (−$72/acre), are the economic measure of profit from this budget because all costs have been charged (land, operator's labor, and management). This budget shows that catfish production is not profitable under the production assumptions and parameter values used in this budget.

Some individuals, in the short term, prefer to include only cash expenses. The results are biased because noncash costs are important for long-term viability of the business. Nevertheless, in the very short term, considering net returns above cash costs alone can provide some additional information for planning purposes. For this farm, net returns above cash costs are $124,594 ($487/acre).

Breakeven prices above variable costs are $0.57/lb and $0.72/lb above total costs. This means that fish prices can fall to $0.72/lb and the farm would still break even, considering all costs. At $0.57/lb the farm could still cover variable costs, but not fixed costs. Breakeven yields are 3,641 lb/acre above total costs

Table 10.3. Annual Costs and Returns for a 256-acre Catfish Farm (Stocking 5,690 5-inch Fingerlings/acre; Feed Fed at 4.78 tons/acre/year; Yield of 4,500 lb/acre; Fingerlings Purchased Off Farm; Pond Owned by Farmers).

Item	Description	Unit	Quantity	Price/cost	Total
Gross receipts	Catfish foodfish	lb	1,152,000	0.70	*806,400*
Variable costs					
Feed	32% protein floating	Ton	1,223.68	227.4	278,265
Fingerlings	5-inch	Inch	7,283,200	0.010	72,832
Labor	Year-round, full-time	FTE*	2	20,280	40,560
	Seasonal or part-time	FTE	0.5	20,280	10,140
Plankton control	Empirical average	Acre	256	14.40	3,686
Gas and diesel	Empirical average	Acre	256	130	33,280
Electricity	Empirical average	Acre	256	289	73,984
Repairs and maint.	Empirical average	Acre	256	97	24,832
Bird depredation supplies		Acre	256	6.25	1,600
Seining and hauling	Catfish foodfish	lb	1,152,000	0.05	57,600
Telephone	Empirical average	Acre	256	10.50	2,688
Office supplies	Empirical average	Acre	256	11.00	2,816
Interest on operating capital		$	501,902†	0.10	50,190
Total variable costs	Per farm				*652,473*
	Per acre				2,549
Income above variable costs per farm					*153,927*
Fixed costs					
Farm insurance	Empirical average	Acre	256	25.3	6,477
Legal/accounting	Empirical average	Acre	256	6.1	1,562
Investment					
Land	Empirical average	$	210,432‡	0.1	21,043
Wells	Empirical average	$	48,000§	0.1	4,800
Pond construction	Empirical average	$	357,888¶	0.1	35,789
Equipment	Empirical average	$	387,570	0.1	38,757
Annual depreciation					
Equipment	Empirical average	Acre	1	42,707	42,707
Total fixed costs	Per farm				*151,135*
	Per acre				590
Total costs	Per farm				*803,608*
	Per acre				3,139
Net returns to operator's labor, management, and risk per farm					2,792
	Per acre				11
Opportunity costs					
Operator's labor	Family	Each		1	10,140
Operator's management	Family	Each		1	11,154
Total opportunity costs of family labor and management					21,294
Total costs					*824,902*

(Continued)

Table 10.3. (*Continued*)

Item	Description	Unit	Quantity	Price/cost	Total
Net returns to operator's risk per farm					*−18,502*
	Per acre				−72
Noncash costs					143,096
Net returns above cash (to operator's risk) per farm					*124,594*
	Per acre				487
Breakeven price	Above variable costs				*0.57*
	Above total costs				*0.72*
Breakeven yield	Above variable costs per farm				*932,104*
	Per acre				3,641
	Above total costs per farm				1,178,431
	Per acre				4,603

*FTE, full time equivalent. One person working one 10-hour day is 1 FTE. Two people working 5-hour days is 1 FTE.
†Operating capital was assumed to be used for 10 months of the year.
‡Land values = $822/acre.
§Six wells at $8,000 each.
¶Pond construction costs = $1,398/acre.

and 4,484 lb/acre above variable costs. This indicates that yields could fall to 4,484 lb/acre and still cover all costs or 3,641 lb/acre and still cover all variable costs.

SENSITIVITY ANALYSES

One of the key limitations of enterprise budgets is that the profitability may or may not hold as the assumed parameter values change. Changes in specific prices of important parameters can have substantial effects on the profitability of the enterprise. For many types of aquaculture enterprises, profits are sensitive to the yield of marketable product and to the price of feed. Table 10.4 presents a simple sensitivity analysis in which marketable yield is varied from 4,000 to 5,500 lb/acre/year at nine different feed prices and breakeven prices (costs of production) recorded. It is clear that increasing yield decreases the cost of production and that increasing feed prices increases the cost of production.

THE PARTIAL BUDGET

Enterprise budgets require careful, conservative, and realistic estimates for all capital investment, equipment, and operating costs for the enterprise under analysis. A thorough, accurate, and useful enterprise budget takes a great deal of thought. Moreover, it can be time-consuming to track down good estimates of costs. However, the enterprise budget considers only one enterprise even if the farm has a variety of different types of enterprises. A partial budget can assess changes that involve interactions between enterprises.

Farm managers must make frequent decisions. Many times, these decisions involve only a relatively

Table 10.4. Sensitivity Analysis that Shows Breakeven Prices above Total Costs for Various Marketable Yields and Feed Prices.

Feed price $/ton	Yield			
	4,000	4,500	5,000	5,500
225	0.80	0.71	0.65	0.59
250	0.83	0.74	0.67	0.62
275	0.86	0.77	0.70	0.64
300	0.89	0.80	0.73	0.66
325	0.93	0.83	0.75	0.69
350	0.96	0.86	0.78	0.71
375	0.99	0.89	0.80	0.73
400	1.02	0.91	0.83	0.76
425	1.05	0.94	0.85	0.78

small change in its operation (e.g., changing the stocking rate, adding a pond with a different species, adding a fingerling pond). If the change is a relatively small one, the majority of costs across the enterprise will remain the same with and without the change. These changes could be evaluated by comparing two whole-farm budgets, but this can consume time that is not warranted by the small size of the change.

A *partial budget* is an analytical technique designed to evaluate the effect of the changes in benefits, costs, and net returns from making such a small change in the farm operation. A partial budget is a standardized and formal method to calculate the change in profit from a possible change. It is important to note that a partial budget does not mean arbitrarily choosing some but not other costs. In a partial budget analysis, all changes in costs and benefits must be considered, but the analysis considers only the changes that will occur.

A partial budget answers the following question: what will be the net economic effect on the farm from making a relatively small change to the operation? A partial budget analysis is limited in that it can be used to analyze only two alternatives at a time. These typically include the current situation as the base and one proposed alternative. Different partial budgets are needed to analyze a series of changes. Partial budgets are useful for existing enterprises that are considering a management change. This might consist of changing stocking densities and stocking sizes, increasing the aeration level, changing to a different type of feed, or beginning to buy fry to produce fingerlings on farm. However, partial budgets can also be used to consider adding investment items such as whether to purchase an in-pond fish grader, add a hatchery building to the fish enterprise, or purchase new aerators. A partial budget is not appropriate for a new enterprise for which comprehensive estimates of the costs and benefits do not exist.

The bottom line of the partial budget is called the change in net benefit. This is the measure of whether net returns would increase or decrease by making the proposed change and by how much. If the change in net benefit is positive, then making this change on the farm will increase net returns. However, if negative, then the change considered would reduce net returns. The magnitude of the values indicates the extent of the change.

The unit of analysis for a partial budget should be the same as for the enterprise budget that serves as the basis for the analysis. If the unit of the base enterprise budget is one 10-acre pond, then the partial budget should be for a 10-acre pond. However, if the base enterprise budget is for an entire farm, then the partial budget should be for the entire farm.

The two broad categories that form the framework of the partial budget are the benefits and costs that result from the proposed change. All items that are expected to change are allocated to the appropriate category.

The benefits include additional revenue and reduced costs, while costs include reduced revenue and additional costs. Additional revenue can result if the proposed change results in higher levels of marketable yield or if the change would add a new species that would generate new revenue. The additional revenue category includes the revenue to be earned if the change is adopted on the farm. However, the analysis includes new revenue that is not being earned in the current situation. Additional revenue often results from increased yields, sales from a new enterprise, gaining a higher price, or increasing the size of the enterprise.

Benefits from the change also result from reduction of costs on the farm. Switching to a cheaper feed, reducing the size of the farm, or eliminating a fingerling operation would reduce costs. For example, if perhaps a new feed resulted in an improved feed conversion, the quantity of feed fed for a given level of marketable yield would be reduced and total feed costs would be lower. The cost category includes any reduction in revenue from the farm. Reduced revenue represents revenue currently earned that would not be available after the change is made. A lower price, reduced sales from lower yields or fewer production acres, or elimination of an enterprise would reduce revenue. Another example would be the revenue lost of one species by switching to a different species.

The cost category also includes the new, additional costs associated with the alternative being considered that are not part of the current costs. Higher stocking and feeding rates, purchasing new fingerlings, buying additional ponds for expansion or for raising a new species will all be new costs. Examples of additional costs would be if the new feed or fingerlings considered cost more.

Once all the changes have been itemized and quantified, the additional revenues are added to the reduced costs to obtain the benefits. Reduced revenue is added to additional costs to obtain the costs. Costs are subtracted from benefits to obtain the change in net benefits.

OTHER CONSIDERATIONS IN DEVELOPING PARTIAL BUDGETS

Opportunity costs must also be considered in partial budgets. Additional variable costs require capital that could be invested elsewhere and that value is included as an additional cost. On the other hand, reduced variable costs also represent an opportunity cost that no longer exists.

Sensitivity analyses should be done on key parameters to evaluate how dependent the results are on changing values for important parameters. Sensitivity analyses are conducted by varying the price of the parameter in consistent increments and measuring the change in the net benefits.

Partial budgets are a simple and easy way to evaluate management changes. The data required are usually minimal. However, it can only compare one change at a time. A partial budget is not well suited to analyze a change where the revenues are not constant from year to year. Typical partial budgets also do not account for risk; yet, changes in risks as a result of the change can be important and should be evaluated.

RECORD-KEEPING FOR ENTERPRISE AND PARTIAL BUDGETS

Records are fundamental to the development of enterprise and partial budgets. For an enterprise budget, records must be maintained on yields, typically in lb/acre or kg/ha for ponds, lb or kg/m³ for cages or tank systems, and lb or kg/gallon of water flow for raceways. Sales records must be maintained. Records of purchase of supplies, including fingerlings, feed, fuel, and repairs and maintenance are necessary.

Labor records must be maintained, divided into salaried personnel and hourly workers. Benefits provided to salaried personnel must be itemized. This may include use of pickup truck for personal use, housing provided on the farm, or other benefits. Hourly workers need to be itemized by full-time seasonal or part-time workers.

It is best, if possible, to maintain records of the hours of use of machinery. This is particularly useful for tractors and aerators. Purchase and sale of equipment as well as service records need to be maintained.

PRACTICAL APPLICATION OF A PARTIAL BUDGET

Table 10.5 illustrates the application of a partial budget to the 256-acre catfish farm situation. The change

Table 10.5. Partial Budget of Adding a Hybrid Striped Bass Enterprise (40 acre) on a 256-acre Catfish Farm.

Category	Value or cost
Benefits	
Additional revenue	
40 acres in hybrid striped bass production, producing 4,500 lb/acre/year, at $1.50/lb	$270,000
Reduced costs	
$3,139/acre for catfish costs, for 40 acre	$125,560
Total Benefits	$359,560
Costs	
Additional costs	
Feed: 191.2 tons for 40 acre, $50/ton more for feed	$9,560
Fingerlings: $0.20/fingerling	$45,520
Interest on operating capital	$4,590
Marketing	$75,000
Reduced revenue	
40 acres at 4,500 lb/acre catfish × $0.70/lb	$126,000
Total Additional Costs	$260,670
Net Benefit	$98,890

being considered is to take 40 acres of water currently in catfish production and stock those acres with hybrid striped bass for foodfish growout.

Additional revenue would consist of the sales of hybrid striped bass foodfish. Sales were estimated to be based on 40 acres in production, with a yield of 4,500 lb/acre/day, at a price of $1.50/lb, for additional revenue of $270,000 (Table 10.5). Costs would be reduced for the catfish no longer raised in those 40 acres of water. Because Table 10.3 shows total costs of $3,139/acre for catfish production on this size of farm, the reduced costs would be $125,560 for the 40 acres taken out of production. Total net benefits (sum of additional revenue and reduced costs) are $395,560.

Hybrid striped bass fingerlings are more expensive than channel catfish fingerlings, and the feed for hybrid striped bass is also more expensive. Moreover, because hybrid striped bass are sold either live or on ice, there would be additional marketing costs estimated at $75,000. The catfish not sold from these 40 acres constitute reduced revenue of $126,000. The total additional costs (sum of additional costs and reduced revenue) are $386,228. The net benefit from switching

40 acres from catfish to hybrid striped bass is $98,890. Because this number is positive, the analysis demonstrates that the change is feasible. However, it should be noted that this analysis has not considered additional risks associated with producing and marketing hybrid striped bass.

OTHER APPLICATIONS IN AQUACULTURE

Dey et al. (2005) used field survey data collected by the WorldFish Center and partners to develop estimates of net returns for a variety of production systems, species, and production technologies. Net returns were estimated for seven different countries (Bangladesh, China, India, Indonesia, Philippines, Thailand, and Vietnam). Species analyzed included carps, tilapia, prawns, snakehead, and walking and striped catfish. Production systems included cages and ponds stocked in both mono- and polyculture.

The analysis indicated that fish farming of freshwater species in the selected Asian countries was generally profitable. Profits tended to be higher with monoculture of higher-valued carnivorous species such as prawns, snakehead, and walking catfish. However, resource-poor fish farmers could also raise omnivorous and herbivorous species profitably in both mono- and polyculture. Semi-intensive levels of production were shown to produce higher rates of return than very extensive or intensive production levels. The extensive production technologies were inefficient in that resources were underutilized while the very intensive production systems had much higher costs of variable inputs.

SUMMARY

This chapter provided a detailed description of enterprise and partial budgets. These budgeting tools provide a mechanism to assess whether or not a proposed enterprise is profitable, how profitable it is expected to be (enterprise budget), and whether a proposed change on the farm is expected to be profitable (partial budget). Care must be taken to include all relevant costs and to use conservative estimates of all cost and revenue items. Prices expected to be received must be based on long-run averages as should be the costs of various input items.

Sensitivity analyses should be run on all parameters for which values are either unknown or those that are highly variable. Breakeven analysis further shows the prices and quantities that will need to be obtained to cover all costs.

REVIEW QUESTIONS

1. Explain how breakeven prices and yields can be used to determine the profitability of the enterprise.

2. How do sensitivity analyses evaluate the risk of the enterprise?

3. What would be an appropriate unit of analysis for an enterprise budget for a trout raceway farm?

4. What would be the greatest single cost for each of the following aquaculture enterprises:
   a. Salmon netpen farm
   b. Trout raceway farm
   c. Subsistence tilapia pond farm
   d. Shrimp pond production farm

5. Describe alternative ways to assign a value to submarketable-sized fish in a pond.

6. List the major limitations to an enterprise budget analysis and how its interpretation should be properly qualified.

7. Explain why there is a charge for interest on operating capital in an enterprise budget.

8. Explain the difference between net returns to operator's labor, management, and risk and net returns to risk.

9. What advice would you give someone when the enterprise budget shows positive income above variable costs but negative net returns?

10. Explain what types of employee charges would be properly charged as fixed costs and which would be categorized as variable costs.

REFERENCES

Dey, Madan M., Mohammad A. Rab, Ferdinand J. Paraguas, Ramachandra Bhatta, Mohammad F. Alam, Sonny Koeshendrajana, and Mahfuzuddin Ahmed. 2005. Status and economics of freshwater aquaculture in selected countries of Asia. *Aquaculture Economics and Management* 9:11–37.

Engle, Carole R. 2007. *Arkansas Catfish Production Budgets*. Cooperative Extension Program MP 466. Pine Bluff, AR: University of Arkansas.

Kay, Ronald D., William M. Edwards, and Patricia A. Duffy. 2008. *Farm Management*. Sixth Ed. Boston: McGraw-Hill.

National Agricultural Statistics Service, U.S. Department of Agriculture. www.usda.mannlib.cornell. edu/MannUSDA/viewDocumentInfo.do?document ID=1375. Accessed November 13, 2009.

United States Bureau of Labor Statistics. www.bls. gov. Accessed November 13, 2009.

United States Department of Energy. www.eia. doe.gov. Accessed November 13, 2009.

11
Financial Statements: Balance Sheet and Income Statement in Aquaculture

Financial management and control of a business for efficient operation requires regular and periodic documentation and review of financial performance and progress. The purpose of financial analysis is to identify problems and opportunities, potential solutions, and then develop plans to take advantage of the opportunities. It summarizes the financial position and condition of the business at a specific point in time. The two main instruments used for periodic financial management and control are the balance sheet and the income statement. This chapter explains what each of these instruments is, how each is structured, and what each tells the financial analyst or farm owner.

The *balance sheet* is a financial statement that itemizes assets and liabilities for the business and calculates net worth as its bottom line. Net worth measures the level of wealth of the business while the relationship of the assets to the liabilities provides a clear view of the financial position of the business. Financial position refers to the total amount of resources that a business controls as well as the financial claims against those resources. Many lenders consider the balance sheet to be the most important financial statement of all because it can be used to determine whether the business or individual can tolerate additional debt.

The income statement measures the revenue and expenses of the business for the specified time period. An income statement resembles an enterprise budget and is sometimes confused with it. The difference between the two is that the enterprise budget is developed for a typical, average, generalized farm business while the income statement is developed for a specific business for a specified time period, typically, a year.

Thus, the income statement documents the financial performance of the business for that particular year, but not for the average conditions facing an industry. The income statement measures the profit or loss for the business for that particular year and is also known as a profit and loss statement.

Many financial transactions affect both the balance sheet and the income statement, and some values necessary from the income statement are obtained from the balance sheet. Thus, it is essential that balance sheets and income statements be prepared for the same time period and at the same time. They must be interpreted together for a thorough understanding of business performance. Any time that revenue and expenses change, the balance sheet and income statements may change. Accurate interpretation of these financial statements requires that the time period of analysis be specified because revenue and expenses are different for different time periods.

The greatest value of the balance sheet and income statement is derived from using them over time to interpret the overall financial performance and position of the business. From year to year, a viable business will show a generally increasing net worth from the balance sheet and increasing net farm income from the income statement.

THE BALANCE SHEET

The balance sheet is a key financial statement used by businesses to monitor financial performance and progress. It is also called the statement of financial condition. The balance sheet measures the financial position of the business for that specific time period. It organizes all that is owned by the business (assets) and the debts that are owed (liabilities) in a systematic

Aquaculture Economics and Financing: Management and Analysis, Carole R. Engle, © 2010 Carole R. Engle.

manner and is prepared for a given point in time. If always prepared at the same time each year, the balance sheet can be used to compare the financial progress of the business from year to year and the growth of the business can be charted over time.

The bottom line of the balance sheet is net worth. Net worth is also referred to as owner's equity and measures the level of capital that the owners have invested in the firm. Assets, liabilities, and owner equity are interrelated and "balanced" in the balance sheet. Owner equity is the difference between assets and liabilities. This relationship can be expressed mathematically as:

Net worth (owner equity) = assets − liabilities

This same relationship can be expressed as:

Assets = owner equity + liabilities

Over time, a successful business is one in which the net worth increases over time. Increases in net worth mean that the owner is accumulating wealth from the business.

The balance sheet is developed for the entire farm business for the fiscal year used by the business. It can be developed to show the capital position at any time, but the most useful balance sheets are developed annually, corresponding to the beginning and ending of the financial year. The ending balance sheet of any one year is identical to the beginning balance sheet for the next year. Balance sheets commonly are prepared for the end of the calendar year, as of December 31. This statement then serves as both the end-of-year statement for that year and the beginning statement for the next year (January 1).

Careful thought should be given to the choice of financial accounting periods. The fiscal year can be the calendar year or it can be some other period of time. The calendar year is convenient from a financial perspective because many people prepare their income taxes for the calendar year. This is also a time when many farms located in temperate areas have time to analyze their finances carefully. However, other financial accounting periods can be used. There is some advantage to matching the production year with the financial reporting year. For a catfish farm, for example, fingerlings typically are stocked in February or March. April is one of the first good feeding months. Thus, using a financial reporting year of April through March of the following year tracks the production cycle more closely. Fish produced that year typically are sold over the winter months and revenue can be matched more closely to expenses when the financial account-

ing period is aligned closely with the production cycle. Alternatively, some farmers maintain financial records developed for both reporting periods.

Regardless of the financial accounting period selected, the balance sheet should be developed at least once a year to be able to monitor financial progress over time. The same time period should be used over the years to maintain consistent accounting practices. Any changes observed, then, will be due to changing financial position of the business and not an artifact of the accounting system. Managers may want to examine quarterly balance sheets to more closely monitor the financial position of the business. This may be especially important during difficult financial and economic times.

The structure of the balance sheet includes major categories of "assets" and "liabilities." All assets and liabilities for the firm are listed on the balance sheet. Assets include anything that is owned by the business while liabilities are the value of debts owed by the business. Any item of value that is owned by the business is considered an asset. Items are considered as assets if they have a value either because they can be sold to generate cash revenue or because they can be used to produce something else that will be sold at a later time. All debts and financial obligations owed to someone else constitute liabilities. Liabilities represent a claim by someone else, an outside entity, on the assets of the business.

Assets and liabilities are further divided into subcategories referred to as "current" and "noncurrent" assets and liabilities and refer to the period of time within which the specific asset or liability will be used. These classifications are based on what is termed "liquidity." *Liquidity* is the extent to which an asset can generate cash quickly and efficiently to meet the business's financial commitments. *Highly liquid assets* are those that can be converted to cash with little or no delay or loss in the net value of the firm. *Current assets* are those that are highly liquid and include assets that are expected to be converted into cash (sold) within a period of 1 year. In other words, current assets are short-term liquid assets.

For example, inventories of catfish that are expected to be sold within a year would be categorized as a current asset while equipment such as tractors and aerators would be classified as noncurrent assets. Current assets include the value of raw materials and supplies on the farm at the time the balance sheet is developed. Stocks of herbicides, copper sulfate, feed, or other inputs should be included. These types of current assets typically are valued at their original purchase cost. Assets commonly included as current assets are cash on hand and balances in checking and savings accounts. These clearly are very liquid types of assets. Accounts receivable (payments not yet received from processing plants for fish already delivered) can also be important current assets.

Any asset that is expected to be used in the business for more than a year is classified as a *noncurrent asset*. Noncurrent assets typically have a useful life greater than 1 year, are usually not purchased for resale, and are used to produce products or services that will be sold. Noncurrent assets typically are those that the business owns because they are used to produce the products sold to generate revenue. Thus, equipment, broodstock, ponds, and land are classified as noncurrent assets. If aerators or tractors would be sold, the fish farm would likely not be able to produce and sell as much product in the future. These noncurrent assets are less liquid (illiquid). They are more difficult to sell quickly and easily at full market value. Broodstock are another example of a noncurrent asset because broodstock are generally considered to be fixed assets. Because broodstock frequently are produced on farm or may have been purchased some time ago, assigning a value can be more difficult. Broodstock can be valued at: (1) cost of production; (2) 75% of reasonable market cost; or (3) anticipated market value after deducting storage, transportation, and marketing costs. The most common way to assign a value to broodstock is the anticipated market value. Costs associated with storage and transportation need to be subtracted out.

Estimating the inventory of aquatic crops can be difficult, depending on the production system. Net pens, cages, tanks, and raceways are more amenable to frequent inventory estimations than open-pond systems. Estimating the quantity of fish available in a 10-acre pond that has been in continuous production for many years is difficult. For many species, there is no very accurate and precise manner to obtain such an inventory estimate. Errors in the estimate of the inventory can create errors in the balance sheet and income statements. Experienced farmers with excellent records will have better estimates of the quantity and sizes of fish in the pond.

Current liabilities are short-term debt obligations. Current liabilities include debt-servicing payments due within the year and accounts payable to feed mills or supply companies. Short-term loans require complete payment of principal and interest within 1 year. Those payments due within the coming year are classified as current liabilities.

Noncurrent liabilities consist of the remaining balances on loans (other than the current year's payments) and other financial obligations that will be incurred at a point in time later than the current year. These include operating loans and lines of credit. Loans taken out for the purchase of noncurrent assets, such as land, pond construction, and equipment, will typically have terms of lending of more than 5 years (see Chapter 14 for more details on commonly used types of loans for aquaculture). Thus, payments on principal and interest due are made over a period of years. The remaining balance of principal on the loans is classified as a noncurrent liability. When calculating noncurrent liabilities, it is important to be certain to deduct the payments due in the coming year. The payments due in the coming year are listed as current liabilities, and need to be subtracted from the noncurrent liabilities.

On the balance sheet, the values of the current assets are summed; the values of the noncurrent assets are also summed. The total value of the current assets is then summed with the total value of the noncurrent assets to obtain total assets. Similarly, all the values of the current liabilities are summed to obtain "total current liabilities" and the values of the noncurrent liabilities are summed to obtain "total noncurrent liabilities." Total current liabilities are added to the total noncurrent liabilities to obtain "total liabilities."

Net worth is calculated on the balance sheet by subtracting the total value of all liabilities from the total value of all assets of the business (total assets − total liabilities). Net worth is also referred to as owner equity. It is the amount of money left for the owner of the business if the business were to be sold and all liabilities paid in full. Net worth (owner equity) is considered to be a claim on the assets by the owner(s) of the business. Owner equity changes over time. The goal of the business is for net worth to increase over time.

Profits generated by the business are used to pay down liabilities and increase net worth as liabilities decrease. Alternatively, if profits are used to acquire additional assets, net worth will increase if the total amount of liabilities remains constant. However, net worth will also change if the value of an asset changes. Thus, if land values decline, the total value of assets will decline. For fish farm balance sheets, the value of the swimming inventory can be a substantial portion of current assets. If fish prices decline, current assets will go down in proportion to the decline in fish prices. It should be noted that if assets increase but are financed through a loan, the liabilities will increase by the same amount as the increase in assets. In such a case, net worth will remain the same. Net worth changes when the owner puts additional personal capital into the business, withdraws capital from the business, or the business shows as a profit or loss.

NOTES ON VALUING ASSETS

There are two principal methods for assigning a value to assets: cost-basis or market-basis. The cost basis value is also referred to as the "book value." Basic accounting principles specify use of a cost basis. With a cost basis, assets are valued using the cost, cost less depreciation, or farm production cost methods. The advantage of a cost-based balance sheet is that it is directly comparable to other balance sheets prepared with standard accounting principles and is a conservative estimation method. Moreover, the net worth will not vary because of market fluctuations. The advantage of a market-based valuation approach is that it more accurately indicates the value of collateral and the current financial position of the business.

Typically, depreciable assets are valued as cost less depreciation. Submarketable fish and broodstock typically are valued at farm production cost. Inventories of market-sized fish, however, are valued at market price even on a cost-based balance sheet. The difference between the cost basis and market basis values is typically larger for assets purchased several years before the preparation of the financial statement. For example, the current market value of ponds may be very different from the original construction cost 20 years ago. The balance sheet should indicate the basis of accounting (cash- or market-based), the nature of the operation (number of ponds, acres), and the depreciation method used.

RECORD-KEEPING FOR THE BALANCE SHEET

Records required to prepare a balance sheet include complete inventories of both current and noncurrent assets. A spreadsheet can be used to itemize all checking and savings accounts for the business. The year-end balance in each serves as the beginning value for the next year's pro forma balance sheet.

To be able to prepare a balance sheet, the farm business must maintain records of all assets and liabilities. The farm should maintain a table or spreadsheet that includes line items for land and each building on the farm. For land, the current value per acre for similar land in that area should be recorded, along with the number of acres and the total value. Each building should be listed separately, with the original purchase price or construction cost, the estimated years of useful life, the annual depreciation, and the current value. Similarly, there should be a separate table or spreadsheet that includes each equipment item for the farm, its initial purchase price, the estimated years of useful life, the annual depreciation, and the current value.

Farm records should include a depreciation schedule for all the equipment on the farm. This table should include the original purchase price, the date of the original purchase, the depreciation method used (i.e., straight-line, double-declining balance), the estimated years of useful life, and the current value.

A similar schedule should be maintained for the ponds, tanks, grading sheds, office, and other buildings on the farm. This table should also include columns to record the original purchase price, the date of the original purchase, the depreciation method used (i.e., straight-line, double-declining balance, etc.), the estimated years of useful life, and the current value.

Additional tables are needed to maintain records of the inventory of the aquatic crops and of materials and supplies in hand at the beginning of the year and at the end of the year. The inventory sheet for aquatic crops should list these by species and by size group per pond. The estimated inventory of fish should include an estimate of the quantity of fish in each pond by size group of fish. This can be estimated from a Fishy file (Fishy is a record-keeping software program used in the U.S. catfish industry.) (Killcreas 2009), by sampling, or by careful examination of the stocking, feeding, and harvesting records for the farm. The estimated number and individual weight of each should be recorded along with the total value. Market-sized fish or shrimp should be valued at the market price while submarketable fish or shrimp should be valued at the estimated production cost.

Supplies on hand and their total value (purchase cost) at the beginning and end of the year should also be itemized with total quantities recorded. The change in value for each (crops and supplies) is calculated for later use in the income statement.

Each farm business should maintain careful records of loans. A table or spreadsheet should be maintained that includes the following information for each year:

1. Initial and ending dates of loan
2. Dates when payments due
3. Amount of payment, separated into interest and principal portions

4. Balance remaining on the principal at the end of financial period

These should be updated annually.

PRACTICAL APPLICATION: 256-ACRE CATFISH FARM

Table 11.1 presents an example of a balance sheet for the 256-acre catfish farm described in Chapter 10. Assets are listed first and then liabilities. Assets and liabilities are shown in order of liquidity. In other words, the more liquid assets (current assets) are listed first. The current assets include the cash available in the farm business ($15,849). The cash available consists of cash on deposit in a bank account. In addition to the cash available, there is an inventory of catfish in the ponds that is expected to reach market size and be sold during the current year. There are 4,500 submarketable fish/acre (based on 75% survival of the

Table 11.1. Balance Sheet for a 256-acre Catfish Farm, December 31.

Item	Total value
Assets	
1. Current assets	
Cash on deposit	$15,849
Checking account balance	0
Accounts receivable	0
Fish inventory*	$266,112
Total current assets	*$281,961*
2. Noncurrent assets	
Equipment	$387,570
Ponds	$357,888
Wells	$48,000
Land	$210,432
Total noncurrent assets	*$1,003,890*
3. Total assets	*$1,285,851*
Liabilities	
4. Current liabilities	
Payments on debt due and payable over next year	
Equipment	$72,368
Real estate	$38,216
Total current liabilities	$110,584
5. Noncurrent liabilities	
Equipment loan	$168,858
Real estate loan	$254,772
Total noncurrent liabilities	$423,630
6. Total liabilities	*$534,214*
7. Net worth (3–6)	*$751,637*

*4,500 submarketable fish/acre at 0.33 lb each @ $0.70/lb.

6,000/acre fingerlings stocked). At 0.33 lb each and a price of $0.70/lb, the total value of fish in inventory is $266,112. Total current assets for the 256-acre farm are $281,961.

Noncurrent assets for the 256-acre catfish farm include equipment and farm real estate, including the land and ponds constructed on the land as well as the necessary wells. The total value of the equipment is $387,570, the ponds are valued at $357,888, the wells at $48,000, and the total land value is $210,432. The total value of noncurrent assets is $1,003,890. Total assets (sum of current and noncurrent assets) are $1,285,851.

On the liabilities side of the balance sheet, current liabilities are listed first. Current liabilities include the payments due over the next year for the catfish farm. The catfish farm has loans for equipment and real estate and, thus, has payments due over the course of the year. The equipment loan payment for the current year is $72,368 and the real estate loan payment for the current year is $38,216. Total current liabilities are $110,584.

Noncurrent liabilities include the value of the principal that remains to be paid on the equipment and pond construction loans. The remaining principal on the equipment loan is $168,858 and the remaining principal on the pond construction loan is $254,772. Total noncurrent liabilities are $423,630. Total liabilities (sum of current and noncurrent liabilities) are $534,214.

Total owner equity (net worth) (total assets − total liabilities) for the 256-acre catfish farm is $751,637. Over time, the net worth should increase as the liabilities decrease and assets increase through equity gained with payments of principal.

INCOME STATEMENT

The income statement is a companion financial statement to the balance sheet and is also key to monitor financial performance and progress. The income statement measures the profit or loss generated by the business for the specific time period used for financial analyses. It summarizes the financial transactions that affect revenue and expenses for the period of time specified in the income statement and measures the difference between revenue and expenses. Thus, its primary value is to determine whether the business was profitable for that time period, but it can also be used to monitor profits from one year to the next. Moreover, it can also be used to identify what portion of profits is generated from different resources used in the production process. (For more detail, see Chapter 4 on monitoring economic and financial performance of aquaculture businesses.)

The bottom line of the income statement is net farm income from operations. This indicator shows whether the farm business resulted in a profit or loss for that particular year. If net farm income is positive, then the farm was profitable that year, and if net farm income is negative, then it suffered losses.

The income statement is calculated for the same time period as the balance sheet. Complete financial analysis requires that both these statements be calculated at the same time and for the same time period. The income statement and balance sheets should be prepared at least annually. However, some managers may wish to examine quarterly statements to keep a closer eye on the business, especially during difficult financial times.

The structure of the income statement is somewhat similar to that of an enterprise budget. The primary difference is that the income statement includes revenue and expenses actually incurred over the time period specified whereas the enterprise budget includes long-run average revenue and expenses. Other differences are that the income statement sorts out interest expenses from the other expenses.

The first category in the income statement is revenue. All revenue received by the farm business over the past financial period is included in the income statement and is itemized by the source of revenue. Thus, the value of sales of foodfish for the past year would be included in the revenue category of the annual income statement. If the farm also sold fingerlings, these would be indicated on a separate line. Any government payments received (such as disaster payments) would also be included in this area. Revenue that results from the sale of a capital asset such as land or another piece of equipment is also included as revenue. However, for a capital asset, the value entered into the income statement is the difference between the price the asset is sold for and the initial cost of purchase. Revenue resulting from the sale of a depreciable asset, such as a tractor, is valued as the difference between the price the tractor is sold for and the initial cost of purchase after subtracting out the accumulated depreciation. For accrual-based income statements (see section on Cash versus Accrual-Based Income Statements), the change in value of the inventory of fish and inputs should also be included in the revenue portion of the income statement. Any change in the accounts receivable (accounts receivable represent payments for product delivered that have not yet been received) must also be

included. For example, if a fish or shrimp farmer delivers a load of product to a processing plant, but has not yet received payment at the time of developing the financial analysis, the value of that payment is an "account receivable." These values are obtained from the differences in the beginning and ending balance sheets.

The next major category in the income statement is that of cash farm expenses. Those operating expenses that were paid during the accounting period, with the exception of interest payments, are itemized in this section. Typically, the value of fingerlings, feed, and labor are included in this category. Repair and utility bills paid and seining and hauling costs are also included. Payments on insurance and taxes are fixed costs in an enterprise budget, but are cash operating expenses and listed as such in the income statement. The value of all the cash operating costs is summed to obtain total cash farm expenses.

Depreciation is a noncash cost and is included as a separate line item in the income statement. The value of depreciation is the sum of the loss in value for all depreciable assets for the financial accounting period being analyzed. Depreciation on all capital items should be included, particularly equipment, ponds, and any buildings. Broodstock are frequently considered to be fixed or capital assets. As fixed assets, the value of broodstock enters the income statement only as depreciation or net replacement cost.

Interest expenses are separated from other cash expenses. The Farm Financial Standards Council considers interest expense a result of financing, not production activities. Thus, interest expenses are not included with the direct cash operating expenses of producing fish and other aquatic crops. Loan principal payments are not considered expenses because the capital borrowed is simply being returned to the original owner. The "payment" or cost is the interest paid. Thus, principal payments are not included in the income statement. (Note that the payment of the principal amounts of loans is an important financial consideration that is treated in Chapter 12 that discusses cash flow.) The principal was an amount borrowed that does have to be returned, but is not an expense.

However, the income statement includes depreciation costs to account for the replacement costs of capital goods. Including a principal payment on loans for capital goods, in a sense, would "double count" the cost of acquiring/replacing capital goods.

Net farm income is the measure of profit or loss of the farm business. If net farm income is positive, then the farm business was profitable for that period of time; if negative, it lost money for that period of time. The Farm Financial Standards Council recommends including income taxes in the income statement. If this is done, net farm income becomes an after-tax net farm income.

CASH VERSUS ACCRUAL-BASED INCOME STATEMENTS

There are two fundamental accounting approaches used in agriculture: cash-based and accrual-based accounting. In cash-based accounting, revenue is recorded when cash is received regardless of when the product was produced or the service provided. Thus, with cash accounting, revenue can be recorded in an accounting period other than the one in which the product was produced. For example, if a crop of catfish is produced in 2010, but fish were off-flavor and not sold until 2011, then the revenue is recorded in the 2011 income statement if cash-based accounting procedures are used.

Expenses similarly are recorded in the accounting period when they are paid, or when the cash is spent with cash-based accounting. Items that are paid for late in one year and not used until the next are charged to the first year's income statement. Alternatively, if fertilizer is charged to an account payable in 2010, but the account payable is not paid until 2011, then the expense is included in the 2011 income statement.

The advantage of the cash accounting method is that it is simple and easily understood. It is also the most commonly used method in fish farming. The disadvantage is that neither revenue nor expense is directly related to the production obtained in the particular financial reporting year. Thus, the profit or loss estimated in the income statement does not truly represent the profit or loss from that particular year's production.

The Farm Financial Standards Council recommends the use of accrual-based, not cash-based accounting. In accrual accounting, the value of products produced and services provided are recorded for the year in which produced, regardless of when the cash revenue was actually received or the expenses actually paid. There is no difference between cash-based and accrual-based accounting when all revenues are received and expenses incurred in the same year as that of the production that resulted. However, when revenues are received in the year following the one in which products are produced, or expenses paid in a different year from the actual production, results of income statements developed based on accrual accounting will differ from those obtained with cash-based accounting.

The major difference in the two methods lies in the way inventories are handled. In accrual accounting, the value of changes in inventories is expressed as revenue in the income statement. For example, on a catfish farm, expenses for feed, fingerlings, utilities, and labor are incurred throughout the production year. However, if the fish are off-flavor at the end of the year, there will be no revenue that year. Cash revenue will be received the next year when the catfish come back on-flavor, are sold, and revenue received. In accrual accounting, the value of the catfish produced is included as revenue in the year during which the production occurred. The way this is handled in the accrual-based income statement is to add a line in the revenue section for the change in inventory of fish. The fish produced that year but not sold are measured by the increase in total weight of fish in the ponds over the course of the year. This increase in the weight of fish on the farm is recorded as a positive change in the inventory. Because these are market-sized fish, the value used would be the market price.

Expenses in accrual-based accounting are handled in a similar fashion. Those expenses incurred in producing the revenue in that year are recorded in that same year. If supplies are purchased and paid for in the same year, the expense section of the income statement will be the same for cash-based and accrual-based accounting statements. However, if items are purchased a year before that when revenue is produced or are not paid for until the year after they are used, then the results will be different for cash-based and accrual-based accounting. If not yet paid for, these enter the accrual-based income statement as a change in inventory. If interest is accrued in one year but the cash payment is not due for several more months, differences will be incurred.

The main advantage of accrual-based accounting is that it provides a more accurate estimate of profit. The information provided for financial analysis and management is more accurate. The disadvantage is that accrual accounting procedures are more difficult and require more time. Also, accrual accounting may not be the most beneficial for income tax reporting. The Farm Financial Standards Council recommends that anyone using cash-based accounting should also calculate an accrual-adjusted net farm income.

RECORD-KEEPING FOR THE INCOME STATEMENT

Farms must maintain adequate records from which the income statement is calculated. Basic records for revenue required for the income statement begin with the delivery tickets from the processing plant. The tickets from the plant will itemize whether dockages were incurred due to: (1) out-of-size fish; (2) diseased and deformed finfish; (3) dead-on-arrival fish; (4) trash fish; or (5) some other reason. This ticket will provide numbers for the weight and number of fish delivered, as well as the losses suffered due to dockages. The net price received can also be recorded after adjusting for the dockages charged.

Purchases of supplies typically involve: (1) obtaining a price quotation; (2) placement of an order; (3) delivery bill; (4) receipt of invoice or bill; (5) receipt of statement of account at end of month; and (6) receipts for payments made. Thus, the records available are the delivery ticket, invoice, monthly statement, and canceled check for payment. These records can be checked for accuracy or errors by comparing: (1) the order against the invoice; (2) the invoice against the delivery ticket; and (3) the delivery ticket against goods supplied.

PRACTICAL APPLICATION: 256-ACRE CATFISH FARM

Table 11.2 illustrates an income statement for the 256-acre catfish farm. The farm sold 1,152,000 lb of food-sized catfish. At $0.70/lb, the total revenue for the year is $806,400.

Cash operating expenses include those costs listed as variable costs on the enterprise budget as well as cash expenses of legal/accounting fees and insurance costs that are listed as "fixed costs" on the enterprise budget. These operating expenses include feed ($278,265), fingerlings ($72,832), labor ($40,560 for a full-time year-round employee and $10,140 for seasonal, part-time labor), plankton control ($3,686), gas, diesel fuel, and oil ($33,280), electricity ($73,984), repairs and maintenance ($24,832), bird depredation supplies ($1,600), seining and hauling costs ($57,600), telephone ($2,688), office supplies ($2,816), legal/accounting fees ($1,562), and insurance expenses ($6,477). Total cash farm expenses sum to $610,322.

Depreciation costs for losses in value of equipment and buildings for this farm for the year totaled $42,707. This value is added to total cash farm expenses to obtain total catfish farm expenses of $653,029.

Interest paid is placed under a separate category in the income statement. This facilitates calculation of financial ratios that are described in Chapter 4. On the

Table 11.2. Income Statement for a 256-acre Catfish Farm, December 31.

Item	Total value
Catfish farm revenue	
Cash catfish sales	*$806,400*
Accounts receivable	*0*
Change in market livestock inventory	*0*
Total catfish farm revenue	*$806,400*
Catfish farm expenses	
Cash operating expenses	
Feed	$278,265
Fingerlings	$72,832
Labor	$40,560
	$10,140
Plankton control	$3,686
Gas, fuel, and oil	$33,280
Electricity	$73,984
Repairs and maintenance	$24,832
Bird depredation supplies	$1,600
Seining and hauling	$57,600
Telephone	$2,688
Office supplies	$2,816
Legal/accounting	$1,562
Insurance	$6,477
Total cash farm expenses	$610,322
Accounts payable	0
Prepaid expenses	0
Depreciation	$42,707
Total operating expenses	*$653,029*
Cash interest paid	
Interest on operating line of credit	*$50,190*
Interest paid on long-term loans	
Land	*$21,043*
Wells	*$4,800*
Pond construction	*$35,789*
Equipment	*$38,757*
Total interest paid	*$150,579*
Total expenses	*$803,608*
Net farm income from operations	*$2,792*

256-acre catfish farm, there are loans for the operating capital as well as long-term loans for the land, wells, pond construction, and equipment. Annual interest payment on the operating line of credit for this farm is $50,190 and annual interest payments on the long-term loans are $21,043 for the loan used to purchase the land, $4,800 for the loan used to dig the wells, $35,789 for the loan used to pay for construction of the ponds, and $38,757 on the equipment loan. All interest expenses are summed to obtain a total of the cash interest paid ($150,579). This is added to the total catfish farm expenses to obtain total expenses for the year of $803,608.

Net farm income from operations is calculated by subtracting total expenses from total farm revenue, or $2,792 for the 256-acre catfish farm. Thus, the farm generated a profit for the year, even though the total amount of profit was not great.

PRACTICAL APPLICATION: CASH VERSUS ACCRUAL ACCOUNTING

Tables 11.3 and 11.4 compare cost-based and accrual-based income statements for the farm situation presented in Table 11.2, but with one difference. In Table 11.2, all the catfish produced in 2010 were sold in 2010, and all expenses for the 2010 crop were paid for in 2010. Thus, there is no difference in the income statement regardless of whether the farm used cash-based or accrual-based accounting.

Farms rarely are able to sell all their fish by December. Off-flavor or market conditions typically result in only part of the fish produced being sold in that same year. Tables 11.3 and 11.4 present income statements showing the profit or loss for the farm if only 50% of the catfish produced are sold in 2010. Table 11.3 uses cash-based accounting and Table 11.4 uses accrual-based accounting. Both tables show results for the years 2010 and 2011.

With cash-based accounting, when the farm sells only 50% of its 2010 crop in 2010, the revenue is half ($403,200) (Table 11.3) of what it was when all fish were sold in 2010 ($806,400) (Table 11.2). This results in a negative net farm income from operations in 2010 of −$400,408. In 2011, assuming that all the off-flavor fish of 2010 were sold as they came on-flavor from January to March, and the 2011 crop was all sold in 2011, the revenue in 2011 would be $1,209,600. This is much higher than that shown in Table 11.2 and results in a higher net farm income from operations of $405,992. Imagine a banker looking at the 2010 income statement; he or she would see a substantial loss on the farm and might begin to question the financial viability of the business.

In the accrual-based income statement in Table 11.4, the increase in the weight of catfish on the farm over the 2010 production season is recorded as a change in inventory. The market-sized inventory that was produced but not sold is valued at its market price. Actual sales in 2010 were 50% of the sales shown in Table 11.2 because only 50% of the fish were sold. Thus,

the total farm revenue for 2010 was $806,400. This is equal to the revenue shown in Table 11.2.

The accrual-based income statement for 2011 shown in Table 11.4 shows increased cash sales of catfish, because the market-sized catfish from 2010 were sold in addition to the 2011 crop. However, at the end of 2011, because the 2011 crop was all sold, there would be a decrease in the total weight of fish on the farm (as compared to the beginning of the year). Thus, the change in inventory line item under revenue shows a decreased value of inventory on the farm. Total revenue for 2011, then, is the same as in 2010, and net farm income from operations is the same as in 2010. The banker reviewing this farm's financial statements sees a consistently profitable business. The profits may be low, but it is a business that is making all its payments and is a viable operation.

The examples shown in Tables 11.3 and 11.4 were simplified to illustrate the differences in cash-based and accrual-based accounting. For a real business, catfish prices would not likely be constant from year to year nor would expenses. The proportion of market-sized fish not sold in December is likewise not constant and varies with a number of factors. Nevertheless, these examples show the benefit of using accrual-based accounting. If the farm uses cash-based accounting for income tax purposes, then an accrual-adjusted statement should also be developed. The difficulties of estimating inventories in open-pond systems were noted earlier in the chapter. Nevertheless, well-managed farms must strive to maintain adequate records and estimates of fish inventories.

RELATIONSHIP BETWEEN THE BALANCE SHEET AND THE INCOME STATEMENT

The balance sheet and the income statement are related. For example, for the balance sheet to show increasing net worth over time, the business must generate profits. Thus, the profit or loss measured by the income statement will affect the financial position of the business, as measured by the balance sheet. Consistent losses from year to year will make it difficult to make loan payments, will affect crop inventories on the farm, or may require the sale of capital assets that reduce the total value of assets to the business.

Changes in values from the beginning and ending balance sheets for livestock, broodstock, and accounts receivable are entered directly into the revenue section of the income statement. Changes in supplies, prepaid

Table 11.3. Income Statement (Cash-Based Accounting) for a 256-acre Catfish Farm with Off-flavor Restricting Sales in 2010, December 31.

Item	2010	2011
Catfish farm revenue		
Cash catfish sales	*$403,200*	*$1,209,600*
Accounts receivable	*0*	*0*
Change in market livestock inventory	*0*	*0*
Total catfish farm revenue	*$403,200*	*$1,209,600*
Cash farm expenses		
Cash operating expenses		
Feed	$278,265	$278,265
Fingerlings	$72,832	$72,832
Labor	$40,560	$40,560
	$10,140	$10,140
Plankton control	$3,686	$3,686
Gas, fuel, and oil	$33,280	$33,280
Electricity	$73,984	$73,984
Repairs and maintenance	$24,832	$24,832
Bird depredation supplies	$1,600	$1,600
Seining and hauling	$57,600	$57,600
Telephone	$2,688	$2,688
Office supplies	$2,816	$2,816
Legal/accounting	$1,562	$1,562
Insurance	$6,477	$6,477
Total cash farm expenses	$610,322	$610,322
Accounts payable	0	0
Prepaid expenses	0	0
Depreciation	$42,707	$42,707
Total operating expenses	*$653,029*	*$653,029*
Cash interest paid		
Interest on operating line of credit	$50,190	$50,190
Interest paid on long-term loans		
Land	$21,043	$21,043
Wells	$4,800	$4,800
Pond construction	$35,789	$35,789
Equipment	$38,757	$38,757
Total interest paid	$150,579	$150,579
Total expenses	*$803,608*	*$803,608*
Net farm income from operations	−*$400,408*	*$405,992*

expenses, accrued expenses, and accounts payable between beginning and ending balance sheets are entered into the income statement.

The change of value of depreciable assets from the beginning to ending balance sheets for a given time period are reflected in the charges for depreciation in the expense component of the income statement. Income taxes due on the closing balance sheet should come directly from the income statement. The change in net worth between the beginning and end of the year in successive balance sheets should correspond to the retained earnings shown on the bottom of the income statement and provide a check on the accuracy of the analysis.

OTHER APPLICATIONS IN AQUACULTURE

Kam et al. (2003) collected data from eight milkfish farmers in Hawaii to develop an economic analysis to

Table 11.4. Income Statement (Accrual-Based Accounting) for a 256-acre Catfish Farm with Off-flavor Restrictions on Sales in 2010.

Item	2010	2011
Catfish farm revenue	*2010*	*2011*
Cash catfish sales	403200	1209600
Accounts receivable	0	0
Change in market livestock inventory	$403,200	−$403,200
Total catfish farm revenue	*$806,400*	*$806,400*
Cash farm expenses		
Cash operating expenses		
Feed	$278,265	$278,265
Fingerlings	$72,832	$72,832
Labor	$40,560	$40,560
	$10,140	$10,140
Plankton control	$3,686	$3,686
Gas, fuel, and oil	$33,280	$33,280
Electricity	$73,984	$73,984
Repairs and maintenance	$24,832	$24,832
Bird depredation supplies	$1,600	$1,600
Seining and hauling	$57,600	$57,600
Telephone	$2,688	$2,688
Office supplies	$2,816	$2,816
Legal/accounting	$1,562	$1,562
Insurance	$6,477	$6,477
Total cash farm expenses	$610,322	$610,322
Accounts payable	0	0
Prepaid expenses	0	0
Depreciation	$42,707	$42,707
Total operating expenses	*$653,029*	*$653,029*
Cash interest paid		
Interest on operating line of credit	$50,190	$50,190
Interest paid on long-term loans		
Land	$21,043	$21,043
Wells	$4,800	$4,800
Pond construction	$35,789	$35,789
Equipment	$38,757	$38,757
Total cash interest paid	$150,579	$150,579
Total expenses	*$803,608*	*$803,608*
Net farm income from operations	*$2,792*	*$2,792*

assess the feasibility of growing milkfish to a smaller market size and of extensive culture of milkfish in Hawaii. Income statements were developed for each scenario evaluated to measure and compare profitability. Sensitivity analyses were conducted to evaluate the effects on profits of varying levels of the price of milkfish, yield (kg/ha), labor cost, feed prices, and stocking costs.

The income statements indicated that the options considered were not economically feasible for milkfish

as a startup business. Further analyses indicated that reducing the share of costs allocated to the milkfish enterprise on the farm will result in positive returns.

SUMMARY

This chapter described balance sheet and income statement, two of the most important financial statements used to analyze the financial performance of a business. The chapter explained how the net worth that is calculated from the balance sheet measures the wealth that is being accumulated in the business. The chapter also explained how the income statement measures the profit or loss for that particular year. The structure and mechanics of each of these statements is explained and practical applications summarized. Issues related to valuation of assets, cash-based and accrual-based accounting, and the relationship between the two statements are discussed. Finally, the records required to develop balance sheets and income statements are described.

REVIEW QUESTIONS

1. Explain what information is obtained from the balance sheet.

2. Explain what information is obtained from the income statement.

3. What is the difference between an enterprise budget and an income statement?

4. What is the difference between cash-based and accrual-based accounting?

5. How are the balance sheet and income statements related?

6. What are the difficulties posed by many aquaculture systems in the use of accrual-based accounting?

7. What is net worth and how can it be used to analyze an aquaculture business?

8. What is the difference between current liabilities and noncurrent liabilities?

9. For real estate loans, how are the payments apportioned between current and noncurrent liabilities on the balance sheet?

10. What is the primary indicator calculated on the income statement and how can it be used to evaluate the farm business?

REFERENCES

Kam, Lotus E., Francisco J. Martínez-Cordero, Ping-Sun Leung, and Anthony C. Ostrowski. 2003. Economics of milkfish (*Chanos chanos*) production in Hawaii. *Aquaculture Economics and Management* 7(1/2):95–123.

Killcreas, Wallace E. 2009. *Fishy 2009*, Version 10.0. Stoneville, MS: Mississippi State University.

12
Cash Flow Analyses in Aquaculture

INTRODUCTION

Most aquaculture businesses are capital-intensive from the perspective of both short-term and long-term capital. Few aquatic farmers have sufficient capital of their own to meet the financial needs of the business. Thus, most farmers borrow capital for short- and long-term capital needs. Short-term capital needs refer to the operating capital needed to purchase feed, fingerlings, fuel, and other operating inputs while long-term capital is required to purchase land, build ponds, and buy equipment. Careful management of these loans requires that close attention be paid to the flow of cash and how the cash flow pattern in the business affects the ability to make loan payments required to service the debt in an adequate manner.

A key consideration in the cash flow budget is the timing of receipt of cash and the timing of expenditures of cash. A critical component of a cash flow analysis is to identify when cash is expected to be received and when it is expected to be spent. Thus, the cash flow budget is a unique type of financial instrument; no other financial statement accounts for the timing of the amounts of cash coming in and going out of the business.

Cash flow analysis results in a picture of the movement of cash into and out of the business. It represents on paper how much cash is available at the beginning of the planning period, when cash will flow into the business from what sources and at what level, and when cash will leave the business for which purposes and in what quantity. Finally, the analysis will show how much cash will be left at the end of the planning period.

The bottom line of the cash flow budget is the cash available for each unit of time included in the analysis. The cash available shows whether there is enough

cash inflow to the business to cover all expenses to be incurred in the business for that particular time period. For example, for a monthly cash flow budget, the cash available for each month shows whether there is enough to meet the expenses anticipated in each month. If the value for that month is negative, then additional operating funds must be either taken from savings or borrowed from a lender to obtain sufficient cash needed for that month to make the payments. This chapter explains the importance of cash flow analysis and the difference between cash flow statements and budgets, describes the structure and mechanics of each, and explains its interpretation.

CASH FLOW STATEMENTS VERSUS CASH FLOW BUDGETS

The terminology related to cash flow analysis differs depending on whether the analysis is done with records from the past year or if it is used to project cash flow into the future. If the analysis is done with records from the past year, it is referred to as a cash flow statement, or a statement of cash flow. If the analysis projects cash revenue and expenses into the future, it is a cash flow budget, sometimes also referred to as a pro forma cash flow budget.

IMPORTANCE OF CASH FLOW ANALYSIS

Cash flow analysis is one of the most useful financial analyses. The balance sheet, income statement, and cash flow analysis combine to provide the fundamental set of financial analyses necessary for understanding the basic financial condition of a business.

One of the most serious financial problems faced by farmers is the control of cash flow. Cash needs are often greater than the amount of cash available during the production season. Sales often tend to occur in groups and do not flow into the farm business evenly

Aquaculture Economics and Financing: Management and Analysis, Carole R. Engle, © 2010 Carole R. Engle.

throughout the year. Many of the receipts from sales occur after production has taken place, following complete harvest of the crop. Such a seasonal pattern of cash inflow creates cash deficits that cause farmers to draw down savings accounts or to borrow additional funds to cover the deficits.

The availability of cash to make necessary payments is as important as profitability to the economic feasibility of a business. Adequate cash flow is a key component of the financial feasibility of a business. Positive cash flow can often make the difference between success and failure of an aquaculture business, especially in the early years of a business startup.

The cash flow budget is a critical component of any new business plan (see Chapter 3 for more detail on developing business plans). It serves as a useful tool for determining the most appropriate time to replace capital assets like aerators or tractors, or to renovate and rebuild ponds. Cash flow budgets can be used to identify the best schedule to make larger, more discretionary purchases of new capital, insurance premiums, or others that can be planned out over time. Extensive cash flow planning may identify opportunities to purchase inputs at discounts, or through more detailed income tax planning.

The cash flow budget is used to evaluate the need for new borrowing. It provides critical insights into whether the business will have adequate cash available when needed to meet its financial obligations. The time periods that exhibit a shortfall in cash are those in which additional new borrowing is needed. The amount of the cash shortfall is the amount that is needed.

A cash flow budget can be used to develop a borrowing and debt repayment plan. Such a plan can prevent excessive borrowing and provide guidance as to the best way to pay down debt quickly. Cash flow budgets are used to develop accurate lines of credit that ensure cash will be available when needed to meet financial obligations.

The cash flow analysis shows whether new loans are feasible and whether there will be cash when needed to make payments on any new loans. Businesses need adequate amounts of capital, but careful attention needs to be paid to the ability of the business to make the necessary payments. If the terms of lending require payments in excess of what the business will have cash for, the business may fail.

Lenders use cash flow budgets to assess the chances of receiving scheduled payments on time from prospective new borrowers. It is a key component in a proper business loan proposal because it shows directly how and when borrowed capital (especially short-term operating loans) will be repaid.

WHAT A CASH FLOW BUDGET IS NOT

Cash flow statements and budgets measure cash flow and cash position, not profits in the business. Cash flow and position are extremely important components of an overall financial analysis of a business, but do not measure profit. Profit for a particular time period is measured by the income statement; enterprise budgets indicate profit in a generalized way for a typical year. The difference is that income statements and enterprise budgets include noncash costs such as depreciation, while cash flow analysis considers movements of cash only. Noncash costs must be included in the calculation of profit because if a business cannot generate enough revenue to replace its equipment when it wears out (cover its noncash depreciation costs), it is not profitable over the longer term and it will not survive. Noncash revenue and expenses are not included in the cash flow budget; so, it does not measure profit, only cash position.

This does not mean that cash flow and position are not important; these areas are as important as the information provided by the other financial statements. For example, a business that shows losses on the income statement for a period of time can survive and recover. However, a business that has inadequate cash resources and cannot pay its bills will not last long. On the other hand, having a positive cash flow by itself is not enough to ensure that a business will be profitable and successful. To fully evaluate the business requires balance sheets and income statements in addition to cash flow statements and budgets.

STRUCTURE AND MECHANICS OF THE CASH FLOW BUDGET

The cash flow budget should be developed for the same enterprise unit as the balance sheet and income statement. If the entire business is for one enterprise, for example, catfish, then the cash flow budget should be developed for the entire farm, as is the case for the balance sheet and income statement.

One of the basic premises of adequate cash flow in a business is that the bills must be paid on time; in other words, there must be cash available when needed to make the payments. If cash revenue in a particular month is less than the cash expenses, then money must be borrowed to pay the bills. However, if there is a cash surplus in one time period, that surplus can roll over to the next time period to be used to make payments that are due later in the year. Or, if not needed for other payments that will become due, it can be used to make additional payments on the principal of a loan to pay it off more quickly and reduce the interest charges. Alternatively, the surplus could be deposited into a savings account or used to purchase other capital assets that might improve productivity of the farm business. Thus, the fundamental premise of cash flow budgeting is that the total cash received, regardless of whether it is received from the sale of product or made available by borrowing capital from a bank, must be at least equal to the total amount of expenses during that time period.

Cash flow budgets can be structured differently depending on the purpose for which the analysis is being developed. For detailed financial planning, monthly cash flow budgets are useful. Monthly cash flow budgets are primarily used to estimate the amount and timing of additional borrowing needed to cover shortfalls of cash in specific time periods, and to assess the ability of the business to make the associated payments on time. Monthly budgets for farm businesses are the most useful in agriculture because of the seasonal nature of most farming businesses. Monthly cash flow budgets allow for development of a detailed analysis of the relationship of the timing and amounts of cash flowing into and out of the business. Quarterly budgets can be used to develop estimates of cash flow needs over a several-year period. Annual cash flow budgets are used in investment analyses to determine cash flow over the life of the investment. (see Chapter 13 on Investment Analysis.)

For maximum benefit to the farm manager, it is best to prepare an end-of-year statement of cash flows as part of the financial analysis for the business. The statement of cash flows is a financial record that records what happened in terms of cash flow over the past year. The statement of cash flows will show the pattern of cash flow over the past year and provides insight into when payments will be needed for the upcoming year. It will also show the time periods in which cash flow is likeliest to be a problem. The cash flow budget is then developed for the upcoming year to project the future needs of the business, plan for new borrowing, and schedule payments to meet debt-servicing requirements.

A cash flow budget summarizes and itemizes cash inflows (cash received into the business) and the cash outflows for the business for the specific time period analyzed. There are certain key principles to keep in mind when constructing a cash flow budget. As trite as it may seem, it is important to keep in mind that only *cash* inflows and outflows are considered. No noncash revenue or noncash expenses are considered. For example, depreciation is not included in the cash flow budget, but the proceeds from sales of any capital assets are included. Inventory values are not included but receipts from sales of product from inventory are.

Principal payments are not included in the income statement, but represent a cash outflow. Thus, payments of both principal and interest are included in the cash flow budget. A primary concern in the cash flow budget is the timing of receipt of revenue and expenses. The cash flow budget is prepared by identifying the timing and magnitude of the expected cash flows into and out of the business. Each type of revenue or expense is charged during the specific period when it is received or incurred. Thus, if a major capital asset is purchased during a given period, its entire cost is charged out in that time period in the cash flow budget. All cash received and expended must be identified and included in the appropriate time period in which it is received and expended.

The cash flow budget is set up with columns that represent specific time periods and rows that include specific line items of cash inflow or outflow. A cash flow budget developed to make management decisions on the farm (typically, a monthly cash flow budget) would include each month in a column in the cash flow budget. The rows would include the specific revenue items, operating expenses, capital expenditures, and scheduled debt payments. The cash available is calculated for each month. If the cash flow budget is developed for an investment analysis, the columns would be years instead of months.

The cash flow budget begins with the sources of cash receipts that are expected to flow into the business. Cash receipts frequently include the cash balance available at the beginning of the time period, cash revenue from sales of farm products sold, cash received from the sale of capital assets (land, equipment, broodstock), and nonbusiness cash receipts from nonfarm cash income, cash gifts, and other sources of cash.

The beginning cash balance or the amount of cash on hand at the beginning of the period is listed first. The beginning cash amount is taken directly from the current portion of the balance sheet for January 1. This is followed by each source of farm cash revenue generated by sales of the crop or of other capital assets. The cash revenue items are summed to generate total cash inflow for the time period.

The easiest way to begin to complete the cash flow budget is to create a total column on the far right of the budget. Annual totals for each line item are recorded in the total column. For example, the total amount of cash inflow from the sale of fish, shrimp, or crabs expected for the coming year is recorded in the total column for the row of product sales for the business. This amount will be the same total value of revenue in the enterprise budget and the total amount of revenue from product sales on the income statement. The next step is to apportion the amount of this revenue to be received for each month. The monthly amounts should add up to the total listed in the total column for each line item.

After cash receipts are itemized by time period, the same is done with cash outflows. Cash outflows typically include farm operating expenses, cash outflows from the purchase of any new capital assets (land, equipment, and broodstock), any nonbusiness expenses such as living expenses, and payments of principal and interest on outstanding loans.

Operating expenses are itemized first. Each type of operating expense is listed. Thus, there is a row for feed expenses, fingerling costs, labor costs, utilities, etc. The total amount for the year is recorded in the total column on the far right of the budget. Then, each month's expense is recorded for the appropriate month. For example, catfish fingerlings typically are purchased in February or March; fingerling expenses would then be listed in either February or March as appropriate. The values of each type of cash outflow for each time period should add to the amount listed in the total column. The amount in the total column should equal that in the enterprise budget and the income statement for each type of operating expense. Comparing the sum of individual months with the amount listed in the total column is a good way to check for errors in the accuracy of the budget.

The next section of cash outflow is to itemize the expenses that result from the purchase of capital assets such as equipment or broodstock. Family living expenses frequently are included in the cash flow budget, particularly when it will be used for applying for loans from banks.

Payments of principal and interest for each loan are itemized in the subsequent section. All loan payments are included, for both current and noncurrent liabilities. It is best to itemize the payments by individual loan. This makes it easier to check the payments and to make adjustments from year to year as the individual loans are paid off.

All expenses are summed to calculate total cash outflow for each month. The difference between total cash inflow and total cash outflow is the cash available,

sometimes called ending cash balance, for that particular time period. If the cash available is negative, this means that there is insufficient cash generated during the period to meet all cash obligations and additional borrowing is needed for that time period. After adding in the new borrowing, the cash balance is obtained. If the cash available is a positive number for a particular month, then there is cash available that can be used for other purposes. An economical use of extra cash is to pay off the principal of existing loans, particularly the operating capital loan. Paying off the principal as quickly as possible will minimize the total interest expenses paid. If not used to pay off loans, this cash amount can be carried over to subsequent time periods for upcoming cash needs in the coming months.

Cash balance becomes the beginning cash available at the start of the next time period. The basic rule for cash balances is that the ending cash balance be at least zero. That means that there is sufficient cash inflow to meet the cash needs for that time period. However, a zero ending cash balance also means that there is no cushion of cash for any unexpected problems or unanticipated expenses. It is prudent for the farm manager to plan for a certain amount of surplus cash to roll over from one time period to the next.

At the bottom of the cash flow budget, it is useful to maintain an accounting of the debt outstanding for each loan. Principal payments made in a time period are subtracted out of the balance owed in the outstanding debt section. Maintaining an accounting of outstanding principal amounts on loans is not necessary for the cash flow budget, but it is a useful practice. It should be adjusted from month to month to reflect additional payments on the principal as well as any new borrowing. These lines of outstanding amounts of principal reflect the overall pattern of debt and how it changes across the annual production cycle and patterns of cash flow. The lines of outstanding debt on operating loans will also indicate whether the farm is approaching any maximum limits on borrowed capital.

COMPARISON OF CASH FLOW BUDGET WITH OTHER FINANCIAL INSTRUMENTS

The cash flow budget represents a flow of cash over a period of time, unlike the balance sheet that represents the financial position at one specific point in time. The income statement also represents a flow of income and expenses, but includes all revenue and expenses, not just cash revenue and expenses. The cash flow statement summarizes the year's transactions and can help to reconcile net income calculated on an accrual basis (see Chapter 11 for a discussion of accrual-based and cash-based accounting for income statements) with the change in cash position.

The cash flow budget contains all cash flows and does not contain any noncash revenue or noncash expenditures. Thus, what is included in a cash flow budget is distinct from the tabulation of revenue and expenses on the enterprise budget and income statement. The income statement includes noncash revenue like positive changes in inventory whereas the cash flow budget does not. Similarly, depreciation is included in an income statement as a noncash expense but because depreciation is not a cash expense, it is not included in the cash flow budget. The total value of a new tractor (capital asset) purchased would not be included in the income statement but is a negative cash flow in the cash flow budget because it is a cash outflow. Cash flow budgets for farms frequently include cash outflow for personal expenses, because these may affect the cash available for the farm business; personal expenses are not included in the income statement.

The cash flow budget does *not* indicate profit. The income statement is the measure of profit in the business. Because the cash flow budget does not include noncash expenses, it does not measure profit. It only measures the flow of cash. It is possible to have a large positive cash flow and low profits from a business. This can occur when capital assets such as equipment or land are sold. The cash proceeds are included as cash inflow, but may decrease profits because fewer crops can be sold from less land. Similarly, it is possible to have negative cash flow and a high profit. If new assets, such as new ponds, are built, or ponds are renovated, these cash expenses will decrease the cash available to the business, but may result in greater profits from the sale of additional crops in the future.

Depreciation and changes in inventory values for many items are included in the income statement but do not enter the cash flow budget. The cash flow budget contains several items that do not enter the income statement, such as the principal payments on debts and capital expenditures. The income statement includes all project expenses (whether paid or not) and the value of items produced (whether sold for cash or not). On the other hand, the cash flow budget identifies the amount and timing of all projected inflows and outflows of cash.

Purchases of assets and loan activity are included in cash flow statements and budgets but are not reflected on the income statement, particularly if it is accrual-based. This is because exchanging an asset for a different asset (selling a tractor to buy aerators) or exchanging an asset for a liability (selling an asset to pay off a debt) do not constitute net income, but these types of transactions do affect the cash position. The income statement is not adequate to show the effects of these transactions on cash position, but the cash analysis does reflect these effects.

The beginning cash of the cash flow budget is the value shown in the beginning balance sheet. Similarly, the ending cash position of the last period in the cash flow budget is the cash on hand for the projected year-end balance sheet. Projected capital sales and capital purchases from the cash flow budget supply the necessary information for determining the ending balance sheet values of capital assets. Information on loans outstanding at the end of the year for the operating line, intermediate-term loans, and long-term loans on the pro forma balance sheet come directly from the cash flow budget.

RECORD-KEEPING FOR CASH FLOW BUDGETS

Inflows and outflows for the cash flow statement can be taken directly from records of receipts and expenditures for each time period. A record-keeping system for cash analysis that has been used effectively is a cash analysis book. In such a book, there is a column to record a description of each transaction. This is followed by two additional columns, one to record the date of each transaction and the other to record the number of the check or receipt. Maintaining the records in a looseleaf notebook provides flexibility and individual sheets can be sent to the accountant.

It is best to have separate columns for receipts and payments, as well as to separate capital receipts and capital payments. The cash analysis book should be updated at least once a month, to keep the receipts and expenses current. The total for each category should be added up each month.

Most cash in a business moves through the checking account of the business. Thus, the checking account statements are a prime source of data for maintaining records for the cash flow budget.

The statement of cash flows is one of the financial records that should be maintained by the business. Careful records of actual cash flows by month, by recording and summarizing cash receipts and cash expenses each month provide a basis for comparison with those budgeted for that month. This monthly comparison provides a basis for early identification of cash flow problems. This allows more time to search for solutions and to make necessary changes.

Inventory records are necessary to develop the cash flow budget (see section on Record-Keeping for the Balance Sheet in Chapter 11). The estimates of inventory should include estimates of the quantity and value of existing crops available for sale during that budget period.

Records of outstanding loans, dates of scheduled payments on each loan, and the amount to be charged to principal and interest will be required.

PRACTICAL APPLICATION: CASH FLOW BUDGET FOR A 256-ACRE CATFISH FARM

Table 12.1 presents a monthly cash flow budget for the 256-acre catfish farm example used in previous chapters. Each month's cash flow is listed as a separate column. Each line item (revenue or expense) is described in the column to the far left. The column on the far right has the total amount. The total amount for each line item should correspond to the amount from the enterprise budget. This serves as a useful check on the accuracy of the budget.

The first line category is revenue. The first item listed under revenue is the "beginning cash." *Beginning cash* is the amount of cash that is available to the business at the beginning of the time period. For the January column, the beginning cash is what the farmer has available to contribute to the farm business, often from a farm savings or checking account. In this example, the amount of beginning cash for the farm at the beginning of January is $578,834.

The next line includes the sales of food-sized catfish, the only product sold by this farm business. Most cash flow budgets include only the dollar value of the products sold by month, but for spreadsheets, it is often easier to track if the actual weight is included as a separate line, as is done in Table 12.1. In this example, the farm sold 46,080 lb of catfish in January. The sales price was $0.70/lb and, thus, the revenue received in January from sales of food-sized catfish was $32,256. The revenue from sales of catfish foodfish is then added to the amount of beginning cash to obtain the total cash inflow for that month, in this case, $611,090 for January.

The next major heading in the cash flow budget is operating cash expenses. Each type of expense is itemized in a separate row. These rows should correspond to the line items in the enterprise budget and in the income statement. Thus, there are separate rows for feed, fingerlings, labor, plankton control, gas, diesel fuel, and oil, electricity, repairs and maintenance expenses, bird depredation costs, seining and hauling charges, telephone expenses, office supplies, farm insurance, and legal/accounting fees. The expected cost of each of these that is anticipated to occur in that month is listed in the appropriate row. There are few feeding days in January for channel catfish, and so only $2,783 is expected to be spent on feed that month. Fingerlings are not planned to be stocked until February and March; thus, there is no charge in January for the purchase of fingerlings. The farm hires two people year-round who feed and check night-time levels of dissolved oxygen during the summer and spend most of their time scaring birds and repairing equipment over the winter. Because they work fairly constant hours across the year, their salary cost is constant. The farm also hires some seasonal or part-time labor over the summer months, so those charges do not appear in the cash flow budget until May and run through September. Plankton control expenses are not incurred until the summer months, either, with no charge in January. Gas, diesel fuel, oil charges, electricity costs, and repairs and maintenance expenses are incurred to some degree over the winter months, but all increase during the primary production period of May through October. On the other hand, bird depredation costs are incurred over the winter months when the migratory birds are resting and feeding on catfish ponds. Seining and hauling costs are incurred in the months when catfish are sold. Because they are charged by the pound of fish sold, this cost can be linked to the pounds of fish sold in each month. Costs of the telephone, office supplies, farm insurance, and legal/accounting fees occur in each month throughout the year.

When the operating expenses are completely recorded by line item for the month of January, these are summed to obtain total monthly cash expenses. For the month of January, these total to $14,940. This is the total amount of cash that is needed to pay the expenses that will be incurred in January.

The next category of items is the debt servicing lines. It is easier to manage and interpret the cash flow budget if each loan is listed separately on the cash flow budget, under the categories of real estate, equipment, and operating capital. It is also useful to separate the principal payments for each loan on a separate line from the interest payments. In Table 12.1, this farm does not have any loans (see Table 12.2 for a cash flow budget with debt-servicing payments) and, thus,

Table 12.1. Monthly Cash Flow Budget for a 256-acre Catfish Farm without Financing.

Item	January	February	March	April	May	June	July	August	September	October	November	December	Total
Beginning cash	578,834	596,150	577,548	557,823	664,921	713,614	727,089	704,967	702,063	698,544	790,834	824,538	824,538
Pounds of catfish sold. lb	46,080	46,080	46,080	207,360	115,200	115,200	69,120	115,200	115,200	195,840	80,640	0	1,152,000
Receipts from catfish sold, $	32,256	32,256	32,256	145,152	80,640	80,640	48,384	80,640	80,640	137,088	56,448	0	806,400
Total cash inflow	611,090	628,406	609,804	702,975	745,561	794,254	775,473	785,607	782,703	835,632	847,282	824,538	8,943,325
Operating cash expenses													
Feed	2,783	5,565	5,565	19,479	13,913	44,522	44,522	55,653	55,653	16,696	8,348	5,565	278,264
Fingerlings	0	36,416	36,416	0	0	0	0	0	0	0	0	0	72,832
Labor													0
Year-round, full-time	3,245	3,245	3,245	3,245	3,245	3,245	3,650	3,245	3,650	3,245	3,650	3,650	40,560
Seasonal, full-time	0	0	0	0	1,521	2,535	2,535	2,535	1,014	0	0	0	10,140
Plankton control	0	0	0	0	922	0	922	0	1,843	0	0	0	3,687
Gas, diesel fuel, and oil	1,137	568	568	568	947	1,516	2,463	2,463	3,031	3,031	1,326	1,326	18,944
Electricity	1,137	1,137	1,705	2,273	2,842	5,683	8,525	9,661	9,661	9,661	2,842	1,705	56,832
Repairs and maintenance	2,980	497	993	993	1,242	2,483	3,228	3,228	2,483	1,242	4,221	4,221	24,832
Bird depredation	240	240	240	160	80	0	0	0	0	160	240	240	1,600
Seining and hauling	2,304	2,304	2,304	10,368	5,760	5,760	3,456	5,760	5,760	9,792	4,032	0	57,600
Telephone	215	215	161	269	269	215	215	215	215	215	215	269	2,688
Office supplies	225	28	141	56	563	563	282	141	141	113	141	422	2,816
Farm insurance	518	518	518	518	518	518	583	518	583	518	583	583	6,476
Legal/accounting	156	125	125	125	125	125	125	125	125	125	125	156	1,562
Total operating expenses	14,940	50,858	51,981	38,054	31,947	67,165	70,506	83,544	84,159	44,798	22,744	18,137	578,833
Debt servicing													

(Continued)

Table 12.1. (*Continued*)

Item	January	February	March	April	May	June	July	August	September	October	November	December	Total
Real estate													
Interest	0	0	0	0	0	0	0	0	0	0	0	0	0
Principal	0	0	0	0	0	0	0	0	0	0	0	0	0
Subtotal	0	0	0	0	0	0	0	0	0	0	0	0	0
Equipment													
Interest	0	0	0	0	0	0	0	0	0	0	0	0	0
Principal	0	0	0	0	0	0	0	0	0	0	0	0	0
Subtotal	0	0	0	0	0	0	0	0	0	0	0	0	0
Operating													
Interest	0	0	0	0	0	0	0	0	0	0	0	0	0
Principal	0	0	0	0	0	0	0	0	0	0	0	0	0
Subtotal	0	0	0	0	0	0	0	0	0	0	0	0	0
Total debt servicing	0	0	0	0	0	0	0	0	0	0	0	0	
Total cash outflow	14,940	50,858	51,981	38,054	31,947	67,165	70,506	83,544	84,159	44,798	22,744	18,137	
Cash available	596,150	577,548	557,823	664,921	713,614	727,089	704,967	702,063	698,544	790,834	824,538	806,401	
New borrowing	0	0	0	0	0	0	0	0	0	0	0	0	
Ending cash balance	596,150	577,548	557,823	664,921	713,614	727,089	704,967	702,063	698,544	790,834	824,538	806,401	
Summary of debt outstanding													
Real estate													
Equipment													
Operating													

Table 12.2. Monthly Cash Flow Budget for a 256-acre Catfish Farm with Financing of 30% of All Capital.

Item	January	February	March	April	May	June	July	August	September	October	November	December	Total
Beginning cash	173,650	183,325	157,082	129,716	202,431	232,022	226,395	192,812	170,806	148,185	208,002	126,094	
Pounds of catfish sold, lb	46,080	46,080	46,080	207,360	115,200	115,200	69,120	115,200	115,200	195,840	80,640	0	1,152,000
Receipts from catfish sold, $	32,256	32,256	32,256	145,152	80,640	80,640	48,384	80,640	80,640	137,088	56,448	0	806,400
Total cash inflow	205,906	215,581	189,338	274,868	283,071	312,662	274,779	273,452	251,446	285,273	264,450	126,094	806,400
Operating cash expenses													
Feed	2,783	5,565	5,565	19,479	13,913	44,522	44,522	55,653	55,653	16,696	8,348	5,565	278,264
Fingerlings	0	36,416	36,416	0	0	0	0	0	0	0	0	0	72,832
Labor													0
Year-round, full-time	3,245	3,245	3,245	3,245	3,245	3,245	3,650	3,245	3,650	3,245	3,650	3,650	40,560
Seasonal, full-time	0	0	0	0	1,521	2,535	2,535	2,535	1,014	0	0	0	10,140
Plankton control	0	0	0	0	922	0	922	0	1,843	0	0	0	3,687
Gas, diesel fuel, and oil	1,137	568	568	568	947	1,516	2,463	2,463	3,031	3,031	1,326	1,326	18,944
Electricity	1,137	1,137	1,705	2,273	2,842	5,683	8,525	9,661	9,661	9,661	2,842	1,705	56,832
Repairs and maintenance	2,980	497	993	993	1,242	2,483	3,228	3,228	2,483	1,242	1,242	4,221	24,832
Bird depredation	240	240	240	160	80	0	0	0	0	160	240	240	1,600
Seining & hauling	2,304	2,304	2,304	10,368	5,760	5,760	3,456	5,760	5,760	9,792	4,032	0	57,600
Telephone	215	215	161	269	269	215	215	215	215	215	215	269	2,688
Office supplies	225	28	141	56	563	563	282	141	141	113	141	422	2,816
Farm insurance	518	518	518	518	518	518	583	518	583	518	583	583	6,476
Legal/accounting	156	125	125	125	125	125	125	125	125	125	125	156	1,562
Total operating expenses	14,940	50,858	51,981	38,054	31,947	67,165	70,506	83,544	84,159	44,798	22,744	18,137	578,833
Debt servicing													

(*Continued*)

Table 12.2. (*Continued*)

Item	January	February	March	April	May	June	July	August	September	October	November	December	Total
Real estate													
Interest	0	0	0	0	0	0	0	0	0	0	0	44,835	44,835
Principal	0	0	0	0	0	0	0	0	0	0	0	20,544	20,544
Subtotal	0	0	0	0	0	0	0	0	0	0	0	65,379	65,379
Equipment													
Interest	0	0	0	0	0	0	0	0	0	0	24,726	0	24,726
Principal	0	0	0	0	0	0	0	0	0	0	77,514	0	77,514
Subtotal	0	0	0	0	0	0	0	0	0	0	102,240	0	102,240
Operating													
Interest	695	695	695	3,126	1,737	1,737	1,042	1,737	1,737	2,952	1,216	0	17,369
Principal	6,946	6,946	6,946	31,257	17,365	17,365	10,419	17,365	17,365	29,521	12,156	0	173,651
Subtotal	7,641	7,641	7,641	34,383	19,102	19,102	11,461	19,102	19,102	32,473	13,372	0	191,020
Total debt servicing	7,641	7,641	7,641	34,383	19,102	19,102	11,461	19,102	19,102	32,473	115,612	65,379	358,639
Total cash outflow	22,581	58,499	59,622	72,437	51,049	86,267	81,967	102,646	103,261	77,271	138,356	83,516	
Cash available	183,325	157,082	129,716	202,431	232,022	226,395	192,812	170,806	148,185	208,002	126,094	42,578	
New borrowing	0	0	0	0	0	0	0	0	0	0	0	0	
Ending cash balance	183,325	157,082	129,716	202,431	232,022	226,395	192,812	170,806	148,185	208,002	126,094	42,578	
Summary of debt outstanding													
Real estate													
Equipment													
Operating													

there are no payments on interest or principal. The total debt-servicing payments are summed and added to the total monthly cash expenses to obtain the total cash outflow, $14,940, for the month of January.

The total cash outflow is subtracted from the total cash inflow to obtain the cash available for the month of January. In this example, $596,150 is available after paying the expenses incurred during the month. Thus, there is adequate cash flow for this time period. If the cash available were negative, that would indicate that there is not sufficient cash flow in January and that additional operating capital would need to be borrowed. Any new borrowing to cover the shortfall in cash for January is listed in the next line, "new borrowing." Then, the negative cash flow amount is subtracted from the new borrowing amount to obtain the amount for the cash balance row. This cash balance is the amount of cash then that is available to begin the next period. The cash balance amount is copied to the beginning cash amount for February. The process continues with filling in the appropriate revenue, expense, and debt-servicing amounts in each line item for each subsequent month.

With no financing charges in Table 12.1, the cash available increases over the year. Thus, without financing, this farm shows a positive cash position.

The final lines on the cash flow budget do not reflect the flow of cash in the business, but rather keep a running tab on the outstanding balances on each loan. These rows are not essential for the cash flow budget, but are useful to have a continuous view of the amount of principal remaining to be paid for each loan. This is particularly useful for the operating loan. Over the course of the production period, the business must generate enough cash to pay back the operating loan. If the principal owed continues to increase and does not get paid down, this is an indication of financial problems. If the operating loan increases for a period of time and then begins to decrease, then the budget will show when it will be paid in full.

Table 12.2 illustrates a cash flow budget for the same 256-acre catfish farm. However, this particular farm has financed 30% of all its capital. It still has an outstanding loan that was originally used to construct the ponds, took out equipment loans to purchase several tractors, and has an operating loan. In Table 12.2, all the revenue and expense items are identical to those in Table 12.1. However, this farm had to borrow its operating capital, not having access to the $578,834 assumed to be available in Table 12.2. This farm, borrowing 30% of its capital, then has the amount of its

operating loan, $173,650, received from the bank as beginning cash at the beginning of January.

The difference between the budget in Table 12.2 and that in Table 12.1 is in the debt-servicing payments. The payments on the real estate loan are scheduled to be paid in December and the equipment loan payments in November. These are convenient payment periods for a catfish farm because the majority of the fish are market-size during this time period and cash frequently is more readily available at the end of the year than at the beginning.

In Table 12.2, with 30% financing, the cash available decreases throughout the year. While the cash position is still positive, the loss of cash throughout the year indicates that there may be problems of cash flow the next year.

Table 12.3 shows the effect of 50% of the market-sized fish being off-flavor. Revenue from catfish sales decreases. This results in negative cash available in July, which necessitates additional borrowing to maintain a cash balance of at least $10,000. Additional borrowing is needed in November and December to meet equipment and real estate payments, respectively. At the end of the year, the operating loan outstanding is higher than at the beginning of the year. Given the amount of cash available in January through June, the operating loan could be restricted to borrow less early on.

Table 12.4 shows that, with careful management, the cash flow deficit is less, although it is not favorable, and still decreases across the year.

OTHER APPLICATIONS IN AQUACULTURE

Kam et al. (2002) included a cash flow analysis as part of an analysis of size economies of the Pacific threadfin *Polydactylus sexfilis*. The 5-year annual cash flows included cash outflows from operating, investment, and financing activities. From these, an internal rate of return was calculated to compare various sizes of production units. The financial analysis indicated that transferring the nursery stage to growout farmers instead of the hatchery reduces the cost per fry.

SUMMARY

Cash flow analysis is the third component of a complete financial analysis. While the balance sheet indicates financial position, strength, and wealth, and the income statement shows profits, the cash flow analysis indicates the cash position of the business. While

Table 12.3. Monthly Cash Flow Budget for a 256-acre Catfish Farm with Financing of 30% of All Capital and Restricted Sales Due to Off-Flavor.

Item	January	February	March	April	May	June	July	August	September	October	November	December	Total
Beginning cash	173,650	168,349	127,130	84,788	90,111	82,262	39,195	10,000	10,000	10,000	6,169	10,000	
Pounds of catfish sold, lb	23,040	23,040	23,040	103,680	57,600	57,600	34,560	57,600	57,600	97,920	40,320	0	576,000
Receipts from catfish sold, $	16,128	16,128	16,128	72,576	40,320	40,320	24,192	40,320	40,320	68,544	28,224	0	403,200
Total cash inflow	189,778	184,477	143,258	157,364	130,431	122,582	63,387	50,320	50,320	78,544	34,393	10,000	1,214,854
Operating cash expenses													
Feed	2,783	5,565	5,565	19,479	13,913	44,522	44,522	55,653	55,653	16,696	8,348	5,565	278,264
Fingerlings	0	36,416	36,416	0	0	0	0	0	0	0	0	0	72,832
Labor													0
Year-round, full-time	3,245	3,245	3,245	3,245	3,245	3,245	3,650	3,245	3,650	3,245	3,650	3,650	40,560
Seasonal, full-time	0	0	0	0	1,521	2,535	2,535	2,535	1,014	0	0	0	10,140
Plankton control	0	0	0	0	922	0	922	0	1,843	0	0	0	3,687
Gas, diesel fuel, and oil	1,137	568	568	568	947	1,516	2,463	2,463	3,031	3,031	1,326	1,326	18,944
Electricity	1,137	1,137	1,705	2,273	2,842	5,683	8,525	9,661	9,661	9,661	2,842	1,705	56,832
Repairs and maintenance	2,980	497	993	993	1,242	2,483	3,228	3,228	2,483	1,242	1,242	4,221	24,832
Bird depredation	240	240	240	160	80	0	0	0	0	160	240	240	1,600
Seining and hauling	1,152	1,152	1,152	5,184	2,880	2,880	1,728	2,880	2,880	4,896	2,016	0	28,800
Telephone	215	215	161	269	269	215	215	215	215	215	215	269	2,688
Office supplies	225	28	141	56	563	563	282	141	141	113	141	422	2,816
Farm insurance	518	518	518	518	518	518	583	518	583	518	583	583	6,476
Legal/accounting	156	125	125	125	125	125	125	125	125	125	125	156	1,562
Total operating expenses	13,788	49,706	50,829	32,870	29,067	64,285	68,778	80,664	81,279	39,902	20,728	18,137	550,033
Debt servicing													

Real estate												
Interest	0	0	0	0	0	0	0	0	0	0	0	44,835
Principal	0	0	0	0	0	0	0	0	0	0	0	20,544
Subtotal	0	0	0	0	0	0	0	0	0	0	0	65,379
Equipment												
Interest	0	0	0	0	0	0	0	0	0	0	24,726	24,726
Principal	0	0	0	0	0	0	0	0	0	0	77,514	77,514
Subtotal	0	0	0	0	0	0	0	0	0	0	102,240	102,240
Operating												
Interest	695	695	3,126	1,737	1,737	1,042	1,737	1,737	1,737	2,952	1,216	17,369
Principal	6,946	6,946	31,257	17,365	17,365	10,419	17,365	17,365	17,365	29,521	12,156	173,651
Subtotal	7,641	7,641	34,383	19,102	19,102	11,461	19,102	19,102	19,102	32,473	13,372	191,020
Total debt servicing	7,641	7,641	34,383	19,102	19,102	11,461	19,102	19,102	19,102	32,473	115,612	358,639
Total cash outflow	21,429	57,347	58,470	67,253	48,169	83,387	80,239	99,766	100,381	72,375	136,340	83,516
Cash available	168,349	127,130	84,788	90,111	82,262	39,195	−16,852	−49,446	−50,061	6,169	−101,947	−73,516
New borrowing	0	0	0	0	0	0	26,852	59,446	60,061	0	111,947	83,516
Ending cash balance	168,349	127,130	84,788	90,111	82,262	39,195	10,000	10,000	10,000	6,169	10,000	10,000
Summary of debt outstanding												
Real estate												
Equipment												
Operating	166,705	159,759	152,813	121,556	104,191	86,826	103,259	145,340	188,036	158,515	258,306	341,822

Table 12.4. Monthly Cash Flow Budget for a 256-acre Catfish Farm with Financing of 30% of All Capital, after Restructuring.

Item	January	February	March	April	May	June	July	August	September	October	November	December	Total
Beginning cash	15,429	10,694	10,000	10,000	10,812	10,381	10,000	10,000	10,000	10,000	10,630	10,000	10,000
Pounds of catfish sold, lb	23,040	23,040	23,040	103,680	57,600	57,600	34,560	57,600	57,600	97,920	40,320	0	576,000
Receipts from catfish sold, $	16,128	16,128	16,128	72,576	40,320	40,320	24,192	40,320	40,320	68,544	28,224	0	403,200
Total cash inflow	31,557	26,822	26,128	82,576	51,132	50,701	34,192	50,320	50,320	78,544	38,854	10,000	531,147
Operating cash expenses													
Feed	2,783	5,565	5,565	19,479	13,913	44,522	44,522	55,653	55,653	16,696	8,348	5,565	278,264
Fingerlings	0	36,416	36,416	0	0	0	0	0	0	0	0	0	72,832
Labor													0
Year-round, full-time	3,245	3,245	3,245	3,245	3,245	3,245	3,650	3,245	3,650	3,245	3,650	3,650	40,560
Seasonal, full-time	0	0	0	0	1,521	2,535	2,535	2,535	1,014	0	0	0	10,140
Plankton control	0	0	0	0	922	0	922	0	1,843	0	0	0	3,687
Gas, diesel fuel, and oil	1,137	568	568	568	947	1,516	2,463	2,463	3,031	3,031	1,326	1,326	18,944
Electricity	1,137	1,137	1,705	2,273	2,842	5,683	8,525	9,661	9,661	9,661	2,842	1,705	56,832
Repairs and maintenance	2,980	497	993	993	1,242	2,483	3,228	3,228	2,483	1,242	1,242	4,221	24,832
Bird depredation	240	240	240	160	80	0	0	0	0	160	240	240	1,600
Seining and hauling	1,152	1,152	1,152	5,184	2,880	2,880	1,728	2,880	2,880	4,896	2,016	0	28,800
Telephone	215	215	161	269	269	215	215	215	215	215	215	269	2,688
Office supplies	225	28	141	56	563	563	282	141	141	113	141	422	2,816
Farm insurance	518	518	518	518	518	518	583	518	583	518	583	583	6,476
Legal/accounting	156	125	125	125	125	125	125	125	125	125	125	156	1,562
Total operating expenses	13,788	49,706	50,829	32,870	29,067	64,285	68,778	80,664	81,279	39,902	20,728	18,137	550,033
Debt servicing													

													Total
Real estate													
Interest	0	0	0	0	0	0	0	0	0	0	0	0	44,835
Principal	0	0	0	0	0	0	0	0	0	0	0	0	20,544
Subtotal	0	0	0	0	0	0	0	0	0	0	0	0	65,379
Equipment													
Interest	0	0	0	0	0	0	0	0	0	0	0	0	24,726
Principal	0	0	0	0	0	0	0	0	0	0	0	0	77,514
Subtotal	0	0	0	0	0	0	0	0	0	0	0	0	102,240
Operating													
Interest	129	71	345	637	319	224	422	797	1,140	1,491	1,270	2,065	8,910
Principal	6,946	6,946	6,946	38,257	17,365	17,365	10,419	17,365	17,365	29,521	12,156	0	180,651
Subtotal	7,075	7,017	7,291	38,894	17,684	17,589	10,841	18,162	18,505	31,012	13,426	2,065	189,561
Total debt servicing	7,075	7,017	7,291	38,894	17,684	17,589	10,841	18,162	18,505	31,012	13,426	2,065	357,180
Total cash outflow	20,863	56,723	58,120	71,764	46,751	81,874	79,619	98,826	99,784	70,914	136,394	85,581	
Cash available	10,694	−29,900	−31,993	10,812	4,381	−31,172	−45,428	−48,506	−49,464	7,630	−97,540	−75,581	
New borrowing	0	39,900	41,993	0	6,000	41,172	55,428	58,506	59,464	3,000	107,540	85,581	
Ending cash balance	10,694	10,000	10,000	10,812	10,382	10,000	10,000	10,000	10,000	10,630	10,000	10,000	
Summary of debt outstanding													
Real estate													
Equipment													
Operating	8,483	41,437	76,484	38,227	26,862	50,669	95,678	136,819	152,397	247,781	333,362		

the business must generate profit over time to continue in business, it will not survive in the short run without adequate cash flow. Cash flow budgets are critical for planning to meet the cash needs of the business but are also required for credit applications to demonstrate adequate repayment capacity. Monthly cash flow budgets are the most useful for planning purposes. The cash flow budget tracks only cash inflow and outflows.

REVIEW QUESTIONS

1. List five uses of a cash flow budget.

2. What is the difference between a cash flow statement and a cash flow budget?

3. Outline the basic structure of a cash flow budget.

4. How does one determine the beginning cash to start a cash flow budget? How is the beginning cash of subsequent time periods determined?

5. Explain the difference between handling capital assets in the cash flow budget and in the income statement.

6. Explain why cash flow budgets cannot be used as a measure of profit.

7. Explain how to set up a cash flow budget to be able to identify an appropriate borrowing pattern for an operating line of credit.

8. Explain how depreciation is handled in a cash flow budget.

9. What are the most important indicators of cash flow in the cash flow budget? What portion of the budget gives the most rapid view of the cash position of the business?

10. Give an example of how a cash flow budget can be used to improve overall farm management.

REFERENCE

Kam, Lotus E., PingSun Leung, Anthony C. Ostrowski, and Augustin Molnar. 2002. Size economies of a Pacific threadfin *Polydactylus sexfilis* hatchery in Hawaii. *Journal of the World Aquaculture Society* 33(4):410–424.

13

Investment Analysis (Capital Budgeting) in Aquaculture

INTRODUCTION

Previous analyses presented in this book have discussed techniques to analyze the profitability of an aquaculture enterprise for a particular point in time. Financial analyses such as the enterprise budget, income statements, and monthly cash flow budgets tend to focus more on operating capital required for production inputs like feed, fingerlings, and labor. However, the land, ponds, tanks, raceways, aerators, hatcheries, and other buildings are just as important to the aquaculture business. Moreover, the farm manager has a variety of options related to these types of capital goods. For example, equipment can be bought or sold but it can also be rented from someone else or leased to another farmer to generate revenue. Thus, management of long-term capital goods is more complex in some regards than the management of operating inputs and capital.

This chapter provides an overview of various types of investment analyses. It presents a summary of the time value of money and how this relates to investment decisions. Results from various investment analyses are compared and contrasted.

OVERVIEW OF INVESTMENT ANALYSIS

Capital is often the most limiting of resources used in agriculture over the long term, and the efficiency of its use in the business is an important consideration. Over time, any resource can be acquired with capital, if sufficient capital resources are available. Theoretically, a farm could be tripled in size by buying more land and

purchasing other farms, or it could be cut in half by selling off half its ponds. However, the scale of operation can be altered similarly by adding equipment to intensify and produce more per acre as opposed to expanding acreage. It is critical for the manager to evaluate the best uses of the capital available to the farm.

There are two broad categories of capital in any business: operating and investment capital. Operating capital is the capital used to cover the annual operating expenses while investment capital is the long-term capital used to acquire "capital goods." Capital goods are those that will be used over time, typically for a period of more than a year. Capital assets such as land, equipment, and ponds typically involve a large initial expense with resulting returns spread over a number of future periods. Time is not a major issue for examining annual operating inputs, but it is of importance when analyzing the profitability of capital assets. Capital assets often involve large amounts of money and the returns are received over many years and at different time periods, sometimes at irregular intervals. Thus, careful consideration of the size and timing of the annual cash flows over the years of useful life of the investment is important.

Aquaculture businesses tend to require higher levels of capital investment than do other forms of agriculture. This is because the structures to hold the water for the aquatic crops being raised, whether outdoor ponds or indoor tanks, are relatively expensive. Moreover, aquaculture farms require some unique structures and equipment, such as systems to maintain oxygen concentrations in water, that are not required for other livestock businesses. These structures frequently require substantial amounts of investment capital to construct along with backup systems to ensure continual operation. Given this high degree of capital intensity in

Aquaculture Economics and Financing: Management and Analysis, Carole R. Engle, © 2010 Carole R. Engle.

many aquaculture businesses, understanding the profitability and management of an aquaculture investment over the course of its business life is critical.

Analyses that estimate the profitability of the uses of capital invested in an aquaculture business over its business life are referred to as "investment analyses," or sometimes as "capital budgeting." An investment is an addition of a durable or capital good to a business. Thus, investments add noncurrent assets to the business. These types of assets have long-lasting consequences. Investment analyses determine the profitability of spending funds on capital assets (those that will be used for more than 1 year).

Investment capital has characteristics and properties that differ substantially from operating capital. Thus, the analytical methods used to evaluate investment capital decisions differ from those used to evaluate operating capital.

The differences in investment capital as compared to operating capital arise primarily from the extended time period between when the initial expense or investment was made and when returns are received. Investment capital is expended at one point in time to build ponds or purchase equipment, but the returns from use of the investment capital are received over a period of years. Investment capital assets often require a large initial outlay of capital that is then followed by additional operating expenses and returns that are spread over a number of years into the future. Operating capital typically is used to purchase production inputs that are used within a period of 1 year or less. Capital investments are assets that last a long time and decisions related to their acquisition cannot be changed quickly or within one production cycle.

The bottom line measure obtained from an investment analysis is the profitability of the investment over its useful life. However, there are a number of different types of methodologies for investment analysis and each results in the calculation of a different indicator as its bottom line measure. The most common types of investment analyses are: (1) payback period; (2) simple rate of return; (3) net present value (NPV); and (4) internal rate of return (IRR). Each of these is named for the indicator that is calculated as the bottom line of the analysis. The payback period results in a calculation of the number of years required to "pay back" the investment or to recover an amount of revenue that equals the amount of the initial investment. The simple rate of return calculation results in a percentage that indicates what percent of the investment is received from the average annual net returns. The NPV anal-

ysis calculates an indicator called NPV as its bottom line. Similarly, the IRR is the bottom line of the IRR analysis.

The unit of analysis for an investment analysis is the investment under consideration. Thus, if a farm is considering investing in additional ponds, the unit of analysis is the investment cost of the additional ponds being proposed and not the investment in the entire existing farm. If the investment analysis is considering a new piece of equipment, then the unit of analysis is the investment cost of that particular piece of equipment. However, if the investment decision were the purchase of an entire farm, then the appropriate unit of analysis would be the entire farm.

In aquaculture, there are many decisions for which a formal investment analysis should be made. Purchase of an aquaculture business is a major investment expense and its profitability over time should be analyzed thoroughly. However, building additional ponds or renovating ponds also represent investment capital decisions that would appropriately be analyzed with investment analysis methods. However, even equipment purchases and technology upgrades in the business should be evaluated with investment analyses prior to incurring the expense.

UNDERLYING CONCEPT OF INVESTMENT ANALYSIS: TIME VALUE OF MONEY

The primary concept that underlies investment analysis is that a dollar received today is worth more than a dollar to be received at some point in the future. This concept is best understood by considering compound interest of a sum of money invested. Interest accrues on the investment with the accrued interest added back into the amount of the value of the investment. The value of the funds invested increases as the compounded interest continues to be added into the original amount. Thus, in 5 years, the future value (FV) will be greater than the original value of what was invested (present value). The extent of the difference is determined by the interest rate.

Figure 13.1 illustrates this concept. Present value (PV) represents the amount invested initially when the ponds were built or the equipment purchased. However, PV also is the value today of amounts of money to be received in the future. FV, on the other hand, represents the amount of money to be received in the future or the amount of money that the PV will become in the future when invested at a certain interest rate.

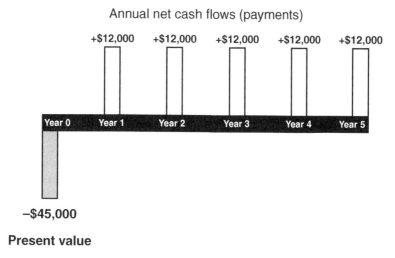

Figure 13.1. Stream of net cash flows from an initial investment.

The payment (PMT) is the amount of money that will be paid at the end of each time period. The interest rate (i) is also called the discount rate and is the rate that is used to calculate the PVs and FVs. The term "interest rate" is used when compounding PVs into FVs and the term "discount rate" is used for the rate used to calculate the PV from future receipts. Both interest and discount rates represent the opportunity cost of capital and serve as pricing mechanisms for the time value of money. Interest rates equate present and future claims on financial assets that will yield revenue at different points in time.

The interest rate (i) is typically expressed as a percentage. Thus, a 10% interest rate is expressed as 0.10. The number of time periods is expressed as n. These time periods are the number of periods during which either a PMT is made, interest is calculated, or revenues are received. For example, if funds invested earn interest that is compounded monthly, there would be 24 time periods over a 2-year period (12/year). However, if PMTs are made only once a year, there would only be two periods in a 2-year period. If the time periods are less than a year, the interest rate must be adjusted accordingly by calculating first a monthly rate and then the rate for the appropriate amount of time.

The calculations of PV and FV can go backwards in time or forward in time, depending upon the question that is being asked. If the question is to identify what an investment will be worth at some point in the future, then one starts with a PV, compounded into the future to reach an FV with compound interest. The interest earned by the investment is added (compounded) into the principal. The compounded amount then earns interest through the next time period. Thus, the principal amount increases through time due to compounding and so do the interest PMTs. The information needed to do the calculation is the PV, the interest rate, and the number of time periods during which PMTs will be made.

Table 13.1 illustrates how this happens. It shows an investment of $5,000 that is invested at a simple annual interest rate of 4%. The value at the end of the first year would be $5,200. In Year 2, this $5,200 would earn annual interest of 4% ($208), for a total value at the end of Year 2 of $5,408. Similarly, in Year 3, this $5,408 would earn $216 and be valued at $5,624 at the end of Year 3. Thus, money that someone has today could be invested to earn interest that will result

Table 13.1. Increase in Value of an Investment Initially Made of $5,000 at an Annual Simple Interest Rate of 4% for 3 Years.

Year	Value at beginning of year	Annual simple interest rate	Annual interest earned	Value of investment at end of year
1	$5,000	4%	$200	$5,200
2	$5,200	4%	$208	$5,408
3	$5,408	4%	$216	$5,624

in an increase in its value to that amount plus interest by some future date. The formula used is:

$$FV = PV(1 + i)^n \qquad (13.1)$$

Table 13.1 is a simple example that can be calculated easily for several years. However, when there are a large number of time periods, the calculation can become quite time consuming. Fortunately, spreadsheet software programs now include functions that make these calculations quite simple. For example, Microsoft® Excel can be used to calculate the FV of an amount saved that will earn an annual simple rate of interest that is compounded annually with the FVSCHEDULE function For the example illustrated in Table 13.1, a spreadsheet is set up that indicates the PV of $5,000 and the annual interest of 4% (entered as 0.04) for each of the 3 years. In the cell where the FV is to appear, the "=" sign is clicked and a dialogue box will appear. Clicking next to "Principal" either the number $5,000 can be entered, or the cell where that amount is can be clicked. Then, the column of values for the interest rates for each year (0.04) must be highlighted. After clicking "enter," the answer will then appear in the cell.

In areas where computers may not be available, tables of values for $(1 + i)^n$ can be used to facilitate this calculation (Table 13.6). The tables show the FV of $1 for various combinations of interest rates and time periods. Thus, for the example in Table 13.1, the value from the table for an interest rate of 4% for 3 time periods (3 years with a single PMT each year) is 1.1249. Multiplying the $5,000 by this value of 1.1249 results in $5,624, the same value as that at the end of the 3-year time period.

When an FV is known, its corresponding PV can be calculated through a similar, but reverse process, using discounting. The FV of an investment is "discounted" back to the PV with calculations that are inverse to those used in compounding. Discounting is used to compare an investment that is made today with the future income that will be earned from that investment. PVs are compounded to find FVs and FVs are discounted to identify PVs. The formula for calculating PV is:

$$PV = \frac{FV}{(1 + i)^n} \qquad (13.2)$$

Spreadsheet software programs can also be used to facilitate these calculations. In this case, the PV function can be used. It is applied in the same manner as that described above for the FVSCHEDULE function

that calculates FV. Tables can also be used to calculate this value for a PV from FVs (Table 13.7). The value for the example in Table 13.1 is 0.889. Thus, 0.889 times $5,624 is the PV of $5,000.

Investment analysis, then, uses the discounting calculations to estimate the profit to be earned from an investment. The revenues to be earned over time as a result of the investment in a new piece of equipment, or additional land and ponds are "discounted" back to the present for each time period and then added together.

The discounting process accounts for the fact that money available today could have been invested in such a way that a rate of interest would have been earned over time. The interest that the investment capital would have earned if it had been invested in some other way is an "opportunity cost," or the value of what would have been earned from some other use of the investment capital. Investment analysis accounts for the opportunity cost of the capital invested.

Table 13.2 illustrates the cash flow for three different investment alternatives. Each investment option requires an initial investment of $45,000, which is shown as a negative value in Year 0. Investment Alternative A shows the greatest overall revenue, $70,000. However, is Investment Alternative A really the best investment? Investment Alternatives A, B, and C have very different cash flows. Investment Alternative A generates little revenue in the early years of the investment, while Investment Alternative C provides a great deal of positive revenue in the early years, but that revenue tapers off in Year 4. Investment Alternative B provides constant net revenue across all 5 years. In this chapter, the payback period, simple rate of return, NPV, and IRR will be calculated for this example and the advantages and disadvantages of each discussed before answering the question of which investment option is

Table 13.2. Net Cash Flows that would Likely Result from Three Different Types of Investments that could be Made on a Farm.

Year	Investment A	Investment B	Investment C
0	−$45,000	−$45,000	−$45,000
1	$8,000	$12,000	$18,000
2	$11,000	$12,000	$17,000
3	$14,000	$12,000	$14,000
4	$17,000	$12,000	$9,000
5	$20,000	$12,000	$3,000
Total	$70,000	$60,000	$64,000

the most profitable. The chapter concludes with an example of a practical application of investment analysis in aquaculture.

INVESTMENT ANALYSIS

The information needed for an investment analysis includes the initial cost of the investment, the annual net cash revenues to be received, the terminal value of the investment, and the interest or discount rate. The initial cost should be the actual expenditure for the piece of equipment or other capital good. Net cash revenues need to be estimated for each time period by subtracting net cash expenditures from cash receipts. Only cash receipts are included. In addition, the family living expenses often included in a monthly cash flow budget should be excluded from the annual cash flow budget used for investment analysis. This is because the investment analysis measures the profitability of the use of capital to be invested. Family living expenses are not part of the costs of the new proposed investment. Noncash expenses like depreciation are already accounted for with the initial cost of the investment and terminal cost. The terminal value is estimated in the same way as the salvage value for a depreciable asset. The discount rate is the opportunity cost of capital. It represents the rate of return that the investor could have earned by investing the funds in some other type of investment.

PAYBACK PERIOD

The payback period is the number of years it would take for an investment to return its original cost through the annual net cash revenues it generates. It is the amount of time that it takes to recoup the initial investment from the net cash flow resulting from that investment. The payback period is expressed as number of years, not as a cash amount. The preferred investment alternative is the one with the shortest payback period. If net cash revenues are constant each year, the payback period can be calculated as follows:

$$P = \frac{I}{E} \qquad (13.3)$$

where P = payback period in years; I = amount of investment; and E = expected annual net revenue.

If annual net cash revenues are not equal, the revenues are added year-by-year to find the year in which the total of the annual net cash revenue is equal to the amount of the investment.

The investment alternatives and their cash flows presented in Table 13.2 can be compared using the payback period. In this table, only Investment Alternative B has equal annual net cash revenue. Because it does, the formula indicated above can be used. The amount of the investment is $45,000 and the annual net cash flow is $12,000. Thus, the payback period is $45,000 divided by $12,000, or 3.75 years. Thus, for Investment Alternative B, the payback period is 3.75 years.

The annual net cash revenues are not equal for investment options A and C. Thus, to calculate the payback period, the annual net cash flows are summed year to year to determine how many years will be required to pay back the original amount of the investment. For Investment Alternative A, the annual net cash revenues of $8,000, $11,000, and $14,000 are summed to $33,000. Thus, it will take 3 years and 8.5 months (an annual net cash flow of $17,000 is equivalent to $1,417/month; it will take 8.5 months into Year 4 to generate the additional $12,000 to pay back the investment) to generate sufficient net revenue to pay back the investment. According to the payback period, Investment Alternative B is preferable to A because it has a slightly shorter payback period. For Investment Alternative C, the payback period is found by adding $20,000 and $18,000 to get $38,000. The additional $7,000 would be received during the 7th month of Year 3. Thus, the payback period for Alternative C is 2.6 years. According to the payback period, then, Alternative C is preferred because it has a shorter payback period in spite of its lower overall net revenue across the 5-year period.

The payback period can be used to rank investments by choosing the investment option with a shorter payback period. The payback period has the advantage of being easy to use, and it quickly identifies investments with the most immediate cash returns. The example from the net cash flows in Table 13.2 shows the ease with which the payback period is calculated.

Disadvantages of the payback period are that it ignores any cash flows occurring after the end of the payback period. For example, it ignores the very large net cash flows in Year 5 of Alternative A. It also does not account for the timing of the cash flows during the payback period. Thus, the payback period measures more the investment's contribution to liquidity than its contribution to profitability. It is not the best measure to use because it is not a true measure of profitability. However, it is useful for firms with low liquidity that must concentrate on quick cash recovery because it measures the speed of recovery of the initial investment.

SIMPLE RATE OF RETURN

The simple rate of return expresses average annual net revenue as a percentage of the investment. Net revenue is found by subtracting the average annual depreciation from the average annual net cash revenue.

The formula used is:

$$\frac{\text{Average annual net revenue}}{\text{Cost of the investment}} \times 100 \qquad (13.4)$$

For the three investment options in Table 13.2, then, the simple rate of return is:

Investment A $14,000 divided by $45,000, or 31%
Investment B $12,000 divided by $45,000, or 27%
Investment C $12,600 divided by $45,000, or 28%.

Thus, according to the simple rate of return, the preferable option is A, with the highest rate of return, followed by C and then B. Alternatives B and C earn a similar rate of return because the average annual net cash flow is similar. This method does not account for the substantial difference in the pattern of cash flow.

The simple rate of return is an improvement over the payback period in terms of measuring profitability because it considers the earnings of an investment over the entire life of the investment. However, there is a major disadvantage in that it fails to consider the size and timing of the annual earnings and can therefore cause errors in selecting investments. This is especially true when net revenues are increasing or decreasing.

NET PRESENT VALUE

NPV is also known as the discounted cash flow method. NPV is equal to the sum of the PVs for each year's net cash flow less the initial cost of the investment. PV is equal to the current value of a sum of money to be received or paid in the future. It is found with discounting and is equal to the sum of PVs for each year's net cash flow less the initial cost of the investment. It can also be viewed as that sum of money which would have to be invested now at the given interest rate to equal the FV on the same date. Figure 13.1 illustrates the stream of net cash flow for Investment Alternative B.

Mathematically, NPV is calculated as follows:

$$NPV = \frac{P_1}{(1 + I)^1} + \frac{P_2}{(1 + I)^2} + \cdots \frac{P_n}{(1 + I)^n} - C \qquad (13.5)$$

where NPV = net present value; P_n = net cash flow in year, n; I = discount rate; and c = initial cost of investment.

Investments with a positive NPV would be accepted; those with a negative NPV rejected, and a zero value makes the investor indifferent. With a positive NPV, the rate of return of the investment is higher than the discount rate used, and it is greater than the opportunity cost of capital used as the discount rate. Projects with a greater NPV are the most favored. The limitations of the NPV analysis are that it depends on the discount rate and that it does not determine the actual rate of return.

NPV can be calculated with a hand calculator or spreadsheet based software, such as Microsoft® Excel. The NPV for the three investment alternatives presented in Table 13.2 were calculated using an Excel spreadsheet with the NPV function. When the NPV function is selected, a dialogue box opens that asks for the discount rate. If the prevailing cost of capital (interest rate) is 10%, for example, then the discount rate is expressed as 0.10. Once the discount rate is entered, the NPV function dialogue box then asks for the values. The first value in the row should be the total investment amount, expressed as a negative number (Figure 13.1). The second value is the net cash flow for Year 1, followed by the net cash flow for Year 2.

The NPV that is calculated is expressed as a dollar value. If it is positive, then the investment is profitable because the flow of returns, after accounting for the time value of money at the discount rate (interest rate) selected, is above 0. If the NPV is negative, then the investment earns less than its opportunity cost (discount rate or interest rate that would have been earned by the investment) and is not profitable.

For the three investment options listed in Table 13.2, Alternative C has the highest NPV ($5,987), followed by Alternative A ($5,374), and then Alternative B ($445) (Table 13.3). Strictly speaking, NPV values, technically, should not be compared to rank investments. Each of these options has a positive NPV and, thus, all are profitable investments.

The limitation of the NPV is that the value calculated depends upon the discount rate selected. Varying discount rates will change the NPV. Moreover, the NPV does not calculate the rate of return actually earned by the investment. The other drawback of the NPV is that the size of the NPV is affected by the size of the investment. To overcome this limitation, a profitability index can be calculated by dividing the total PV of the

Table 13.3. Net Cash Flows, Net Present Value (NPV), and Internal Rate of Return (IRR) Calculation with Excel Dialogue Boxes.

| Investment option | Year | | | | | | NPV* | IRR† |
	0	1	2	3	4	5		
A	($45,000)	$8,000	$11,000	$14,000	$17,000	$20,000	$5,374	14%
B	($45,000)	$12,000	$12,000	$12,000	$12,000	$12,000	$445	10%
C	($45,000)	$20,000	$18,000	$14,000	$9,000	$3,000	$5,987	17%

*Excel formula for NPV is NPV(0.1, B13:G13). The range of cells is specific to the particular Excel file the analyst is working on. In the formula, 0.1 is the interest rate of 10% and B13:G13 is the row of cells with the net cash flows, with B13 being the initial investment, that is negative.
†Excel formula for IRR is IRR(B13:G13,0.1). The range of cells is specific to the particular Excel file the analyst is working on. In the formula, B13:G13 is the row of cells with the net cash flows, with B13 being the cell with the initial investment, that is expressed as a negative value. The value of 0.1 in the formula is an estimate of what the IRR will be; this facilitates the analysis.

cash flows by the total PV of the investment required. The higher the profitability index, the better.

INTERNAL RATE OF RETURN

The IRR is the actual rate of return on the investment with proper accounting for the time value of money. It is also called the marginal efficiency of capital, yield on investment, or discounted yield. The equation used is that for the NPV, but the equation is solved for i, the interest rate when NPV = 0. This equation is difficult to solve. It requires trial and error, but it can be solved through Microsoft Excel and other programs. Its interpretation is that any investment with an IRR greater than the opportunity cost of capital is profitable. Some investors select an arbitrary cutoff point. Unlike the NPV, it can be used to rank investments which have different initial costs and lives. The limitation of the IRR is that it implicitly assumes that annual net returns or cash flows can be reinvested to earn a return equal to the IRR. If IRR is fairly high, this may not be possible and the IRR may overestimate the actual rate of return.

The IRR calculated for the three options in Table 13.2 were calculated with the IRR function in Excel (Table 13.3). The dialogue box asks for the array of values that represent first the investment, again expressed as a negative number, and then the series of annual net cash flows. The same spreadsheet is used to calculate both the NPV and IRR. For the examples in Table 13.2, the IRR was 14% for Alternative A, 10% for Alternative B, and 17% for Alternative C.

One of the limitations of the IRR is that when the cash flow of a project changes from positive to negative during the life of the project, the IRR can produce more than one solution. Moreover, it does not reflect the size of the project. However, it is readily understood by businesses and lenders.

COMPARISON OF INVESTMENT ANALYSIS INDICATORS

Table 13.4 compares the results for the three investment options described in Table 13.2. Based strictly on the payback period, Alternative C is preferred. The simple rate of return shows Alternative A as the best. The IRR calculations show both Alternatives A and C with the highest rate of return, but the NPV selects Alternative C. Overall, Alternative C would be selected as the best choice in spite of not having the highest overall net returns over time. It has the shortest payback period, hence, contributes the most to the business' liquidity. It generates the same IRR as A, but has a higher NPV. This is because it returns revenues faster and earlier than A; these revenues are worth more over time because they could then be invested in other types of revenue-generating activities.

Table 13.4. Comparison of Investment Analysis Indicators for Three Investment Options (Described in Table 13.2).

| Investment analysis | Investment option | | |
Indicator	A	B	C
Payback period	4.2 years	4.5 years	2.6 years
Simple rate of return	31%	27%	28%
Net present value	$5,374	$445	$5,987
Internal rate of return	14%	10%	17%

RELATIONSHIP BETWEEN NPV AND IRR

NPV and IRR are closely related. This is because they both are based on net cash flows that use the same discounting method. The difference is that the calculation of NPV requires that a discount (interest) rate be specified at the outset. IRR, on the other hand, computes the interest rate that yields a zero NPV. The IRR and NPV typically result in the same rankings of projects, although there are some exceptions.

DATA REQUIREMENTS FOR INVESTMENT ANALYSIS (CAPITAL BUDGETING)

The amount of the initial investment must first be determined. This is the amount of funds that the investor or owner is willing to commit to the project or investment. The total amount of investment must include all costs associated with the investment. These include any taxes, transportation charges, installation costs associated with roads, or electrical supply in addition to the purchase cost of the asset. If there is a trade-in value, it should be subtracted from the purchase amount.

Net cash flows measure the expected future returns for the investment alternatives. Net cash flows are obtained by subtracting out cash outflows from cash receipts for each time period. Net cash flows are not profits because noncash items are not included. At the end of the planning horizon (10 years for a 10-year cash flow budget), whatever remaining value exists in capital goods is included as a terminal or residual value.

The planning horizon for the analysis must be specified. One factor that is important in deciding on the length of the planning horizon is the useful life of the asset. If the analysis were for an investment in a new piece of equipment, an appropriate planning horizon would be the useful life of the capital good.

RECORD-KEEPING

Investment analysis combines records maintained for other purposes on the farm. Essentially, enterprise budget-style cost analyses must be developed. The types of information and records required for a proper enterprise budget can be found in Chapter 10.

Once an investment analysis has been developed, the Excel files associated with it should be maintained. These can be used as templates for future analyses or to develop sensitivity analyses related to variations in parameter values to evaluate other conditions.

PRACTICAL APPLICATION: DOUBLING THE NUMBER OF AERATORS ON A 256-ACRE CATFISH FARM

An example of an investment decision on a fish farm would be the decision to expand aeration capacity from 1 hp/acre to 2 hp/acre. For a 10-acre pond, then, this would require adding 1 additional 10-hp aerator to each pond to supplement the existing 10-hp aerator. For the 256-acre farm that has 25 10-acre ponds, an additional 25 aerators would need to be purchased. At a cost of $4,500, the additional investment would be $112,500 for the farm.

The benefit of expanding aeration capacity would be the ability to maintain higher levels of dissolved oxygen in the pond. Higher levels of oxygen would be expected to result in higher yields (lb/acre) of catfish. Thus, if yield increased by 500 lb/acre and catfish price was $0.70/lb, the additional revenue would be $89,600. The additional weight of fish sold would require additional harvesting and seining costs. Electricity costs would also increase. The total increase in use of electricity was assumed to be 90%, instead of a doubling of use, because aerators would be turned on sequentially as oxygen begins to fall. This example analysis assumed that a 5-year loan was taken out at a 10% interest rate, to finance the purchase of the aerators.

Table 13.5 shows the results of the investment analysis. The payback period would be 3.3 years with a NPV of $16,038 (10% discount rate) and an IRR of 16%. Thus, under the assumptions used in this analysis, the expansion of aeration capacity is a profitable investment. Additional yields above 500 lb/acre, catfish prices above $0.70/lb, electricity usage less than that assumed, or lower electric rates would result in greater profit levels. Conversely, if yield remained the same, this investment would not be profitable. If additional yields were less than 500 lb/acre, or if catfish prices were lower or more electricity would be used, the profitability of this investment would be reduced.

OTHER APPLICATIONS IN AQUACULTURE

Trimpey and Engle (2005) used an investment analysis to evaluate the profitability of purchasing an in-pond mechanical grader for food-sized channel catfish. The cost of a commercial-scale version of the grader was estimated. Benefits were estimated on the basis of trials of the efficiency of the grader in both experimental

Table 13.5. Annual Cash Flow Budget for Investment Analysis of Adding 1 hp/acre of Additional Aeration to a 256-acre Farm.

Item		Year 0	Year 1	Year 2	Year 3	Year 4	Year 5
Cash receipts			89,600	89,600	89,600	89,600	89,600
Cash inflow			89,600	89,600	89,600	89,600	89,600
Operating expenses							
	Fingerlings		—	—	—	—	—
	Feed		—	—	—	—	—
	Fertilizer		—	—	—	—	—
	Chemicals		—	—	—	—	—
	Labor		—	—	—	—	—
	Electricity		44,390	44,390	44,390	44,390	44,390
	Diesel		—	—	—	—	—
	Gasoline		—	—	—	—	—
	Repairs and maintenance		—	—	—	—	—
	Levee repairs		—	—	—	—	—
	Postharvest handling		6,400	6,400	6,400	6,400	6,400
	Marketing		—	—	—	—	—
	Insurance		—	—	—	—	—
Total			50,790	50,790	50,790	50,790	50,790
Capital expenditures			—	—	—	—	—
Other expenses			—	—	—	—	—
Scheduled debt payments							
	Principal—real estate		—	—	—	—	—
	Interest		—	—	—	—	—
	Principal—equipment		22,500	22,500	22,500	22,500	22,500
	Interest		4,479	4,479	4,479	4,479	4,479
	Principal—operating		—	—	—	—	—
	Interest		—	—	—	—	—
Total cash outflow			55,269	55,269	55,269	55,269	55,269
Cash available			34,331	34,331	34,331	34,331	34,331
New borrowing			—	—	—	—	—
Cash balance			34,331	34,331	34,331	34,331	34,331
Debt outstanding							
	Real estate		—	—	—	—	—
	Equipment		90,000	67,500	45,000	22,500	—
	Operating		—	—	—	—	—
Net cash flow		(112,500)	34,331	34,331	34,331	34,331	34,331
Net present value (NPV)	Discount rate		0.10				
	NPV		16,038				
Internal rate of return (IRR)	IRR		*16%*				

and commercial catfish ponds. The trials provided data on the size distributions of fish harvested from the traditional live car passive grading system and the active mechanical grader. Various scenarios of deductions from processing plants for out-of-size fish were used to compare the relative benefits for the size distributions of fish from the traditional, passive technology to the active mechanical technology.

The results showed payback periods from 0.1 to 2.0 years. The mechanical grader was shown to be profitable with positive NPVs and internal rates of return that were greater than the current opportunity cost of capital. Both NPV and internal rates of return increased with farm size. The analysis showed that it was economically feasible to adopt the in-pond mechanical grader.

Table 13.6. Future Value of a $1 Investment.

Years	4%	4.5%	5%	5.5%	6%	6.5%	7%	7.5%	8%	8.5%	9%	9.5%	10%
1	1.0400	1.0450	1.0500	1.0550	1.0600	1.0650	1.0700	1.0750	1.0800	1.0850	1.0900	1.0950	1.1000
2	1.0816	1.0920	1.1025	1.1130	1.1236	1.1342	1.1449	1.1556	1.1664	1.1772	1.1881	1.1990	1.2100
3	1.1249	1.1412	1.1576	1.1742	1.1910	1.2079	1.2250	1.2423	1.2597	1.2773	1.2950	1.3129	1.3310
4	1.1699	1.1925	1.2155	1.2388	1.2625	1.2865	1.3108	1.3355	1.3605	1.3859	1.4116	1.4377	1.4641
5	1.2167	1.2462	1.2763	1.3070	1.3382	1.3701	1.4026	1.4356	1.4693	1.5037	1.5386	1.5742	1.6105
6	1.2653	1.3023	1.3401	1.3788	1.4185	1.4591	1.5007	1.5433	1.5869	1.6315	1.6771	1.7238	1.7716
7	1.3159	1.3609	1.4071	1.4547	1.5036	1.5540	1.6058	1.6590	1.7138	1.7701	1.8280	1.8876	1.9487
8	1.3686	1.4221	1.4775	1.5347	1.5938	1.6550	1.7182	1.7835	1.8509	1.9206	1.9926	2.0669	2.1436
9	1.4233	1.4861	1.5513	1.6191	1.6895	1.7626	1.8385	1.9172	1.9990	2.0839	2.1719	2.2632	2.3579
10	1.4802	1.5530	1.6289	1.7081	1.7908	1.8771	1.9672	2.0610	2.1589	2.2610	2.3674	2.4782	2.5937
11	1.5395	1.6229	1.7103	1.8021	1.8983	1.9992	2.1049	2.2156	2.3316	2.4532	2.5804	2.7137	2.8531
12	1.6010	1.6959	1.7959	1.9012	2.0122	2.1291	2.2522	2.3818	2.5182	2.6617	2.8127	2.9715	3.1384
13	1.6651	1.7722	1.8856	2.0058	2.1329	2.2675	2.4098	2.5604	2.7196	2.8879	3.0658	3.2537	3.4523
14	1.7317	1.8519	1.9799	2.1161	2.2609	2.4149	2.5785	2.7524	2.9372	3.1334	3.3417	3.5629	3.7975
15	1.8009	1.9353	2.0789	2.2325	2.3966	2.5718	2.7590	2.9589	3.1722	3.3997	3.6425	3.9013	4.1772
16	1.8730	2.0224	2.1829	2.3553	2.5404	2.7390	2.9522	3.1808	3.4259	3.6887	3.9703	4.2719	4.5950
17	1.9479	2.1134	2.2920	2.4848	2.6928	2.9170	3.1588	3.4194	3.7000	4.0023	4.3276	4.6778	5.0545
18	2.0258	2.2085	2.4066	2.6215	2.8543	3.1067	3.3799	3.6758	3.9960	4.3425	4.7171	5.1222	5.5599
19	2.1068	2.3079	2.5270	2.7656	3.0256	3.3086	3.6165	3.9515	4.3157	4.7116	5.1417	5.6088	6.1159
20	2.1911	2.4117	2.6533	2.9178	3.2071	3.5236	3.8697	4.2479	4.6610	5.1120	5.6044	6.1416	6.7275
21	2.2788	2.5202	2.7860	3.0782	3.3996	3.7527	4.1406	4.5664	5.0338	5.5466	6.1088	6.7251	7.4002
22	2.3699	2.6337	2.9253	3.2475	3.6035	3.9966	4.4304	4.9089	5.4365	6.0180	6.6586	7.3639	8.1403
23	2.4647	2.7522	3.0715	3.4262	3.8197	4.2564	4.7405	5.2771	5.8715	6.5296	7.2579	8.0635	8.9543
24	2.5633	2.8760	3.2251	3.6146	4.0489	4.5331	5.0724	5.6729	6.3412	7.0846	7.9111	8.8296	9.8497
25	2.6658	3.0054	3.3864	3.8134	4.2919	4.8277	5.4274	6.0983	6.8485	7.6868	8.6231	9.6684	10.8347
30	3.2434	3.7453	4.3219	4.9840	5.7435	6.6144	7.6123	8.7550	10.0627	11.5583	13.2677	15.2203	17.4494
35	3.9461	4.6673	5.5160	6.5138	7.6861	9.0623	10.6766	12.56889	14.7853	17.3796	20.4140	23.9604	28.1024
40	4.8010	5.8164	7.0400	8.5133	10.2857	12.4161	14.9475	18.0442	21.7245	26.1330	31.4094	37.7194	45.2593

Years	10.5%	11%	11.5%	12%	12.5%	13%	13.5%	14%	14.5%	15%	15.5%	16%	16.5%
1	1.1050	1.1100	1.1150	1.1200	1.1250	1.1300	1.1350	1.1400	1.1450	1.1500	1.1550	1.1600	1.1650
2	1.2210	1.2321	1.2432	1.2544	1.2656	1.2769	1.2882	1.2996	1.3110	1.3225	1.3340	1.3456	1.3572
3	1.3492	1.3676	1.3862	1.4049	1.4238	1.4429	1.4621	1.4815	1.5011	1.5209	1.5408	1.5609	1.5812
4	1.4909	1.5181	1.5456	1.5735	1.6018	1.6305	1.6595	1.6890	1.7188	1.7490	1.7796	1.8106	1.8421
5	1.6474	1.6851	1.7234	1.7623	1.8020	1.8424	1.8836	1.9254	1.9680	2.0114	2.0555	2.1003	2.1460
6	1.8204	1.8704	1.9215	1.9738	2.0273	2.0820	2.1378	2.1950	2.2534	2.3131	2.3741	2.4364	2.5001
7	2.0116	2.0762	2.1425	2.2107	2.2807	2.3526	2.4264	2.5023	2.5801	2.6600	2.7420	2.8262	2.9126
8	2.2228	2.3045	2.3889	2.4760	2.5658	2.6584	2.7540	2.8526	2.9542	3.0590	3.1671	3.2784	3.3932
9	2.4562	2.5580	2.6636	2.7731	2.8865	3.0040	3.1258	3.2519	3.3826	3.5179	3.6580	3.8030	3.9531
10	2.7141	2.8394	2.9699	3.1058	3.2473	3.3946	3.5478	3.7072	3.8731	4.0456	4.2249	4.4114	4.6053
11	2.9991	3.1518	3.3115	3.4785	3.6532	3.8359	4.0267	4.2262	4.4347	4.6524	4.8798	5.1173	5.3652
12	3.3140	3.4985	3.6923	3.8960	4.1099	4.3345	4.5704	4.8179	5.0777	5.3503	5.6362	5.9360	6.2504
13	3.6619	3.8833	4.1169	4.3635	4.6236	4.8980	5.1874	5.4924	5.8140	6.1528	6.5098	6.8858	7.2818
14	4.0464	43104	4.5904	4.8871	5.2016	5.5348	5.8877	6.2613	6.6570	7.0757	7.5188	7.9875	8.4833
15	4.4713	4.7846	5.1183	5.4736	5.8518	6.2543	6.6825	7.1379	7.6222	8.1371	8.6842	9.2655	9.8830
16	4.9408	5.3109	5.7069	6.1304	6.5833	7.0673	7.5846	8.1372	8.7275	9.3576	10.0302	10.7480	11.5137
17	5.4596	5.8951	6.3632	6.8660	7.4062	7.9861	8.6085	9.2765	9.9929	10.7613	11.5849	12.4677	13.4135
18	6.0328	6.5436	7.0949	7.6900	8.3319	9.0243	9.7707	10.5752	11.4419	12.3755	13.3806	14.4625	15.6267
19	6.6663	7.2633	7.9108	8.6128	9.3734	10.1974	11.0897	12.0557	13.1010	14.2318	15.4546	16.7765	18.2051
20	7.3662	8.0623	8.8206	9.6463	10.5451	11.5231	12.5869	13.7435	15.0006	16.3665	17.8501	19.4608	21.2089
21	8.1397	8.9492	9.8350	10.8038	11.8632	13.0211	14.2861	15.6676	17.1757	18.8215	20.6168	22.5745	24.7084
22	8.9944	9.9336	10.9660	12.1003	13.3461	14.7138	16.2147	17.8610	19.6662	21.6447	23.8124	26.1864	28.7853
23	9.9388	11.0263	12.2271	13.5523	15.0144	16.6266	18.4037	20.3616	22.5178	24.8915	27.5034	30.3762	33.5348
24	10.9823	12.2392	13.6332	15.1786	16.8912	18.7881	20.8882	23.2122	25.7829	28.6252	31.7664	35.2364	39.0681
25	12.1355	13.5855	15.2010	17.0001	19.0026	21.2305	23.7081	26.4619	29.5214	32.9190	36.6902	40.8742	45.5143
30	19.9926	22.8923	26.1967	29.9599	34.2433	39.1159	44.6556	50.9502	58.0985	66.2118	75.4153	85.8499	97.6737
35	32.9367	38.5749	45.1461	52.7996	61.7075	72.0685	84.1115	98.1002	114.3384	133.1755	155.0135	180.3141	209.6078
40	54.2614	65.0009	77.8027	93.0510	111.1990	132.7816	158.4289	188.8835	225.0191	267.8635	318.6246	378.7212	449.8182

Table 13.7. Present Value of a $1 Lump Sum.

Years	4%	4.5%	5%	5.5%	6%	6.5%	7%	7.5%	8%	8.5%	9%	9.5%	10%
1	0.96154	0.95694	0.95238	0.94787	0.94340	0.93897	0.93458	0.93023	0.92593	0.92166	0.91743	0.91324	0.90909
2	0.92456	0.91573	0.90703	0.89845	0.89000	0.88166	0.87344	03.86533	0.85734	0.84946	0.84168	0.83401	0.82645
3	0.88900	0.87630	0.86384	0.85161	0.83962	0.82785	0.81630	0.80496	0.79383	0.78291	0.77218	0.76165	0.75131
4	0.85480	0.83856	0.82270	0.80722	0.79209	0.77732	0.76290	0.74880	0.73503	0.72157	0.70843	0.69557	0.68301
5	0.82193	0.80245	0.78353	0.76513	0.74726	0.72988	0.71299	0.69656	0.68058	0.66505	0.64993	0.63523	0.62092
6	0.79031	0.76790	0.74622	0.72525	0.70496	0.68533	0.66634	0.64796	0.63017	0.61295	0.59627	0.58012	0.56447
7	0.75992	0.73483	0.71068	0.68744	0.66506	0.64351	0.62275	0.60275	0.58349	0.56493	0.54703	0.52979	0.51316
8	0.73069	0.70319	0.67684	0.65160	0.62741	0.60423	0.58201	0.56070	0.54027	0.52067	0.50187	0.48382	0.46651
9	0.70259	0.67290	0.64461	0.61763	0.59190	0.56735	0.54393	0.52158	0.50025	0.47988	0.46043	0.44185	0.42410
10	0.67556	0.64393	0.61391	0.58543	0.55839	0.53273	0.50835	0.48519	0.46319	0.44229	0.42214	0.40351	0.38554
11	0.64958	0.61620	058468	0.55491	0.52679	0.50021	0.47509	0.45134	0.42888	0.40764	0.38753	0.36851	0.35049
12	0.62460	0.58966	0.55684	0.52598	0.49697	0.46968	0.44401	0.41985	0.39711	0.37570	0.35553	0.33654	0.31863
13	0.60057	0.56427	0.53032	0.49856	0.46884	0.44102	0.41496	0.39056	0.36770	0.34627	0.32618	0.30734	0.28966
14	0.57748	0.53997	0.50507	0.47257	0.44230	0.41410	0.38782	0.36331	0.34046	0.31914	0.29925	0.28067	0.26333
15	0.55526	0.51672	0.48102	0.44793	0.41727	0.38883	0.36245	0.33797	0.31524	0.29414	0.27454	0.25632	0.23939
16	0.53391	0.49447	0.45811	0.42458	0.39365	0.36510	0.33873	0.31439	0.29189	0.27110	0.25187	0.23409	0.21763
17	0.51337	0.47318	0.43630	0.40245	0.37136	0.34281	0.31657	0.29245	0.27027	0.24986	0.23107	0.21378	0.19784
18	0.49363	0.45280	0.41552	0.38147	0.35034	0.32189	0.29586	0.27205	0.25025	0.23028	0.21199	0.19523	0.17986
19	0.47464	0.43330	0.39573	0.36158	0.33051	0.30224	0.27651	0.25307	0.23171	0.21224	0.19449	0.17829	0.16351
20	0.45639	0.41464	0.37689	0.34273	0.31180	0.28380	0.25842	0.23541	0.21455	0.19562	0.17843	0.16282	0.14864
21	0.43883	0.39679	0.35894	0.32486	0.29416	0.26648	0.24151	0.21899	0.19866	0.18029	0.16370	0.14870	0.13513
22	0.42196	0.37970	0.34185	0.30793	0.27751	0.25021	0.22571	0.20371	0.18394	0.16617	0.15018	0.13580	0.12285
23	0.40573	0.36335	0.32557	0.29187	0.26180	0.23494	0.21095	0.18950	0.17032	0.15315	0.13778	0.12402	0.11168
24	0.39012	0.34770	0.31007	0.27666	0.24698	0.22060	0.19715	0.17628	0.15770	0.14115	0.12640	0.11326	0.10153
25	0.37512	0.33273	0.29530	0.26223	0.23300	0.20714	0.18425	0.16398	0.14602	0.13009	0.11597	0.10343	0.09230
30	0.30832	0.26700	0.23138	0.20064	0.17411	0.15119	0.13137	0.11422	0.09938	0.08652	0.07537	0.06570	0.05731
35	0.25342	0.21425	0.18129	0.15352	0.13011	0.11035	0.09366	0.07956	0.06763	0.05754	0.04899	0.04174	0.03558
40	0.20829	0.17193	0.1420	0.11746	0.09722	0.08054	0.06678	0.05542	0.04603	0.03827	0.03184	0.02651	0.02209

Years	10.5%	11%	11.5%	12%	12.5%	13%	13.5%	14%	14.5%	15%	15.5%	16%	16.5%
1	0.90498	0.90090	0.89686	0.89286	0.88889	0.88496	0.88106	0.87719	0.87336	0.86957	0.86580	0.86207	0.85837
2	0.81898	0.81162	0.80436	0.79719	0.79012	0.78315	0.77626	0.76947	0.76276	0.75614	0.74961	0.74316	0.73680
3	0.74116	0.73119	0.72140	0.71178	0.70233	0.69305	0.68393	0.67497	0.66617	0.65752	0.64901	0.64066	0.63244
4	0.67073	0.65873	0.64699	0.63552	0.62430	0.61332	0.60258	0.59208	0.58181	0.57175	0.56192	0.55229	0.54287
5	0.60700	0.59345	0.58026	0.56743	0.55493	0.54276	0.53091	0.51937	0.50813	0.49718	0.48651	0.47611	0.46598
6	0.54932	0.53464	0.52042	0.50663	0.49327	0.48032	0.46776	0.45559	0.44378	0.43233	0.42122	0.41044	0.39999
7	0.49712	0.48166	0.46674	0.45235	0.43846	0.42506	0.41213	0.39964	0.38758	0.37594	0.36469	0.35383	0.34334
8	0.44989	0.43393	0.41860	0.40388	0.38974	0.37616	0.36311	0.35056	0.33850	0.32690	0.31575	0.30503	0.29471
9	0.40714	0.39092	0.37543	0.36061	0.34644	0.33288	0.31992	0.30751	0.29563	0.28426	0.27338	0.26295	0.25297
10	0.36845	0.35218	0.33671	0.32197	0.30795	0.29459	0.28187	0.26974	0.25819	0.24718	0.23669	0.22668	0.21714
11	0.33344	0.31728	0.30198	0.28748	0.27373	0.26070	0.24834	0.23662	0.22550	0.21494	0.20493	0.19542	0.18639
12	0.30175	0.28584	0.27083	0.25668	0.24332	0.23071	0.21880	0.20756	0.19694	0.18691	0.17743	0.16846	0.15999
13	0.27308	0.25751	0.24290	0.22917	0.21628	0.20416	0.19278	0.18207	0.17200	0.16253	0.15362	0.14523	0.13733
14	0.24713	0.23199	0.21785	0.20462	0.19225	0.18068	0.16985	0.15971	0.15022	0.14133	0.13300	0.12520	0.11788
15	0.22365	0.20900	0.19538	0.18270	0.17089	0.15989	0.14964	0.14010	0.13120	0.12289	0.11515	0.10793	0.10118
16	0.20240	0.18829	0.17523	0.16312	0.15190	0.14150	0.13185	0.12289	0.11458	0.10686	0.09970	0.09304	0.08685
17	0.18316	0.16963	0.15715	0.14564	0.13502	0.12522	0.11616	0.10780	0.10007	0.09293	0.08632	0.08021	0.07455
18	0.16576	0.15282	0.14095	0.13004	0.12002	0.11081	0.10235	0.09456	0.08740	0.08081	0.07474	0.06914	0.06399
19	0.15001	0.13768	0.12641	0.11611	0.10668	0.09806	0.09017	0.08295	0.07633	0.07027	0.06471	0.05961	0.05493
20	0.13575	0.12403	0.11337	0.10367	0.09483	0.08678	0.07945	0.07276	0.06666	0.06110	0.05602	0.05139	0.04715
21	0.12285	0.11174	0.10168	0.09256	0.08429	0.07680	0.07000	0.06383	0.05822	0.05313	0.04850	0.04430	0.04047
22	0.11118	0.10067	0.09119	0.08264	0.07493	0.06796	0.06167	0.05599	0.05085	0.04620	0.04199	0.03819	0.03474
23	0.10062	0.09069	0.08179	0.07379	0.06660	0.06014	0.05434	0.04911	0.04441	0.04017	0.03636	0.03292	0.02982
24	0.09106	0.08170	0.07335	0.06588	0.05920	0.05323	0.04787	0.04308	0.03879	0.03493	0.03148	0.02838	0.02560
25	0.08240	0.07361	0.06579	0.05882	0.05262	0.04710	0.04218	0.03779	0.03387	0.03038	0.02726	0.02447	0.02197
30	0.05002	0.04368	0.03817	0.03338	0.02920	0.02557	0.02239	0.01963	0.01721	0.01510	0.01326	0.01165	0.01024
35	0.03036	0.02592	0.02215	0.01594	0.01621	0.01388	0.01189	0.01019	0.00875	0.00751	0.00645	0.00555	0.004477
40	0.01843	0.01538	0.01285	0.01075	0.00899	0.00753	0.00631	0.00529	0.00444	0.00373	0.00314	0.00264	0.00222

SUMMARY

Investments in assets that will be used over a period of years must be evaluated differently from operating inputs because of the time periods involved. Many aquaculture businesses are capital-intensive and require a great deal of capital investment. Thus, understanding how to evaluate and manage these investments from a financial perspective is important.

The time value of money is central to understanding how to analyze these investments. Investment capital has an initial PV that can be invested at a particular interest rate to compound over time into a FV. Conversely, the returns generated over a period of years in the future (FV) can be discounted back to calculate their PV.

The most commonly used types of investment analyses are the payback period, NPV, and the internal rate or return. The payback period is simple to calculate and provides a measure of the contribution to liquidity of an investment by indicating how many years it will take to recover the amount of the investment. However, the payback period does not consider revenue generated beyond that point in time and does not account for the timing of cash flows. The NPV is the sum of the PVs of the revenue generated in each year. It accounts for the timing of revenue and considers all revenue over the life of the investment. However, the NPV depends upon the discount rate selected. The IRR uses the same annual cash flow and calculations, but results in a determination of the interest rate generated by the investment.

REVIEW QUESTIONS

1. What are the differences between operating and investment capital?

2. Explain how compounding and discounting are similar and how they are different.

3. Define the following investment indicators:
   a. Payback period
   b. Simple rate of return
   c. NPV
   d. IRR

4. Compare and contrast NPV and IRR.

5. Explain why revenue earned in the future does not have the same values as capital in hand at the moment.

6. Explain how PVs and FVs are related.

7. Contrast the data requirements for calculating the payback period, the simple rate of return, the NPV, and the IRR.

8. Develop a table that indicates the limitations of each of the four major investment indicators.

9. What is the most important information to be obtained from each of the four major investment indicators?

10. What specific information can be obtained from each of the four investment indicators that can be used to make good investment decisions?

REFERENCE

Trimpey, Jeremy and Carole Engle. 2005. The economic feasibility of adoption of a new in-pond mechanical grader for food-sized channel catfish (*Ictalurus punctatus*). *Aquacultural Engineering* 32:411–423.

14
Lending in Aquaculture

INTRODUCTION

Aquaculture is a capital-intensive business. As shown in Chapter 10, aquaculture businesses frequently require over $1,400/acre to construct pond facilities and over $3,000/acre to operate the business. Few individuals have sufficient capital to be able to provide all the capital required from their own resources. Thus, most aquaculture entrepreneurs must borrow capital to have adequate amounts for a successful business.

Capital is classified into terms that reflect its source. Capital provided by the owners, including any partners, is referred to as equity capital. Equity capital has been an important source of capital for agriculture in general and aquaculture specifically. Historically, the primary source of equity capital in agriculture has been from retained earnings. Capital that is borrowed on credit is referred to as debt capital. This chapter focuses on considerations related to borrowing debt capital from lenders.

Credit is used in a variety of ways in an aquaculture business. For example, it can be used to create and maintain an adequate size of business. This is especially true for those aquaculture businesses for which economies of scale exist (see Chapter 7 for more on economies of scale). It can also be used to increase the efficiency of operation in the business. For example, older ponds that have silted in may be very shallow with reduced capacity to store oxygen. Thus, credit that provides capital to renovate ponds may result in deeper ponds with improved oxygen profiles and better fish yields.

Borrowing capital may also be necessary to adjust to changing economic conditions. Increased commodity prices in 2007–2008 resulted in dramatic increases in the price of feed. In such a situation, additional borrowing may be needed to continue to purchase the feed necessary to produce an adequate crop of fish.

Credit can also be used to meet seasonal fluctuations in income and expenditures. These fluctuations may result in cash shortfalls at certain times of the year, and it may be necessary to borrow additional operating capital to cover these cash shortfalls. Good financial planning includes taking steps necessary to protect the business against adverse conditions. Maintaining a credit reserve will provide a ready source of capital that can be drawn upon when market prices and revenue fall or when input costs and expenses increase. Credit can also provide continuity in the business, particularly in times of inheritance, and transfers of estates.

Borrowing debt capital creates financial risk for the aquaculture business. Loans and the resulting debt result in fixed financial obligations. Moreover, the loan becomes a claim against the assets of the business, the land, ponds, buildings, and equipment.

The financial plan for the business must carefully match up the expected cash flow, repayment of the loans taken out to provide necessary capital, and the anticipated financial performance of the aquaculture business. These must be coordinated carefully. The structure of the loans affects the business's cash flow. Ideally, each loan is structured to match the length of the payoff periods with the asset's useful life, and the earnings pattern generated by the assets acquired with the loan. These are referred to as self-liquidating loans, loans that are used to acquire assets that generate enough revenue over the life of the loan. With self-liquidating loans, the asset being financed generates more revenue than the size of the loan taken out and the payments are made from the cash produced from that asset. Thus, the loan does not decrease the business's liquidity. A maturity period of a loan that is too short to generate the revenue required will result in liquidity and cash flow problems.

Aquaculture Economics and Financing: Management and Analysis, Carole R. Engle, © 2010 Carole R. Engle.

The first step in preparing to borrow capital is to carefully evaluate the credit capacity of the borrower. It is essential to determine the ability of the business to bear additional financial risk. The balance sheet (see Chapter 11 for more detail on preparing balance sheets) can be used to measure whether the farm can withstand financial losses without being forced into liquidation. The next step is to carefully assess whether the new loan will add to potential profits or not, to understand what are the effects on the business's net returns. Finally, the firm's repayment capacity must be evaluated. The lender will want to be repaid in cash. The business's ability to do so will be influenced by the liquidity of fish, cash flow, and the income-generating capacity of the business.

This chapter discusses various types of loans, types of security and collateral, loan repayment, amortization calculations, and repayment schedules.

TYPES OF LOANS

While there are different ways to classify different types of loans, the most common types of loans are: operating, equipment, and real estate loans. Operating loans typically are short-term loans that are repaid in the same year. Operating loans typically are used to purchase production inputs such as fingerlings, feed, fuel, and electricity that are used in production and are not carried over to another year. These assets cannot be recovered in their original form; they are recovered when the crop is sold. Repayment frequently is tied to the sale of the crop by placing a lien on the fish crop.

The need for operating capital is more urgent in some respects than it is for other types of capital. The business must have adequate quantities of capital available to acquire the necessary types and quantities of inputs to operate at a profitable level. It is also critical to have sufficient credit reserves to withstand unexpected losses or adverse conditions (mortalities due to a disease outbreak, for example). Operating loans are listed as current liabilities on the balance sheet. Equipment loans are most often used to purchase tractors, aerators, generators, and hauling trucks. Equipment loans are intermediate-term loans that typically are repaid over a period of about 5–7 years. Equipment loans frequently have 1–2 payments per year. Because the equipment purchased with an equipment loan will be used for a number of years in the aquaculture business, it would be financed over its useful life. Equipment loans are listed as noncurrent liabilities on the balance sheet.

Real estate loans are long-term loans that are typically repaid over a period of 15–30 years. Real estate loans typically are used to purchase land, build ponds, dig wells, and to construct building and grading sheds. Real estate loans are listed as noncurrent liabilities.

Operating loans can be set up in different ways. With single-payment loans, principal is payable in one lump sum when the loan is due, along with accrued interest. Simple interest is charged on single-payment loans. The disadvantage of a single-payment loan is that the farmer may end up borrowing more money than what is needed at that one point in time.

With a line of credit, loan funds are transferred into a farm account as needed, up to an approved maximum amount. The aquaculture owner, then, can borrow freely up to the approved maximum amount. Payments are made when income is received. Accumulated (accrued) interest is paid first, and then payments are made on the principal. Approval of a line of credit by a bank typically requires a comprehensive cash flow budget.

Financing noncurrent assets (equipment, ponds, buildings) requires different types of loans because these types of assets are used for more than 1 year. The cost of these types of assets is charged as depreciation to the farm business (see Inset 10.1 in Chapter 10 for details). Loans for depreciable assets should be structured to use the capital set aside for depreciation losses for repayment of the loan. If the loan is structured in this manner, the loan becomes a self-liquidating loan because it is paid off over the useful life of the asset, the period of time when it is generating revenue for the farm business. Thus, for a piece of equipment that is expected to be used for 7 years, the loan should be structured to be repaid within 7 years. A loan that is repaid over a shorter period of time will often result in cash being diverted from other uses to pay down the loan.

Land is a noncurrent asset but is not depreciable because it does not have a definable useful life. Funds used to pay off farm real-estate loans typically are taken from retained earnings (profits minus income taxes) for the business.

SECURITY AND COLLATERAL

Lenders require borrowers to commit assets to provide assurance that the loan will be repaid. Assets committed to a loan are referred to as security or collateral. In the event that the borrower cannot make the payments necessary on the loan, the lender can take possession of the assets pledged to the loan. The lender typically will sell these assets to recover the amount of the loan outstanding.

Loans with collateral pledged are referred to as secured loans whereas a borrower with outstanding credit may be granted an unsecured loan, one without specific collateral committed to it. Unsecured loans are also referred to as signature loans. There are few, if any, unsecured loans granted for aquaculture businesses. Lenders prefer secured loans because the collateral pledged reduces the risk that the lender will lose the principal amount loaned to the borrower. Collateral used for equipment and real estate loans may include land or a home. Collateral for long-term loans typically is some sort of pledge of farm real estate, frequently, the land and the ponds.

LOAN REPAYMENT

Repayment schedules should be developed primarily on the basis of the borrower's cash flow budget. The purpose of the loan and the type of collateral used may affect the repayment plan. The repayment schedule should be developed jointly between the lender and the borrower.

Total interest paid will be greater for loans with longer terms and repayment periods. Shorter repayment periods will result in lower amounts of total interest paid but higher payments. Interest is calculated with the following equation:

$$I = P \times I \times T$$

Where I is the total amount of interest to be paid, P is the amount that is borrowed (principal), i is the interest rate, and T is the number of payments to be made.

A single-payment loan is the most simple type of loan. There is only one payment made that includes the total amount of the principal and interest paid in one lump sum. For this type of loan, interest is charged for 1 year on the total amount of the principal. Because the entire amount plus interest is paid in one payment,

careful cash flow budgeting is critical to be certain that adequate cash is available to make that payment.

As an example, if $30,000 is borrowed at 10% annually for 1 year, the single payment would be $30,000 × 10% × 1 year. The interest would be $3,000 and the total payment (principal plus interest) would be $33,000. This interest payment is referred to as simple interest. It should be noted that, if this loan were repaid in 6 months instead of a year, then the total interest would only be $1,500 because the "$T$," or time period would be 0.5 (half a year) instead of 1 for 1 year.

A single-payment operating loan is the simplest type of loan, but it is generally not adequate for fish farming operating capital. A single-payment loan must be renegotiated each year, with new paperwork, applications, and approvals. The paperwork takes time and often results in operating capital not being available when needed. Single-payment loans also often result in borrowing more money than what is needed at one time.

The most common type of operating loan for aquaculture is a line of credit. With a line of credit, loan funds are transferred into the farm account as needed. The line of credit is established with an approved maximum amount or cap. Once the borrowing cap is reached, the farmer cannot use any additional funds from the operating line of credit. Payments on the operating line of credit typically are tied to receipt of revenue from crop sales. The accumulated interest on the loan is paid first with the rest applied to the principal amount. While there is no absolute repayment schedule, the timing of the distribution of funds typically is determined by a cash flow budget (see Chapter 12 for details on cash flow budgeting).

Table 14.1 demonstrates how an operating loan is structured and used. In this example, $15,000 is borrowed on February 1 and $15,000 more borrowed on April 1 with another $5,000 borrowed on September 1. Total borrowing, then, on September 1 is $35,000. On October 1, following sales of fish, a payment is made of $26,542. Of this, $1,542 was

Table 14.1. Line of Credit (Interest Rate = 10%).

Date	Amount borrowed	Amount repaid	Interest paid	Principal paid	Outstanding balance
February 1	$15,000	0	0	0	$15,000
April 1	$15,000	0	0	0	$30,000
September 1	$5,000	0	0	0	$35,000
October 1		$26,542	$1,542	$25,000	$10,000
December 1		$10,167	$167	$10,000	$0
Total	$35,000	$36,709	$1,709	$35,000	—

paid on the accumulated interest and $25,000 paid on the principal. This leaves an outstanding balance of $10,000 on the loan. On December 1, a final payment is made of $10,167 that includes an interest payment of $167 and a payment on the principal of $10,000.

Lines of credit are preferable to single-payment operating loans because funds are borrowed only when needed. Thus, the total amount of interest paid is less than with a single-payment loan. However, a line of credit requires good financial management and careful cash flow budgeting and monitoring to avoid borrowing more than can be repaid.

LOAN AMORTIZATION AND SCHEDULES

Loans can also be "amortized" or spread across the length of the loan. Amortization of a loan results in a table that shows the total payment due on each payment date, the amount of the payment that is applied to the principal, and the amount that is applied to the interest. This table is referred to as an "amortization schedule." There are different techniques and types of amortization schedules. This chapter discusses the most common types of amortization schedules for the most common types of loans. With amortized loans, the original amount borrowed (principal), the annual interest rate, the length of the loan, and the number of payments per year are negotiated between the borrower and the lender. An amortized loan has periodic interest and principal payments. It is also referred to as an "installment loan." As the principal is repaid and the loan balance declines, interest payments decline.

There are two main types of amortized loans: equal principal payments and equal total payment loans. With equal principal payments, each payment includes an equal amount of payment on the principal. For example, Table 14.2, shows that an equal principal payment amortized loan for a $50,000 loan to be repaid over 5 years at an interest rate of 6%, will have an equal principal payment of $10,000 per year over the 5 years. With an equal principal loan, the amount of interest paid declines each year. Moreover, the total payment made declines, as the amount of interest paid declines while the principal portion of the payment remains the same. However, the payments in the early years are quite high and may be difficult for borrowers to make. In Table 14.2, the payment in Year 1 is $1,800 higher than the payment in Year 4. This can pose a particularly difficult payment schedule for a new or startup business that does not generate much revenue in the early years.

Table 14.2. Equal Principal Payment Amortized Loan for a $50,000 Loan to be Repaid over 5 Years at an Annual Interest Rate of 6%.

Year	Principal paid	Interest paid	Total payment	Principal remaining
1	$10,000	$3,000	$13,000	$40,000
2	$10,000	$2,400	$12,400	$30,000
3	$10,000	$1,800	$11,800	$20,000
4	$10,000	$1,200	$11,200	$10,000
5	$10,000	$600	$10,600	$0

With equal total payment amortization schedules, the payment made each time remains the same, but the amount paid on principal and interest will vary over the life of the loan. Thus, the payment in the first year or so is the same as the last payment on the loan. The debt-servicing requirements in the early years of the loan are lower than with an equal principal payment loan.

Table 14.3 illustrates an equal total payment amortized loan. Tables of amortization factors (Tables 14.7 and 14.8) can be used to facilitate development of the amortization schedule for an equal total payment amortized loan. The payment to make for each scheduled payment period is calculated by multiplying the original principal amount borrowed by the amortization factor.

In Table 14.7, an amortization factor of 0.23740 (found in Table 14.7 by looking at the appropriate column for an interest rate of 6% and the row that corresponds to the 5-year term of the loan). The amortization factors are calculated with the following equation:

$$\text{Amortization factor} = \frac{i}{[1 - (1 + i)^{-n}]} \quad (14.1)$$

Table 14.3. Equal Total Payments for a $50,000 Loan over a Period of 5 Years at an Interest Rate of 6%.

Year	Total payment	Interest paid	Principal paid	Principal remaining
1	$11,870	$3,000	$8,870	$41,130
2	$11,870	$2,468	$9,402	$31,728
3	$11,870	$1,904	$9.966	$21,762
4	$11,870	$1,306	$10,564	$11,198
5	$11,870	$672	$11,198	$0

Applying this amortization factor to the principal amount of $50,000 generates the payment amount of $11,870. Spreadsheet-based software programs, like Microsoft® EXCEL include functions to make these calculations. In EXCEL, the function to use is "PMT." It requires specification of the interest rate (as a decimal), the number of payments, and the total principal amount.

In Year 1, the interest owed is 6% of the total principal owed ($50,000), or $3,000. The rest of the payment, then, is used as a principal payment of $8,870. At the end of Year 1, the remaining amount of principal is $41,130. In Year 2, the total payment is the same, at $11,870. Interest paid is 6% of $41,130, or $2,468, with the principal payment of $9,402. The payments on the principal increase over the period of the loan.

BALLOON PAYMENTS

Balloon payments are amortization schedules in which the initial payments are particularly low. Balloon payments result in the principal not being paid by the end of the term of the loan. In some balloon payment loans, the periodic payments include only interest payments and the principal amount is not being paid down. Because there often is a substantial amount of the principal remaining at the end of the loan period, refinancing with another loan may be necessary.

Table 14.4 illustrates a balloon payment for a loan of $50,000 for 5 years at an interest rate of 6%. This example shows an annual payment of $7,429 for each of the first 4 years with a final payment in Year 5 of $32,509.

SOURCES OF CREDIT

Private lenders (commercial banks) have been the primary source of debt capital in the aquaculture industry. Private lenders have provided loans for real estate, equipment, and operating loans. Banks generally hold a large share of farm loans in general. This is due

Table 14.4. Balloon Payment for $50,000 for 5 Years at an Annual Interest Rate of 6%.

Year	Total payment	Interest paid	Principal paid	Principal remaining
1	$7,419	$3,000	$4,419	$45,581
2	$7,419	$2,735	$4,684	$40,897
3	$7,419	$2,454	$4,965	$35,932
4	$7,419	$2,156	$5,263	$30,669
5	$32,509	$1,840	$30,669	$0

mostly to the numbers of banks in rural areas. Many banks that supply a great deal of credit to agriculture are small, rural banks that rely heavily on the local economy for their funds.

The farm credit system has also been an important source of debt capital for aquaculture businesses. The U.S. Congress established the farm credit system in 1916 primarily as a source of capital for farmers. The farm credit system is a private cooperative that is audited and regulated by the farm credit administration. Loan funds are acquired by selling bonds and notes to supply funds to four regional farm credit banks and one agricultural credit bank that provide funds to local associations.

The farmers home administration (FmHA), a part of the U.S. Department of Agriculture, guarantees loans, up to 95% repayment in the event that a borrower defaults on the loan. One of the requirements for a guaranteed loan is that the borrower was turned down by a conventional bank. Often, these are new farmers without the equity necessary to qualify for conventional loans.

FmHA lending includes both direct loans and loan guarantees. The major role of FmHA in lending for aquaculture has been through loan guarantees. With the guaranteed loan program, the loans are made and financed through private banks but the FmHA guarantees up to 90% of the loan to the private bank if the borrower defaults. This reduces the risk of losing capital for the private lender.

Feed companies have been another source of capital, particularly operating capital used to purchase feed. In some cases, these loans have been tied to commitments to supply fish to processing plants owned by the same company.

RECORD-KEEPING

The farm manager should maintain complete records of each loan for the farm. The data that need to be maintained should include the initial date of the loan, the length of the loan (5 years, 10 years, etc.), the number of payments a year, and the dates that those payments are due, the interest rate, whether the interest rate is a fixed or variable rate, and the ending date of the loan. It is best to have a complete amortization schedule such as that illustrated in Table 14.5 for each loan. A summary of the status of each loan should be developed to have available at the beginning of each year. Table 14.5 provides a format that could be used for this table. The payments included in this table can then be entered quickly into the cash flow budget for the year. The

Table 14.5. Format for a Table that Summarizes the Current Status of Each Loan on the Farm to be Available at the Beginning of Each Year.

Name of loan	Lender	Description of item financed	Initial amount	Initial date of loan	Length of loan	Amount of payment		Payment dates for new year	Remaining balance on loan	Ending date of loan
						Principal	Interest			
Real estate										
Ponds										
Shed										
Office										
Equipment										
Tractor										
Aerators										
Operating										

Table 14.6. Practical Application.

Year	15-year loan (amortization factor = 0.10979)			20-year loan (amortization factor = 0.09439)		
	Total payment	Principal	Interest	Total payment	Principal	Interest
1	$20,300	$7,358	$12,942	$17,452	$4,511	$12,941
2	$20,300	$7,873	$12,427	$17,452	$4,826	$12,626
3	$20,300	$8,424	$11,876	$17,452	$5,164	$12,282
4	$20,300	$9,014	$11,286	$17,452	$5,526	$11,926
5	$20,300	$9,645	$10,655	$17,452	$5,913	$11,539
6	$20,300	$10,320	$9,980	$17,452	$6,326	$11,126
7	$20,300	$11,042	$9,258	$17,452	$6,769	$10,683
8	$20,300	$11,815	$8,485	$17,452	$7,242	$10,210
9	$20,300	$12,642	$7,658	$17,452	$7,749	$9,703
10	$20,300	$13,527	$6,773	$17,452	$8,291	$9,161
11	$20,300	$14,474	$5,826	$17,452	$8,872	$8,580
12	$20,300	$15,487	$4,813	$17,452	$9,493	$7,959
13	$20,300	$16,571	$3,729	$17,452	$10,157	$7,295
14	$20,300	$17,731	$2,569	$17,452	$10,868	$6,584
15	$20,300	$18,973	$1,327	$17,452	$11,629	$5,823
16				$17,452	$12,443	$5,009
17				$17,452	$13,314	$4,138
18				$17,452	$14,246	$3,206
19				$17,452	$15,243	$2,209
20				$17,452	$16,312	$1,140
Total	$304,496	$184,896	$119,606	$349,047	$184,896	$164,142

Amortization schedule for a real-estate loan for a 256-acre catfish farm. Total long-term capital required is $616,320 (from Chapter 10 enterprise budget). 30% borrowed is $184,896. 7% interest rate.

principal and interest payments for the upcoming year and the remaining balances on each loan are available for the balance sheet, and the total interest payments are available for the income statement.

PRACTICAL APPLICATION

Table 14.6 illustrates how the terms of lending can affect both cash flow as well as operating costs. It demonstrates the payments required for a long-term loan to be repaid over either 15 or 20 years. This example is for the 256-acre catfish farm that has been used throughout this book for the financial analyses. The total amount of long-term investment capital needed for this farm to construct ponds and wells and to purchase the land is $616,320. If 30% of the required capital is borrowed, then the loan principal would be $184,896. If borrowed over a term of 15 years at an interest rate of 7%, the amortization factor would be 0.10979 (Table 14.7). Thus, multiplying the amortization factor of 0.10979 by the principal amount borrowed ($184,896) indicates

the annual payment to be made of $20,300. Of this, the annual interest amount is $12,942 for the first year (7% of $184,896). The amount of the principal paid with this payment is $7,358 ($20,300−$12,942). As the principal is paid down, the amount of the payment that goes to interest is reduced and greater amounts of the payment are used to pay down the principal. This is the equal total payment method discussed above.

If the loan is structured over a 20-year payment horizon, each annual payment is lower ($17,452, as compared to $20,200). Thus, cash flow obligations created by this loan are smaller when it is repaid over a longer time period. However, the total interest payment will be greater ($164,142 with the 20-year term as compared to $119,606 with the 15-year term).

OTHER APPLICATIONS IN AQUACULTURE

Bacon et al. (1998) conducted a survey of lenders in the Appalachian region of the United States to compare

Table 14.7. Amortization Factors for Amortization Schedules 4–10%.

Years	4%	4.5%	5%	5.5%	6%	6.5%	7%	7.5%	8%	8.5%	9%	9.5%	10%
1	1.04000	1.04500	1.05000	1.05500	1.06000	1.06500	1.07000	1.07500	1.08000	1.08500	1.09000	1.09500	1.10000
2	0.53002	0.53400	0.53780	0.54162	0.54544	0.54926	0.55309	0.55693	0.56077	0.56462	0.56847	0.57233	0.57619
3	0.36035	0.36377	0.36721	0.37065	0.37411	0.37758	0.38105	0.38454	0.38803	0.39154	0.39505	0.39858	0.40211
4	0.27549	0.27874	0.28201	0.28529	0.28859	0.29190	0.29523	0.29857	0.30192	0.30529	0.30867	0.31206	0.31547
5	0.22463	0.22779	0.23097	0.23418	0.23740	0.24063	0.24389	0.24716	0.25046	0.25377	0.25709	0.26044	0.26380
6	0.19076	0.19388	0.19702	0.20018	0.20336	0.20657	0.20980	0.21304	0.21632	0.21961	0.222292	0.22625	0.22961
7	0.16661	0.16970	0.17282	0.17596	0.17914	0.18233	0.18555	0.18880	0.19207	0.19537	0.19869	0.20204	0.20541
8	0.14853	0.15161	0.15472	0.15786	0.16104	0.16424	0.16747	0.17073	0.17401	0.17733	0.18067	0.18405	0.18744
9	0.13449	0.13757	0.14069	0.14384	0.14702	0.15024	0.15349	0.15677	0.16008	0.16342	0.16680	0.17020	0.17364
10	0.12329	0.12638	0.12950	0.13267	0.13587	0.13910	0.14238	0.14569	0.14903	0.15241	0.15582	0.15927	0.16275
11	0.11415	0.11725	0.12039	0.12357	0.12679	0.13006	0.13336	0.13670	0.14008	0.14349	0.14695	0.15044	0.15396
12	0.10655	0.10967	0.11283	0.11603	0.11928	0.12257	0.12590	0.12928	0.13270	0.13615	0.13965	0.14319	0.14676
13	0.10014	0.10328	0.10646	0.10968	0.11296	0.11628	0.11965	0.12306	0.12652	0.13002	0.13357	0.13715	0.14078
14	0.09467	0.09782	0.10102	0.10428	0.10758	0.11094	0.11434	0.11780	0.12130	0.12484	0.12843	0.13207	0.13575
15	0.08994	0.09311	0.09634	0.09963	0.10296	0.10635	0.10979	0.11329	0.11683	0.12042	0.12406	0.12774	0.13147
16	0.08582	0.08902	0.09227	0.09558	0.09895	0.10238	0.10586	0.10939	0.11298	0.11661	0.12030	0.12403	0.12782
17	0.08220	0.08542	0.08870	0.09204	0.09544	0.09891	0.10243	0.10600	0.10963	0.11331	0.11705	0.12083	0.12466
18	0.07899	0.08224	0.08555	0.08892	0.09236	0.09585	0.09941	0.10303	0.10670	0.11043	0.11421	0.11805	0.12193
19	0.07614	0.07941	0.08275	0.08615	0.08962	0.09316	0.09675	0.10041	0.10413	0.10790	0.11173	0.11561	0.11955
20	0.07358	0.07688	0.08024	0.08368	0.08718	0.09076	0.09439	0.09809	0.10185	0.10567	0.10955	0.11348	0.11746
21	0.07128	0.07460	0.07800	0.08146	0.08500	0.08861	0.09229	0.09603	0.09983	0.10370	0.10762	0.11159	0.11562
22	0.06920	0.07255	0.07597	0.07947	0.08305	0.08669	0.09041	0.09419	0.09803	0.10194	0.10590	0.10993	0.11401
23	0.06731	0.07068	0.07414	0.07767	0.08128	0.08496	0.08871	0.09254	0.09642	0.10037	0.10438	0.10845	0.11257
24	0.06559	0.06899	0.07247	0.07604	0.07968	0.08340	0.08719	0.09105	0.09498	0.09897	0.10302	0.10713	0.11130
25	0.06401	0.06744	0.07095	0.07455	0.07823	0.08198	0.08581	0.08971	0.09368	0.09771	0.10181	0.10596	0.11017
30	0.05783	0.06139	0.06505	0.06881	0.07265	0.07658	0.08059	0.08467	0.08883	0.09305	0.09734	0.10168	0.10608
35	0.05358	0.05727	0.06107	0.06497	0.06897	0.07306	0.07723	0.08148	0.08580	0.09019	0.09464	0.09914	0.10369
40	0.05052	0.05434	0.05828	0.06232	0.06646	0.07069	0.07501	0.07940	0.08386	0.08838	0.09296	0.09759	0.10226

Table 14.8. Amortization Factors for Amortization Schedules 10.5–16.5%.

Years	10.5%	11%	11.5%	12%	12.5%	13%	13.5%	14%	14.5%	15%	15.5%	16%	16.5%
1	1.10500	1.11000	1.11500	1.12000	1.12500	1.13000	1.13500	1.14000	1.14500	1.15000	1.15500	1.16000	1.16500
2	0.58006	0.58393	0.58781	0.59170	0.59559	0.59948	0.60338	0.60729	0.61120	0.61512	0.61904	0.62296	0.62689
3	0.40566	0.40921	0.41278	0.41635	0.41993	0.42352	0.42712	0.43073	0.43435	0.43798	0.44161	0.44526	0.44891
4	0.31889	0.32233	0.32577	0.32923	0.33271	0.33619	0.33969	0.34320	0.34673	0.35027	0.35381	0.35738	0.36095
5	0.26718	0.27057	0.27398	0.27741	0.28085	0.28431	0.28779	0.29128	0.29479	0.29832	0.30185	0.30541	0.30898
6	0.23298	0.23638	0.23979	0.24323	0.24668	0.25015	0.25365	0.25716	0.26069	0.26424	0.26780	0.27139	0.27499
7	0.20880	0.21222	0.21566	0.21912	0.22260	0.22611	0.22964	0.23319	0.23677	0.24036	0.24398	0.24761	0.25127
8	0.19087	0.19432	0.19780	0.20130	0.20483	0.20839	0.21197	0.21557	0.21920	0.22285	0.22653	0.23022	0.23395
9	0.17711	0.18060	0.18413	0.18768	0.19126	0.19487	0.19851	0.20217	0.20586	0.20957	0.21332	0.21708	0.22087
10	0.16626	0.16980	0.17338	0.17698	0.18062	0.18429	0.18799	0.19171	0.19547	0.19925	0.20306	0.20690	0.21077
11	0.15752	0.16112	0.16475	0.16842	0.17211	0.17584	0.17960	0.18339	0.18722	0.19107	0.19495	0.19886	0.20280
12	0.15038	0.15403	0.15771	0.16144	0.16519	0.16899	0.17281	0.17667	0.18056	0.18448	0.18843	0.19241	0.19643
13	0.14445	0.14815	0.15190	0.15568	0.15950	0.16335	0.16724	0.17116	0.17512	0.17911	0.18313	0.18718	0.19127
14	0.13947	0.14323	0.14703	0.15087	0.15475	0.15867	0.16262	0.16661	0.17063	0.17469	0.17878	0.18290	0.18705
15	0.13525	0.13907	0.14292	0.14682	0.15076	0.15474	0.15876	0.16281	0.16690	0.17102	0.17517	0.17936	0.18357
16	0.13164	0.13552	0.13943	0.14339	0.14739	0.15143	0.15550	0.15962	0.16376	0.16795	0.17216	0.17641	0.18069
17	0.12854	0.13247	0.13644	0.14046	0.14451	0.14861	0.15274	0.15692	0.16112	0.16537	0.16964	0.17395	0.17829
18	0.12586	0.12984	0.13387	0.13794	0.14205	0.14620	0.15039	0.15462	0.15889	0.16319	0.16752	0.17188	0.17628
19	0.12353	0.12756	0.13164	0.13576	0.13993	0.14413	0.14838	0.15266	0.15698	0.16134	0.16572	0.17014	0.17459
20	0.12149	0.12558	0.12970	0.13388	0.13810	0.14235	0.14665	0.15099	0.15536	0.15976	0.16420	0.16867	0.17316
21	0.11971	0.12384	0.12802	0.13224	0.13651	0.14081	0.14516	0.14954	0.15396	0.15842	0.16290	0.16742	0.17196
22	0.11813	0.12231	0.12654	0.13081	0.13512	0.13948	0.14387	0.14830	0.15277	0.15727	0.16179	0.16635	0.17094
23	0.11675	0.12097	0.12524	0.122956	0.13392	0.13832	0.14276	0.14723	0.15174	0.15628	0.16085	0.16545	0.17007
24	0.11552	0.11979	0.12410	0.12846	0.13287	0.13731	0.14179	0.14630	0.15085	0.15543	0.16004	0.16467	0.16933
25	0.11443	0.11874	0.12310	0.12750	0.13194	0.13644	0.14095	0.14550	0.15008	0.15470	0.15934	0.16401	0.16871
30	0.11053	0.11502	0.11956	0.12414	0.12876	0.13341	0.13809	0.14280	0.14754	0.15230	0.15708	0.16189	0.16671
35	0.10829	0.11293	0.11760	0.12232	0.12706	0.13183	0.13662	0.14144	0.14628	0.15113	0.15601	0.16089	0.16579
40	0.10697	0.11172	0.11650	0.12130	0.12613	0.13099	0.13586	0.14075	0.14565	0.15056	0.15549	0.16042	0.16537

perspectives of lenders toward aquaculture as compared to agricultural loans. The study focused on attitudes toward the risk of insolvency and the characteristics associated with approvals of aquaculture loans. Overall, aquaculture was not viewed as unusually high risk when compared to other types of alternative agricultural enterprises. However, selling through a roadside stand and fee fishing were considered to be higher risk marketing strategies than selling directly to restaurants, supermarkets, or to farmers markets. Lowest marketing risk, as viewed by lenders, was associated with sales to processors or through a growers' cooperative.

Smaller lenders (those with assets of $75 million to $100 million) were 10 times more likely to approve loans to aquaculture than were lenders with assets above $100 million. Lenders with positive attitudes to alternative agriculture were the most likely to approve aquaculture loans. In addition, lenders who required a marketing plan were three times more likely to approve an aquaculture loan than those who did not insist on a marketing plan. The majority of aquaculture loans were provided by banks with at least 75% of their portfolio in agricultural loans.

SUMMARY

This chapter discusses various uses for credit and describes the types of loans commonly used by aquaculture businesses. It also demonstrates how to calculate loan payments, amortize loans, and contrasts the effects on cash flow and expenses of various types of loan structures. It ends with brief descriptions of various sources of loan funds.

REVIEW QUESTIONS

1. What are some of the major uses of credit in aquaculture businesses?

2. What are the principal sources of capital that are available to aquaculture businesses?

3. Contrast the differences between single-payment operating loans and lines of credit.

4. What are typical terms of repayment for operating, equipment, and real estate loans? Why are these so different?

5. Why must long-term loans be amortized?

6. Develop an amortization schedule for an equipment loan to purchase a tractor for $60,000 to be repaid over a 7-year period at an annual interest rate of 8%, with both an equal total payment and an equal principal payment schedule.

7. For the two amortization schedules developed in Question #6, explain the differences between the two in terms of effects on cash flow and the total amount of interest paid.

8. Explain why finance experts do not consider loans with balloon payments to be sound financial practices.

9. List and describe the major sources of funds for loans for aquaculture businesses.

10. What types of securities and collateral are used for aquaculture businesses?

REFERENCE

Bacon, Richard, Conrado M. Gempesaw II, Marten R. Jenkins, and Joe A. Hankins. 1998. Aquaculture markets in the Appalachian region: a lender's perspective. *Aquaculture Economics and Management* 2(2):81–87.

Section III
Research Techniques to Analyze
Farm-Level Decision-Making

INTRODUCTION TO SECTION III: RESEARCH TECHNIQUES TO ANALYZE FARM-LEVEL DECISION-MAKING

The chapters in the following section presents information that should be of especial interest to those who conduct research on aquaculture production systems, strategies, equipment, or management practices. The challenges and common errors are discussed, and detailed approaches to developing comprehensive economic analyses with production data are presented.

Cost analyses are commonly used to analyze the economics of treatments from production research.

Chapter 15 describes common errors made in these analyses and clarifies the appropriate methods. Chapter 16 illustrates one of the simplest ways to account for risk in production economics research analyses. Chapter 17 demonstrates methods to develop analyses at the farm level. Allocation of resources across the farm may influence the choice of production alternatives. What is best for the farm as a whole may be different from what is indicated by a simple budget or cost analysis. The final chapter talks about approaches to managing government policies.

15

Use and Misuse of Enterprise and Partial Budgets

INTRODUCTION

No published enterprise budget will answer specific questions about the most profitable way to raise fish because there is so much variation from farm to farm. Reliable answers are developed by adapting published budgets to the specific situation to be analyzed. This is true for both farmers and research analysts. For farmers, winter is a good time to review farm records, calculate feed conversion and financial ratios, and analyze the end-of-year financial position of the business. From there, plans to improve efficiencies related to yield, feed conversion, aeration, and financial performance can be assessed. Goals for the upcoming year can be evaluated by comparing the costs and returns expected from proposed improvements and changes.

Researchers can adapt published budgets by including results of experimental trials to determine the effect on production costs. Enterprise and partial budget templates can, if used properly, provide an excellent basis for end-of-year analysis and planning for farmers and for researchers to develop an estimate of the effect on annual cost and returns that may be associated with farm adoption of the research recommendations.

Enterprise and partial budgets are also used commonly in the research literature on the economics of aquaculture. However, their frequent use has also resulted in incomplete and inappropriate applications that can result in erroneous conclusions. This chapter reviews some common omissions, errors, and misuses of enterprise and partial budgets for both farm-level planning and in research.

ENTERPRISE BUDGETS

Chapter 10 presents details on the proper way to prepare an enterprise budget and its most common uses. One of the most common problems with enterprise budgets is that they have not been developed using standardized methodologies. A budget that omits cost items, or does not provide sufficient information is misleading because its estimate of profitability is erroneous.

"WISHFUL ENTERPRISE BUDGETS"

Enterprise budgets, when developed in a careful and comprehensive manner, demonstrate whether the production system or technology is profitable, and the magnitude of the profitability. However, at times, a scientist who has devoted a great deal of effort to developing certain aquaculture production systems or technologies, may develop an enterprise budget as a way to justify continuing the work and also to promote the production system or technology. The "wish" to demonstrate that the technology will be feasible at times leads to budgets that are not accurate, are overly optimistic, and in the worst cases, misleading. This type of misuse of enterprise budgets is, unfortunately, fairly common, particularly when a biologist without proper economics training attempts to develop a budget. Economists who develop thorough, comprehensive, and realistic budgets are often criticized as being too "dismal" and too "pessimistic" when, in fact, conservative, realistic estimates of profitability or losses provide the type of accurate guidance that is very much needed by aquaculturists.

Aquaculture Economics and Financing: Management and Analysis, Carole R. Engle, © 2010 Carole R. Engle.

ENTERPRISE BUDGETS DEVELOPED WITHOUT USING STANDARDIZED STRUCTURE

Some publications list "budgets" that include only a dollar amount for each line item and omit a clear description, the quantity selected, and the unit cost or price. With only a dollar value, it is not possible to evaluate the relevance of the scenario budgeted. When these occur, budgets cannot be compared over time or across states and locations, and little guidance is provided.

ECONOMIC ENGINEERING TECHNIQUES AS OPPOSED TO SURVEY DATA FOR ENTERPRISE BUDGETS

The majority of enterprise budgets for aquaculture have been developed on the basis of economic engineering techniques. Economic engineering is considered a subset of microeconomic applications to engineering projects. While originally developed as a method of estimating costs of engineering projects, its use has become common in aquaculture. This is primarily because many aquaculture production systems and technologies are new and experimental. If commercial farmers have not yet adopted a particular production system or technology and implemented it on a commercial scale, there will be no data available from which to estimate commercial-scale costs of production. Yet, even on an experimental basis, it is important to have an idea whether the production system or technology is economically feasible, to have a measure of its profitability, and to have an idea which costs are the largest for that production system or technology. Thus, for such new production systems or technologies, an enterprise budget developed with economic engineering techniques, may be the only type of economic analysis possible.

Enterprise budgets based on economic engineering techniques are frequently based on "recommended management practices." The values selected for the yield, prices, and quantities tend to be those that will optimize the production system or technology under study. This is because the goal of the engineers and scientists working on the systems is to develop a system and refine it to perform at optimal levels.

Commercial production levels frequently fall short of what engineers and researchers define as an optimal level for the system. On commercial-scale operations, farm realities often result in lower levels of production due to adjustments made by farm management to marketing, weather, and practical realities that change the quantities of inputs used, yields, and sometimes prices. Thus, enterprise budgets based on economic engineering techniques tend to be more optimistic and demonstrate more positive returns than can be achieved on a regular basis on a commercial farm.

More realistic budgets are produced when the parameter values selected for the budget are based on farm-level production surveys. The key parameters frequently include the yields of fish sold, the quantities used of various inputs, the prices of product sold, and the costs of inputs. In the United States, crop surveys are done on a national basis by the U.S. Department of Agriculture for many types of crops, but particularly those that have been established for a number of years. National survey data such as these can then be used to develop realistic budgets for the various crops, particularly if the surveys are repeated over time. Such cost-of-production surveys are not done for what are considered to be minor crops, such as the aquaculture crops in the United States.

USING EXPERIMENTAL DATA IN ENTERPRISE BUDGETS

It is fairly common for researchers to take data from an experimental trial and extrapolate the results for use in an enterprise budget. This can result in useful information or it can be problematic, depending on which parameter values are extrapolated and which are maintained intact from farm-level data.

For example, to extrapolate from an aquarium or tank study and imply that the results obtained would likely be replicated across a commercial farm is not a realistic assumption. Experimental pond studies will produce results that are closer to what might be expected on a commercial farm, but how the size of the pond may affect the results is a continuing question. Moreover, data from one given pond does not represent a farm-level average. Farm-wide data will encompass effects of various factors that may include losses due to bird depredation, marketing constraints that affect how and when fish can be harvested and sold, the efficiency of the work force, and others. Thus, while an individual pond can produce fish with a certain yield and feed conversion ratio, it may not be realistic to extrapolate that yield and feed conversion ratio across the acreage of the entire farm. In such a case, the more appropriate unit of analysis may be an individual pond rather than an entire farm.

Part of the explanation for why this should not be done is that not all ponds across a farm can be managed in precisely the same way. A farm may have broodstock or fingerling ponds, for example. Some ponds may be stocked at different densities with different sizes of fish due to variations in the availability, mortality events, or varying growth rates from the previous year. These factors can result in various densities, sizes, and ranges of sizes in different ponds. Moreover, some ponds are more productive than other ponds, because they were renovated more recently, are deeper, or for some unknown reason.

When using experimental data from several treatments in enterprise budgets, care must be taken not to use nominal treatment means if the treatment means were not significantly different from each other. For example, assume that the treatment of an experiment was varying levels of animal protein in a fish diet. Further assume that there were no significant differences in yield or total feed fed across the treatments. In such a case, it would be incorrect to use the means of each treatment in separate enterprise budgets and compare the results. The nominal values of the means would undoubtedly be different from one another and would result in what would appear to be an "economic" difference. This "difference" is false because, upon repeating the experiment, the nominal values of treatment means would likely show slightly different patterns (due to random fluctuations) and the "economic" analysis would change with the random fluctuations. In this example, the appropriate analysis would be to do a simple partial budget because the only significant difference appears to be the difference in cost of the feeds.

Some enterprise budgets extrapolate from one production cycle to an entire year. This may not represent the reality of commercial production systems. For example, a shrimp hatchery may need to dry out tanks in between batches of postlarvae. While the production cycle may be 6 months, it may not be possible to achieve two crops a year due to the time needed to dry and disinfect the tanks. It may also be unrealistic to assume that postlarvae are available for stocking in growout ponds every month of the year. While theoretically it might be possible to achieve two crops a year of shrimp in ponds, if postlarvae are not available year-round, then it would be erroneous to extrapolate to two crops a year in an enterprise budget.

Some budgets have extrapolated yields for an entire year from an average daily rate of gain. However, such an average, if calculated from an experiment conducted in one season of the year may not hold true in another. For example, water temperatures often vary from rainy to dry season in the tropics. In temperate climates, growth rates are reduced drastically in the winter period.

INVESTMENT COSTS

For pond-based aquaculture farms, land is a major resource required for aquaculture production. Farmers who have owned the land for many years, or those who have inherited land from their parents and grandparents, may not wish to account for its cost in an enterprise budget. It may be viewed as a "sunk" cost in that its cost was paid many years previously. This perspective on land frequently leads individuals to ignore its costs in budgets.

However, if a farmer can earn more by using the land for something other than the aquaculture crop, at some point the farmer will decide to use the land for its more profitable use. As an example, rapidly increasing prices of land in the state of Florida in the United States have enticed fish farmers to sell their land to developers and cease farming fish. This type of cost (the value of what would have been earned from the land in some other use) is referred to by economists as an opportunity cost. The opportunity cost of land is equal to the value from the most profitable alternative use of the land.

The cost of land, moreover, is not a depreciable cost. With proper care and attention to its nutrients, the land should be usable for many years without loss in value.

Pond construction cost is another major investment cost in pond-based aquaculture systems that is too frequently ignored. The existence of the pond is assumed in these cases to be another sunk cost for which no charge is required. The relevant costs associated with the pond are both the annual interest on the investment and the annual depreciation of ponds. Annual depreciation is a charge to account for the loss in value as a capital item like a pond or tractor wears out or becomes obsolete. Failure to account for the depreciation of ponds means that the business will not be prepared financially to rework the ponds when necessary. If that happens, the ponds will not continue to be as productive as previously; thus, the value of the ponds in generating revenue for the business will be diminished. For example, the cost of rebuilding ponds has been estimated to range from $500 to $1,500/acre, nearly as high as that of constructing new ponds. Thus, accounting for depreciation as an expense in the financial planning of the business is essential for the long-term survival of the farm.

Early budgets frequently used an estimated earth-moving cost as the basis for the cost of building ponds. However, such an estimate does not include the substantial costs that are incurred in expenses related to establishing a good grass cover on the levees, or gravel to support a road for fish hauling trucks and equipment to reach ponds that need to be harvested. The cost of drilling sufficient wells to have adequate capacity to fill all ponds in a reasonable amount of time similarly is frequently underestimated in many enterprise budgets.

Equipment Costs

One of the most common mistakes in an enterprise budget is the lack of an effort to compile a complete equipment list or to ignore the costs of equipment already owned by the farmer. Both types of mistake result in underestimating the substantial equipment investment and the associated annual fixed costs required for a successful aquaculture business.

A number of budgets developed for cage production frequently ignore the costs of essential structures and equipment. Boardwalks are as necessary to provide access to the cages for feeding and for disease treatment as gravel and levees are to a pond farm. However, few budgets for cage production include the costs of boardwalks. Pumps and aerators for supplemental oxygenation are also frequently ignored in cage budgets. Net pen systems frequently require buffer floats, supply barges, and feeding systems. Annual depreciation must be charged for these items in addition to delineating the total capital investment cost to avoid unanticipated costs. The cost of the cages themselves has been ignored in some budgets! Moreover, many budgets for cage production assume the water to be available at no cost to the farmer. However, there are costs even if the water source is a public water body. Expensive permits may be required. Delays in obtaining permits may delay opening the business, and these delays result in costs to the business.

Equipment already owned by the farmer is also frequently ignored. A farmer who owns a tractor, though, already has uses for that tractor. If the tractor is needed for the fish crop but is in use for other businesses, then the fish crop likely will suffer. The successful farmer must be certain to have enough equipment to fully cover all needs on the farm. This means that sufficient numbers of additional tractors, aerators, and other pieces of equipment are purchased so that the farmer can maintain the crops in healthy condition.

When cost-of-production surveys are used as the basis for developing budgets, there is an opportunity to record all equipment used on each farm. Compiling complete lists of equipment on each farm surveyed is tedious but will ensure that all required equipment is included in the budgets developed.

Aquaculture businesses also need sufficient backup equipment for the farm to operate successfully even during emergency situations. This will require purchases of backup generators, replacement aerators, spare parts, and adequate tools to make repairs. For example, on catfish farms in the United States, power outages or brownouts can result in substantial levels of fish mortality across the farm. Few farms have adequate emergency aeration equipment to keep all fish alive across the entire farm during a power outage that occurs at the peak of the production season. However, many farms have purchased generators to provide additional capacity to survive a power outage. Similarly, aquaculture businesses that operate marketing and grading sheds frequently purchase generators as a backup power source to survive disaster events such as a tornado or ice storm that may damage power lines and interrupt the supply of electricity to the farm. Many budgets developed on the basis of economic engineering principles do not include backup redundancy of equipment to minimize the risk of these types of yield losses. Such budgets do not adequately account for the costs of mitigating the risks associated with these events.

Equipment costs enter into the enterprise budget as annual fixed costs in the form of annual depreciation and of annual interest charges on the investment in equipment. Including annual depreciation in enterprise budgets assures that the business will generate enough revenue to be able to replace the equipment when it wears out. If the business cannot afford to replace equipment when it wears out, it will not be able to stay in business.

A common error in calculating annual depreciation is to use an unrealistic value for the years of useful life for each equipment item. This will result in erroneous estimates of annual depreciation, annual fixed costs, annual total costs, and net returns. Equipment usage on farms differs from that of equipment used on research stations. The environment on a farm and the level and quality of the workforce that maintains and cares for the equipment will vary from that of research facilities. The level of care will affect the useful life of the types of equipment. This difference will result in varying years of useful life that are realistic for commercial farm situations.

ANNUAL COSTS AND RETURNS

Annual returns calculated in enterprise budgets often are based on the current price rather than the long-term average price that should be used (see Chapter 10 for more detail). The problem with using a current price is that, if the budget is prepared in a year when prices are high, the annual net returns estimated will be unrealistically high. Over time, prices will come down, and when that happens, the business will not have prepared to cope with the reduction in revenue. Individuals who estimate returns frequently are reluctant to use longer-term average prices if these are lower than current prices. The argument becomes that the current prices are more realistic. On the other hand, if long-term average prices are higher than current prices, lenders may be reluctant to "believe" the higher, long-term prices even though these are the best to use in the budget.

Yields

Yields selected for enterprise budgets similarly are often overly optimistic. Researchers frequently drop data from studies if results from a particular pond are considered to be an "outlier." In other words, if one pond or tank in a trial has much greater mortality rates than others or if the growth is much lower, there is a tendency to eliminate the data from that pond, tank, or cage study, in the analysis and in the manuscript. The problem is that a farmer cannot "eliminate" that type of substantial mortality or reduced growth from its effect on the farm's revenue. Budgets that are based strictly on research data that may have been purged of outliers may use yield values that are upwardly biased. Such values can result in misleading information on the profitability of the enterprise when used in an enterprise budget.

In some cases, the best possible yield is selected as the yield value for the enterprise budget. When this is the case, the budget demonstrates the maximum amount of profit that can be obtained, under the best possible conditions. The results then apply only to those situations in which the best possible yield is obtained. Unfortunately, budgets that are developed this way frequently do not make it clear that it is a best case, not a typical, scenario. When this is not explained fully, those examining the budget are likely to believe that the values are average, or typical, values, rather than

values that they are unlikely to see or achieve very often on their farms.

In some cases, researchers choose to calculate the yields of fish for enterprise budgets based upon assumed survival and growth rates. Budgets that attempt to do so frequently result in estimations that are unrealistic. This is particularly true for pond aquaculture systems that are in continuous production. There is no known practical technology to accurately inventory values in continuous pond production systems with mixed sizes of fish. Survival and growth will differ among sizes of fish because differing behaviors will affect feed consumption.

Annual Variable Costs

Economically engineered budgets often use an assumed feed conversion ratio as the basis for calculating the amount of feed to include in the enterprise budget. In fishponds that are managed in a continuous production system, ponds can end up with mixed-size populations of fish. In such a situation, a single feed conversion ratio is nearly meaningless. This is because the feed conversion ratio measures the rate at which the individual fish converts feed to flesh. Feed conversion ratios vary with the size and condition of the fish. Fish farmers who manage ponds in continuous production cannot measure feed consumption of individual fish. What the farmer can measure on a commercial farm is the weight of feed fed on the farm divided by the weight of fish sold. This is not the same metric as the biological feed conversion ratio calculated by researchers (the amount of feed that it takes for an individual fish to gain a pound in weight).

The conversion metric calculated by the farmer overestimates the "feed conversion ratio" because feed fed to fish that died or that were eaten by birds is inherent in the calculation. Farmers have no way to separate out the amount of feed that was eaten by fish that did not survive and grow to be sold. Similarly, feed fed to maintain weight of fish that are off-flavor and are held until they can be purged and sold, is also included in the calculation of "feed conversion ratio" calculated on farms. Thus, when a farmer says that their "feed conversion ratio" is 2.5, that does not mean that the fish on the farm are converting individually at 2.5. That simply means that, across the farm, the feed yield ratio (feed fed compared to the marketed yield of fish) was 2.5. Thus, using a biological feed conversion ratio of what an individual fish can do in an enterprise budget will not result in an accurate estimate of what the feed costs across the farm will be. The probability of that individual fish dying or growing poorly is omitted from the calculation. Across the farm, that omission can result in substantial differences. It is more accurate to use farm records to identify the quantity of feed fed across the farm as the quantity of feed to use in the budget. Similarly, it is best to use farm sales records for the estimated yield.

Economically engineered budgets also frequently omit a number of annual cost items. Most budgets do include the obvious costs associated with feed, fingerlings, and labor. However, less frequently included variable costs include items such as: repairs and maintenance, the costs of bird depredation, telephone and office expenses, legal and accounting fees, taxes and insurance, and interest on operating capital. These can be substantial costs on some farms. For example, a farm that spent less investment capital on equipment by purchasing older equipment may have substantial repair and maintenance costs. Similarly, a farm with older ponds will need to spend more to maintain the levees and rework levees to keep them wide enough to allow vehicles to pass.

Bird pressure on fishponds in the United States continues to grow, and fish farmers must spend more and more each year to scare birds. These costs include both the ammunition costs as well as the manpower required to have an effective bird dispersal program on the farm. Telephone expenses have increased over time with the need for farm managers and employees to have cell phones. The need for office supplies has grown as increasing amounts of regulations have targeted aquaculture businesses and increased the required amount of paperwork for the farm to remain in compliance.

Many enterprise budgets neglect to include interest charges on the operating capital used. If the farm does not borrow operating capital, then there is no cash charge for its use. However, there remains an opportunity cost for use of the owner's capital to operate the farm. This is because the owner could have used the capital in some other activity and the profits to be earned from its use in another activity constitute an opportunity cost to the fish farm.

Realistically, there are few aquaculture businesses that rely exclusively on the owner's capital to cover all the operating costs of the farm. The majority of aquaculture businesses borrow operating capital in the form of an operating line of credit or revolving loan. Thus, interest charges must be paid and accounted for in the enterprise budget. It is unrealistic to assume that the capital is available at no cost to the farm owner for the majority of aquaculture farm business owners.

Many enterprise budgets for aquaculture farms are not consistent in the treatment of expenses associated with family labor. Many budgets do not account for the cost of the farmer's own labor. Yet, if the farmer does not make enough money to compensate him or her for the time and effort, they will likely quit farming to do something else with their time. Unpaid family labor is an important opportunity cost and must be valued at more than minimum wage. Moreover, older farmers may not be able to continue doing the physical labor that they could when younger. Farmers tire of working long hours with little or no compensation. Farmers need and want medical insurance to save for retirement and to receive these types of benefits. Management positions in other industries frequently pay as much as 25% or more of the salary to provide these benefits to employees.

If some budgets include unpaid family labor costs as an opportunity cost, but others do not, then the net returns estimated are not comparable. Yet, these types of comparisons are made all too frequently.

Annual Fixed Costs

The annual fixed cost portion of the enterprise budget is perhaps the portion of the budget that is most often misused and underestimated. Annual fixed costs in aquaculture tend to be high and often compose a substantial portion of the total annual costs of the aquaculture business. Costs associated with all fixed resources used in the business should be accounted for.

Many of the fixed costs in an aquaculture business are composed of interest on the investment and the annual depreciation of the ponds, buildings, and equipment owned by the business. The value of the interest on the investment may or may not be a cash expense. If the owner does not have loans for the investment items, then there is no cash interest charge on the amount invested. However, because this capital could have been used in some other way, the value of what would have been earned in another use is a cost to the business. This cost must be accounted for by charging an interest on the investment items as an annual fixed cost in the enterprise budget.

However, many individuals omit such an interest charge in the annual budget if the owner has not taken out loans for that capital. Doing so greatly underestimates the fixed costs and will result in an incorrect

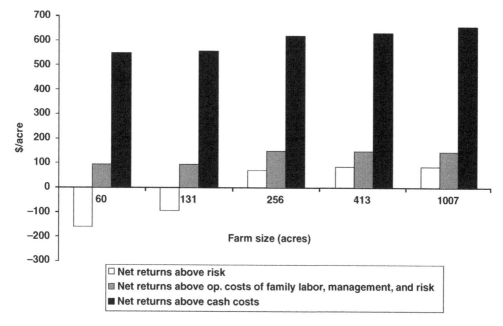

Figure 15.1. Three different measures of net returns.

estimate of the profitability of the business, the breakeven prices and the breakeven yields. Such an omission will result in highly misleading results.

Reducing annual fixed costs by omitting charges that are necessary to reflect the total usage of capital in the business will also obscure the economies of scale of the farm (see Chapter 7 for more discussion of economies of scale and their importance to aquaculture businesses). If the owner of the business does not believe that there are economies of scale, then no effort will be made to expand the business to achieve lower per-unit costs. The farm will continue to operate at a financially inefficient level and will be less competitive as compared to other farms.

Annual depreciation charges also are frequent omissions from enterprise budgets. Without full accounting for the loss in value of the capital goods used in the production of the aquaculture products, there is no accounting for the capital that will be needed to replace these items. This will also result in net returns that are artificially high and artificially low breakeven prices and yields. In all cases, the results of the enterprise budget will be erroneous and misleading, pointing the farm owner to erroneous conclusions and likely poor management decisions.

Annual Net Returns

Some enterprise budgets calculate net returns incorrectly by omitting various combinations of the items discussed above, leaving out unpaid family labor and omitting noncash expenses. However, others calculate a labor charge in their calculation of net returns. Thus, different enterprise budgets may not always be comparable.

Figure 15.1 shows net returns calculated properly accounting for all costs (noncash annual depreciation and opportunity costs of unpaid family labor). These are referred to as net returns above risk because the other costs have been charged out. The figure also shows net returns without the costs of unpaid family labor and management. These are referred to as net returns above opportunity costs of family labor, management, and risk. The figure also shows net returns above cash costs only.

When properly accounting for the use of all resources, the 60-acre- and 131-acre farms were not profitable. The other three farm sizes showed modest profit levels. However, without accounting for the costs of all equipment, labor, other noncash costs, and all farm scenarios appear to be very profitable. Without accounting for the cost of unpaid family labor and management, net returns were $49, $86, $144, $158, and $145 per acre for the 60-acre-, 131-acre-, 256-acre-, 431-acre-, and 1,007-acre farms, respectively. The decline in net returns per acre between the 431- and 1,007-acre farms is due to hiring an overall manager in addition to foremen or managers of each farm unit on this largest farm size. When values for the unpaid family labor and management were included, net returns decreased and became negative on the 60-acre- and 131-acre farms. Across farm sizes, net returns above all costs increased as farm size increased, from −$161/acre to $111/acre on the largest farm size. This set of net returns is the true economic returns because all resources used in the production of catfish are accounted for.

Some enterprise budgets include a return on investment (ROI) calculation, in addition to net returns. This is not appropriate. A properly calculated ROI is not based on the net returns as calculated in an enterprise budget. ROI is calculated from the income statement based on a specific formula that avoids double counting charges related to capital resources. Chapter 4 presents details on how to properly calculate the ROI.

SENSITIVITY ANALYSES

Values used in an enterprise budget are single values that should represent long-term mean values of that particular parameter. However, these values frequently are not well known and there may not be an adequate database to be confident of the value. In such cases, and particularly when the parameter is an important one, sensitivity analyses must be run. If omitted, a false image of the profitability of the enterprise may emerge.

PARTIAL BUDGETS

Partial budgets are used to evaluate the effect of a relatively small change in the farm business. Chapter 10 presents details on developing a partial budget and its interpretation. One of the misuses of partial budgets is to include only some costs and not others, accounting only "partially" for costs. This is a total misuse and misconception related to a partial budget.

A common scenario is one in which a researcher has conducted a study that shows an improvement in production. The researcher wishes others to acknowledge his or her work and feels that he or she has been successful in identifying some new information and advances in his or her particular branch of aquaculture. Because economics is a way of assigning monetary values to parameters, he or she decides to conduct a "partial" budget analysis of the results of their experiment. In developing the partial budget, the researchers decide to consider costs associated only with the treatments that were assigned to the study. Let's assume that the treatment was additional oxygen. In the tank study, the oxygen was supplied by an air blower located in the laboratory where the study was conducted.

In their partial budget, the researchers assigned a market value to the additional weight of fish produced at the greater amounts of oxygen. However, in their system, there was no additional cost for the additional oxygen because of the way it was supplied to the tanks. Because their analysis was only a "partial" one, they were not concerned with working through all details of what a commercial farmer would have to do to achieve the higher oxygen levels on their farm.

In reality, however, the farm would have to purchase additional aerators to supply an equivalent amount of oxygen to that used by the researchers in their tanks. The additional aerators would entail a capital acquisition that would increase both annual depreciation and interest on the investment in the annual fixed cost section of the budget. Moreover, adding additional aerators to the farm would increase the demand for electricity and would increase the electric bill that would be listed under the annual variable costs on the budget. A proper partial budget would include the additional annual depreciation, interest on investment, and electric charges under the category of "Additional Costs."

The partial budget includes only those costs and benefits that change from adopting the new change on the farm. However, it must include all the costs and benefits that will change. To include only some of those changes and not others is a misuse of the analytical method known as a partial budget. This practice will inevitably result in an erroneous conclusion for the analysis.

RECORD-KEEPING

For the farm owner or manager to maintain proper budgets for each enterprise on the farm, he or she will need to maintain complete and comprehensive records. Chapter 10 provides details on the records needed to prepare comprehensive enterprise budgets. However, the key factors are to be certain to account for the annual fixed costs associated with all the capital resources on the farm. These include the annual depreciation and interest on investment on the equipment, ponds, cages, tanks, and all buildings on the farm. The annual depreciation should be maintained in the form

Table 15.1. Comparison of "Net Returns" Using Common Errors of Omission, 256-acre Catfish Farm.

| Category | Net returns | Breakeven price | | Breakeven yield | |
		Above variable cost (VC)	Above total cost (TC)	Above VC	Above TC
	$/acre	$/lb	$/lb	lb/acre	lb/acre
Fully accounting for all costs	−72	0.57	0.72	3,641	4,603
Without opportunity costs of unpaid family labor	11	0.57	0.58	3,641	4,484
Without fixed costs	518	0.57	0.70	3,641	3,760
Without fixed costs or opportunity costs of unpaid family labor	601	0.57	0.57	3,641	3,641

of a depreciation schedule that lists the annual loss in value of each piece of equipment, each building, and each block of ponds, cages, or tanks. The interest on the investment should be calculated from a schedule that shows the initial cost of each capital item (equipment, ponds, cages, tanks, buildings) and the current interest rate.

To be sure that all variable costs are included, all expense records should be checked each year to be certain that none are omitted in the enterprise budget. Special attention should be paid to accounting for repairs and maintenance. It is a good idea to maintain a separate record of the repairs and maintenance required on each piece of equipment. This will provide a good idea of when the repair costs begin to exceed the cost of replacing that piece of equipment.

The final key piece of information is that of the labor and salaried individuals. Hourly labor costs are included as variable costs and contracted salaried employees are fixed costs.

As long as all costs are accounted for properly and the quantities of inputs purchased and fish sold are based on averages calculated from thorough records, the enterprise budget will be a useful tool. Omission of cost categories, overestimation of yields or underestimation of costs will lead to misuse of enterprise budgets.

PRACTICAL APPLICATION

Figure 15.1 contrasts three different measures of net returns (profit) for five catfish farm sizes. These include: (1) properly accounting for the use of all equipment, labor, and other inputs; (2) including only cash expenses; and (3) omitting opportunity costs of unpaid family labor.

Table 15.1 shows that not accounting for all the resources used on the farm will obscure the fact that, in the long-run, the farm as budgeted is not profitable. Net returns for the 256-acre catfish farm, as budgeted, are negative. Thus, in the long run, the business is not profitable. It covers variable costs, but does not cover all its fixed costs. Breakeven prices were $0.57/lb above variable costs and $0.72/lb above total costs. Not accounting for unpaid family labor similarly demonstrates an artificially low breakeven price and yield and artificially high net returns. Net returns were positive and breakeven prices were $0.70/lb above total costs, indicating apparent profitability. Without accounting for annual fixed costs, the breakeven yields above total costs are nearly identical ($0.58/lb) to those of the breakeven prices above variable costs.

OTHER APPLICATIONS IN AQUACULTURE

Engle (2007) discusses the use of enterprise budgets (and other approaches) to assist in thorough business planning. Such uses of well-designed budgets can help to transfer new technologies to the private sector to become economically and financially viable businesses.

Enterprise budgets may be most valuable in the analysis of new technologies and new start-up types of aquaculture businesses. Engle (2007) summarizes studies that analyze rapid advances in new offshore mariculture technologies. Enterprise budgets have been developed for grouper production in cages (Pomeroy et al. 2004), hard clam hatcheries (Adams and Pomeroy 1992), southern bay scallops (Adams et al. 2001), haddock hatcheries (Dalton et al. 2004), red-claw crawfish in ponds (Medley et al. 1994), tilapia in irrigation ditches (Sherif et al. 2002), and catfish and tilapia in the partitioned aquaculture system (Goode et al. 2002).

However, enterprise budgets can be used appropriately in the analysis of existing technologies and businesses. For example, Brummett et al. (2004) evaluated polyculture and monoculture of tilapia, clarias, and kanga production in ponds. Lutz (2000) compared the costs of raising tilapia in raceways, ponds, and recirculating systems with budget-based cost analyses. Valderrama and Engle (2001) used enterprise budgets to evaluate the economics of shrimp production in semi-intensive ponds in Honduras.

SUMMARY

This chapter has discussed some common uses and misuses of enterprise and partial budgets. Most of the misuses of these analytical tools result from a type of wishful thinking on the part of either researchers or farmers. People tend to be optimistic in their projections and too often do not spend the time to carefully think through the details. The result often is that expenses are omitted, prices and yields are set at an unrealistically high level, or unit costs of inputs are set at unrealistically low levels. The end result is erroneous and can lead the farm owner or manager to commit mistakes in management and exercise poor judgment due to faulty economic analyses.

REVIEW QUESTIONS

1. What are the most common types of errors associated with product prices of aquaculture products produced on the farm? What is the correct manner to assign such a price?

2. What is the most common type of mistake with regard to the yield used in the enterprise budget? What is the correct manner to assign a yield in the enterprise budget?

3. What are commonly omitted types of variable costs?

4. What are the types of annual fixed costs most frequently omitted from an enterprise budget?

5. What problems result from omitting annual fixed costs from the budget?

6. What is the most common type of misuse of a partial budget?

7. Why is it not correct to calculate an ROI from an enterprise budget?

8. Why should land be included in an enterprise budget if the farm owner owns the land?

9. What problems can arise when experimental data are used to develop enterprise budgets?

10. What is the difference between economically engineered enterprise budgets and those based on survey data? Why does this difference matter?

REFERENCES

Adams, Charles M., Leslie Sturmer, Don Sweat, Norman Blake, and Robert Degner. 2001. The economic feasibility of small-scale, commercial culture of the southern bay scallop (*Argopecten irradians concentricus*). *Aquaculture Economics and Management* 5(1/2):81–97.

Adams, Charles M. and Robert Pomeroy. 1992. Economies of size and integration in commercial hard clam culture in the southeastern United States. *Journal of Shellfish Research* 11(1):169–176.

Brummett, Randall E., James Gockowski, Jeshma Bakwowi, and Angoni Desire Etaba. 2004. Analysis of aquaculture investments in periurban Yaounde, Cameroon. *Aquaculture Economics and Management* 8(5/6):319–328.

Dalton, Timothy J., Kate M. Waning, and Linda Kling. 2004. Risk efficient juvenile haddock production systems: an ex-ante recursive stochastic approach. *Aquaculture Economics and Management* 8(1/2):41–59.

Engle, Carole R. 2007. Investment and farm modeling for feasibility assessment and decision making in aquaculture. In: P-S Leung, C-S Lee and P. O'Bryen (eds). *Species & System Selection for Sustainable Aquaculture*. Oxford: Blackwell Publishing. pp. 67–84.

Goode, Timothy, Michael Hammig, and David Brune. 2002. Profitability comparison of the partitioned aquaculture system with traditional catfish farms. *Aquaculture Economics and Management* 6(1/2):19–38.

Lutz, Greg. 2000. Production economics and potential competitive dynamics of commercial tilapia culture in the Americas. In: B.A. Costa-Pierce and J.E. Rakocy (eds). *Tilapia Aquaculture in the Americas*. Baton Rouge, LA: The World Aquaculture Society. Volume 2, pp. 119–132.

Medley, Paul B., Robert G. Nelson, Upton Hatch, David B. Rouse, and Gerard F. Pinto. 1994. Economic feasibility and risk analysis of Australian red claw crayfish *Cherax quadricarinatus* aquaculture in the southeastern United States. *Journal of the World Aquaculture Society* 25(1): 135–146.

Pomeroy, Robert S., Romeo Abayani, Marietta Duray, Joebert Toledo, and Gerald Quinitio. 2004. The financial feasibility of small-scale grouper aquaculture in the Philippines. *Aquaculture Economics and Management* 8(1/2):61–83.

Sherif, S.M., R.W. Fox, and O.E. Maughan. 2002. Economic feasibility of introducing pulsed-flow aquaculture into the irrigation system of cotton farms in Arizona. *Aquaculture Economics and Management* 6(5/6):349–361.

Valderrama, Diego and Carole R. Engle. 2001. Risk analysis of shrimp farming in Honduras. *Aquaculture Economics and Management* 5(1/2):49–68.

16
Risk Analysis in Production Aquaculture Research

WHAT IS RISK?

Aquaculture is often considered as a high-risk production activity. While there are many sources of risk, economists frequently characterize risk into the following categories: production (yield or technical), marketing (or price), and financial risk. These can combine to produce income risk. To an economist, risk refers to the measurable fluctuations in values for parameters for each of these categories.

Production or yield risk can occur as a result of losses due to bird depredation, losses due to disease, oxygen problems, equipment failure, power outages, or floods. Figure 16.1 demonstrates production/yield risk that results from fluctuating weather conditions. In Figure 16.1, each curve represents the output that is likely to be produced at different input levels (feed, stocking densities, etc.) under one of three weather conditions (good, normal, or poor) for the crop under consideration. Clearly, yields and output are higher for each input level with good weather. Yields are lower at each input level for normal or poor weather. Weather parameter values can be measured mathematically as probabilities, and these probabilities constitute a measure of risk.

Market or price risk is created by variability in prices. Supply varies with the weather and with production decisions made on the farm. Demand varies as consumer incomes, exchange rates, export policies, strength of the economy, supply of competing products, seasonal or cyclical trends, and input prices vary. Because market price is dictated by the interaction of supply and demand, fluctuations in both result in market (price) risk.

Aquaculture Economics and Financing: Management and Analysis, Carole R. Engle, © 2010 Carole R. Engle.

Financial risk results from fluctuations in those factors that affect loans and loan payments. These include interest rates, the lender's willingness to continue lending, market values of loan collateral, and the ability of the business to generate cash flows needed for debt payments.

The ability to bear risk varies with different farms. Farms with high levels of net worth can withstand larger losses. However, highly leveraged farms (those with high debt levels) can lose net worth quickly and are more vulnerable. Cash flow commitments related to loan payments or outstanding accounts payable affect the ability of the business to remain profitable when faced with negative risks. An individual's willingness to bear risk also varies. Some individuals enjoy the risk often associated with the opportunity to gain high returns (risk lovers). Others, however, shy away from risk and prefer lower returns that are more secure with less chance of losses (risk avoiders or risk averters).

Higher risk is often accompanied by higher profits. Figure 16.2 illustrates the trade-off between risk and expected profits. In many cases, profits increase as the level of risk increases. In such cases, higher profits are possible, but the higher profits are accompanied by higher probabilities of losses.

This chapter explains how risk is measured and describes indicators that can be used to measure risk. It also reviews the literature of risk analyses in aquaculture that include studies of stochastic dominance and efficiency, spreadsheet-based risk simulations, and risk-based programming model analyses.

RISK VERSUS UNCERTAINTY

Risk and uncertainty are distinct concepts in economics. The fundamental distinction between risk and

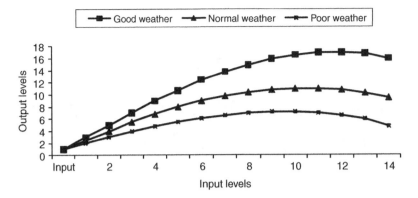

Figure 16.1. Fish yield risk due to various weather conditions.

certainty is whether or not mathematical probabilities of the randomness that a farmer is facing can be estimated (Knight 1921). Risk refers to measurable variations for which decision makers can assign mathematical probabilities while uncertainty refers to fluctuations in variables that cannot be measured and expressed in mathematical probabilities. While many people use the terms risk and uncertainty interchangeably, this distinction is important in economics.

HOW RISK IS MEASURED

There are some simple measures that can be calculated that reflect fluctuations in parameter values. The measures of central tendency (mean/expected value, median, and mode) can be used in combination with measures of spread or variability (standard deviation, variance, the coefficient of variation (CV), the confidence interval, and the probability of key events).

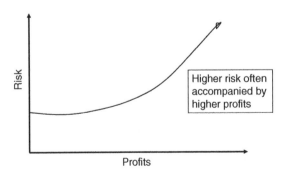

Figure 16.2. Trade-offs in risk and expected profits.

Mean is the average or expected outcome. It is the probability-weighted average of the random variable. For discrete variables, if x is a random variable with N possible values, each with probability p_n, then the mean or expected value of x is:

$$\sum_{i=1}^{N} p_i x_j \tag{16.1}$$

For continuous values, if x is a random variable with a probability density function $f(x)$, then the mean or expected value of x is:

$$\int_x x f(x) \mathrm{d}x \tag{16.2}$$

For example, if a fish crop has three possible yields: low (3,500 lb/acre), medium (5,500 lb/acre), or high (7,500 lb/acre) with probabilities of 0.35 for the low yields, 0.55 for the medium yield, and 0.10 for the high yield, then, given Equation 16.1,

$$\text{Mean} = (0.35 \times 3,500) + (0.55 \times 5,500)$$
$$+ (0.10 \times 7,500) = 1,225 + 3,025$$
$$+ 750 = 5,000 \text{ lb/acre}$$

The mean is not what will happen, but rather, if the random event occurs several times, the mean is the average of the outcomes. For example, if the mean yield of shrimp is 1,250 lb/acre, this does not mean that the yield next year will be 1,250 lb/acre, but rather that, over the next several years, the average shrimp yield will be approximately 1,250 lb/acre.

In addition to the mean, the median and the mode also measure a distribution's central tendency. The median is the middle or half-way point. Thus, half of the

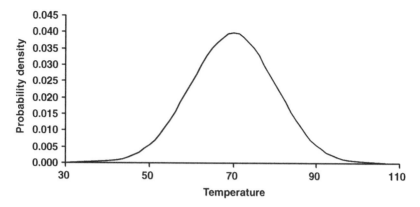

Figure 16.3. Measures of central tendency for various types of distributions.

observations will be less than the median and half of the observations will be greater than the median. The mode is the most common or most likely value.

When the distribution is symmetric, the mean is equal to the median. When the distribution is normally distributed, then the mean is equal to the median and is also equal to the mode. Distributions that are skewed or asymmetric have means, medians, and modes that are all different (Figure 16.3).

Measures of spread or variability can also be used to measure risk. Probability-weighted measures of variability can be calculated. All begin with the deviation of the possible outcomes from the mean. If x_i is an outcome and $E[x] = $ mean, then

Deviation $D_i = x_i - E[x] = x_i - m$

Variance is the probability-weighted average of the squared deviation of each outcome from the mean and can be represented as:

$$\text{Variance} = s^2 = \sum_{i=1}^{N} p_i D_i^2 = \sum_{i=1}^{N} p_i (x_i - E[x_i])^2$$

(16.3)

Why is the deviation squared? This is because it converts all deviations to be positive. Negative and positive deviations of the same size have the same value; in other words, $-4^2 = 16$.

The CV is calculated by dividing the standard deviation by the mean and is expressed as a percent. Thus, if the standard deviation is 150 and the mean is 600, then the CV is 0.25 or 25%. The CV normalizes

the standard deviation by the mean. A CV of 25% indicates that the standard deviation is 25% of the mean.

Risk can also be calculated by developing probability density functions, statistical functions that describe all possible outcomes and associated probabilities. Table 16.1 illustrates the data for a probability density function. The probability of occurrence is specified (0.2–0.4) for each range of fish yield (2,000–6,000 lb/acre). The possible range of fish yields is listed along with the midpoint of the range and the probability of occurrence. The midpoint yield is multiplied by the probability of occurrence to form the expected yield.

If the data are discrete, there are only two options, win or lose, or heads or tails from a coin toss. For continuous data, such as catfish or shrimp prices at harvest or next year's earnings, the functions developed are

Table 16.1. Using Probabilities to Form Expectations.

Possible fish yields (lb/acre)	Midpoint (lb/acre)	Probability	Expected yield*
2,000–3,000	2,500	0.2	500
3,000–4,000	3,500	0.4	1,400
4,000–5,000	4,500	0.3	1,350
5,000–6,000	5,500	0.1	550
Total		1.0	3,800

*Probability X midpoint (lb/acre).

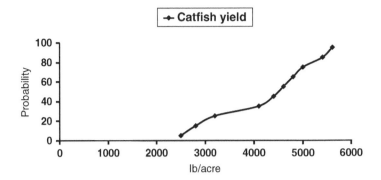

Figure 16.4. Normal distribution with continuous data.

continuous functions. Figure 16.4 illustrates a continuous distribution. In this case, the maximum temperature has a normal distribution with a mean of 70 and a standard deviation of 10.

The formula for the probability density function is:

$$f(x) = \frac{1}{\sqrt{2\pi(10^2)}} \exp\left[-\frac{(x-70)^2}{2(10^2)}\right] \quad (16.4)$$

Distributions can take forms other than that of a normal distribution. For example, a distribution developed for price data may be lognormal, a beta distribution may be used for percent losses or rates of gain, and gamma or Weibull distributions used for yields. Empirical distributions can be developed from raw data.

A cumulative distribution function can be traced that calculates the probability that the random variable takes on a value less than or equal to that of the random variable. Figure 16.5 illustrates a cumulative distribution function that shows all possible levels of catfish yield in response to the weather effects shown

in Figure 16.1. The cumulative distribution function in Figure 16.5 becomes steeper at higher yield levels that correspond to good weather. Thus, the yield response to poor weather is greater than that to good weather.

FINANCIAL INDICATORS THAT PROVIDE INSIGHT INTO RISK

There are other approaches to measuring risk. Several of the indicators presented in Chapter 4 provide a type of measure of risk. For example, the liquidity measures presented in Table 4.5 provide a measure of the ability of the business to meet its short-term cash obligations. For example, the current ratio of 2.55 shows ample current assets to meet current liabilities. This is supported by the amount of working capital for the farm, $171,377. This amount of working capital indicates the extent of the safety margin of cash available to the farm business to cover any unexpected adverse conditions. Similarly, the cash flow coverage ratio and the debt-servicing ratio provide a measure of the safety margin

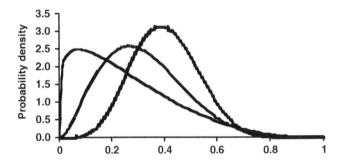

Figure 16.5. Cumulative distribution function.

of available cash above that required for meeting debt-servicing requirements associated with loan payments.

Additional measures of cash flow risk can be calculated from the cash flow budget. These include: (1) percent farm revenue can decline and still meet cash flows; (2) percent farm expenses can increase and still meet cash flows; and (3) percent interest rates can increase and still meet cash flows. These are self-explanatory, but provide a measure of the safety margin for adverse situations with farm revenue, farm expenses, and interest rates. These measures indicate how much of a change can occur in these measures and still be able to make the payments needed.

STOCHASTIC DOMINANCE AND EFFICIENCY ANALYSES

Researchers have developed the theory and methodology to rank various alternatives according to the riskiness of each. Much of this work has been done relative to choices among various financial investments. This work has since been extended to agriculture scenarios that entail risky situations. While there are other frameworks within which to measure and compare risky alternatives, the stochastic dominance framework has been used more frequently in recent years. It provides a quantitative means to rank various alternatives and identify which is "better" or which one "dominates" (stochastic dominance) the others in terms of the relationship between expected returns (profits) and the level of risk. Levy (2006) presents a detailed discussion of stochastic dominance, from development of the underlying theory, stochastic dominance decision rules, relevant algorithms, and presents several empirical studies.

There have been several applications of stochastic dominance analysis in aquaculture. Gempesaw et al. (1992) used stochastic dominance with respect to a function to evaluate pond size and the profitability of growout of hybrid striped bass. The analysis was based on a dynamic whole simulation model developed in a capital budgeting framework (see Chapter 13 for more detail on capital budgeting). Data used were obtained from the literature and were used to develop representative farm scenarios. The analysis showed that the optimal size of pond was in the range of 2.5 to 10 acres.

Kazmierczak and Soto (2001) compared the risk efficiency of various farm sizes. Monte Carlo simulations were run to generate probability distributions of net returns based on stochastic prices, feed prices, and yields of catfish. Net returns were shown to be distributed in a non-normal fashion. Results showed that large farms were the most risk efficient.

Stochastic dominance criteria can also be incorporated into other modeling analyses. For example, Dalton et al. (2004) developed a simulation model for producing juvenile haddock based on multivariate stochastic cost functions. Principal results showed that land-based production of juvenile haddock was viable in spite of the high opportunity cost of capital. The financial risk, however, was shown to pose a constraint to industry development. Four alternative feeding technologies for juvenile haddock were compared. Stochastic dominance criteria demonstrated that early weaning of fry from *Artemia* was not feasible.

SPREADSHEET-BASED RISK SIMULATIONS

There are several spreadsheet-based risk simulation software programs that can be used relatively easily to incorporate elements of risk into basic budget analyses. The two most common forms are @RISK (Palisade Corporation) and Crystal Ball (Microsoft Corporation). These software programs allow the practitioner to insert distributions rather than single-point numbers into spreadsheet cells. The spreadsheet software program then draws a sample from each distribution, calculates the net returns, records it, and then draws another sample from each distribution in the spreadsheet, calculates a net returns, records it, and so on. The process is repeated for as many simulations as are established, frequently 500–1,000 iterations. The results generated are composed of distributions of net returns instead of a single value of net returns. Such a distribution provides a basis for calculating the probability of losing money, or the probability of earning a specific amount of profit in the business.

Spreadsheet-based risk simulations have become fairly popular in aquaculture economic analyses. This is likely due to the riskiness of aquaculture technologies and businesses and the ease of use of spreadsheet-based risk methodologies.

Risk simulations have been used to evaluate the economic feasibility of new species and production systems. For example, Medley et al. (1994) used a spreadsheet-based risk simulation software (@RISK; Palisade Corporation) to evaluate the economic feasibility of red claw (*Cherax quadricarinatus*) culture at three different densities. The risk analysis showed that the economic feasibility is sensitive to the cost of juveniles, the percentage of harvest biomass of the larger size classes, the price of the larger size classes of red claw crayfish, and the length of the production season.

Zucker and Anderson (1999) compared various alternative production and marketing scenarios with a dynamic, stochastic model of a land-based summer flounder (*Paralichthys dentatus*) farming business. The model was developed using Microsoft Excel® (Microsoft Corporation, Washington, USA) software. The simulation model developed included submodels for the production facility, for growth and biological parameters, marketing, and a financial summary. Data used were primarily from literature and expert opinion. Two selling strategies and six marketing strategies were modeled along with various alternative scenarios that included: (1) varying growth rates by 25%; (2) effects of a low interest rate of 2%; (3) advancement in system engineering that results in a 25% increase in density; and (4) change in policy that allows for possession of fish smaller than 0.34 kg. Model results showed that conditions for such a land-based flounder production system to be profitable included a location near a source of salt water, sales that targeted premium market outlets such as sushi markets, and development of adaptive market strategies.

Spreadsheet-based risk analyses have also been used to identify key sources of risk for established aquaculture species and technologies. Several studies used data from shrimp farms in risk analyses. For example, Valderrama and Engle (2001) developed an analysis of the financial risks of shrimp farming in Honduras with a stochastic simulation model using Crystal Ball™ software (Decisioneering, Denver, Colorado). Data used were obtained from a survey of shrimp farms in Honduras that included farms from 10 ha to over 400 ha as well as an alternative semi-intensive technology for raising shrimp. Correlations identified in the survey data for feed prices and farm sizes were used to link six other variables (total seed costs, feed quantity, total full-time labor, total diesel costs, debt payment, and infrastructure depreciation). The primary source of risk in this situation was not price fluctuation but fluctuating yields. Certainty levels of achieving profits on various farm sizes and production technologies were calculated.

Seijo (2004) used Crystal Ball© to develop a bioeconomic model to estimate the risk of exceeding specified limit reference points for shrimp aquaculture production. Particular emphasis was placed on assessing the alternative timing of harvesting decisions. Results showed that risk can be reduced by harvesting in 6–7 months. The model also showed which types of financial indicators are affected the most by harvesting in other periods.

Martinez and Seijo (2001) further developed the model to evaluate the economic effects of risk and uncertainty of various alternative water exchange and aeration rates in semi-intensive shrimp culture systems. Particular sources of risk included: seed price, shrimp growth rate, survival rate, and shrimp prices. Results showed that the traditional production technology was more profitable than reducing water exchange and increasing aeration levels. Nevertheless, the differences were small and, under the assumed risks, the probability of achieving a positive net present value was high.

Martinez and Seijo (2001) developed an Excel-based bioeconomic model with a Bayesian decision analysis approach. The model was used to choose the cycling strategy for shrimp farms in Mexico, either a 6–8 month production cycle or 2-, 3-, or 4-month production cycles with complete harvests. The two-cycle strategy was a better option for the southern areas while the one-cycle strategy was better for the northern areas.

Pomerleau and Engle (2003) used a spreadsheet-based risk simulation analysis to evaluate the effects of stochastic fluctuations of feed cost, feed conversion ratio (FCR), yield, survival, and fish length on breakeven costs of production data. Data used were from an experimental pond study. The lowest risk strategy was the medium stocking density. The risk analysis further showed that the probability of breakeven prices was 0–9% of being above the stocker market price of $2.20/kg.

Hishamunda et al. (1998) developed a spreadsheet-based risk analysis of small-scale fish farming in Rwanda. The add-in software program to Lotus 1-2-3, @Risk, was used to develop a simulation of fish farming by cooperatives and by individual farmers. Labor was the major determinant of economic success on small-scale farms in Rwanda. Overall, the probability of economic success was below 15% and net returns to land and management were negative. Moreover, fish production was riskier than production of beans and cabbages. Farms managed as cooperatives were likelier to fail than were farms that were managed by individual farmers.

RISK PROGRAMMING MODEL-BASED RESEARCH

Chapter 17 describes methods to develop whole-farm models with mathematical programming techniques, including methods of incorporating risk and stochastic elements into the analysis. This section will provide a few examples of applications of risk programming models in aquaculture.

Risk programming models have been used to evaluate new production technologies. For example, Hochmann et al. (1990) developed a stochastic dynamic decision model to evaluate the potential of a round pond technology for shrimp production. The model results provided an optimal stocking and harvesting schedule for a shrimp pond that resulted in doubling net profits from shrimp production.

Other analyses have used risk programming techniques to evaluate management alternatives for established aquaculture industries such as shrimp and catfish. For example, Valderrama and Engle (2002) developed a Target MOTAD risk programming model of shrimp farming in Honduras based on survey data from commercial shrimp farms. Overall risk levels measured were relatively low, primarily due to selection of intermediate and low stocking densities during the rainy season. However, dry season production was associated with a relatively high degree of risk. Alternate farm plans were identified that further reduced risk levels.

Engle and Hatch (1988) developed a Target MO-TAD risk programming model to compare alternative aeration strategies for catfish production. Data used included experimental data available from the literature. Greater amounts of electric aeration on catfish farms allowed farmers to increase stocking densities and feeding rates. However, these more intensive production levels entailed greater risk of losses from power outages or other events that result in losses. The study results identified the profit-maximizing choice of aeration strategy for different levels of risk tolerance of farmers. Continuous nightly aeration was selected most often, but greater levels of concern of financial risk resulted in selecting a less expensive type of electric aerator.

OTHER SIMULATIONS AND RISK ANALYSES

Arnason (1992) developed a dynamic optimization model to develop optimal feeding and harvesting schedules for a variety of different growth response relationships. The analysis demonstrated that the profit-maximizing feeding paths may take a variety of different shapes that depend on the shape of the growth function. The Arnason (1992) study was extended by

Heaps (1993) to demonstrate that, if fish growth is density-independent, then harvesting scheduling need not account for culling fish before.

Jin et al. (2005) developed a simulation model to assess the risk of open-ocean aquaculture. The model is based on an expected value-variance analysis of a firm-level investment–production model to simulate aquaculture growout, estimate the benefit–cost values of the project, and to calculate the risk premium for a risk-averse investor. The analysis included a case study of open-ocean aquaculture of Atlantic cod (*Gadus morhua*) in New England, USA. The analysis showed that the risk level and the risk-aversion parameter were inversely related to the investment level. Under conditions of uncertainty, the scale of aquaculture operations would be smaller than with certainty of conditions.

Tveteras (1999) evaluated risky production technologies on Norwegian salmon farms by examining the risk properties of inputs and patterns of technical change for both deterministic and risky parts of the technology. The analysis further examined how different specifications of farm-specific effects, different functional forms, and different estimators influenced the empirical results. Results showed that increasing feed and fish inputs increased risk while increasing labor inputs decreased risk. A risk preference function was developed for Norwegian salmon farming based on panel data to estimate a Cobb-Douglas production function. Absolute, relative, and downside risks for each producer were calculated from the parameters estimated from the risk preference function. Salmon farms were found to be risk averse and salmon farmers' risk preferences exhibited decreasing absolute risk aversion.

RECORD-KEEPING

Risk analysis requires data on the variability of key parameters. In aquaculture, these key parameters often include the price of the product, the price of feed, and the yield (lb/acre of production). Thus, it is important to maintain access to records of how the values of these parameters fluctuate over time.

Records of prices of some types of aquaculture products can be found in publicly available sources. For example, in the United States, prices paid to farmers of catfish and trout are available online from the U.S. Department of Agriculture. However, there are no long-term datasets of prices of other types of aquaculture species. Even when there is a long-term industry dataset for prices, it is important for the farmer/

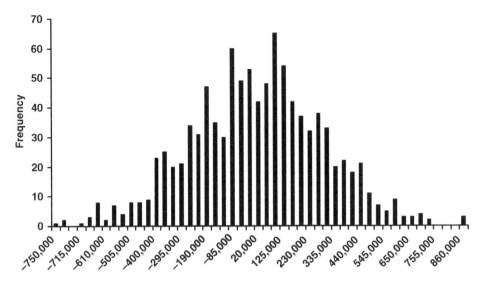

Figure 16.6. Frequency distribution of net returns for 256-acre catfish farm.

manager to maintain a record of the prices received for each shipment of fish.

Net prices are even more important to track. Deductions from the stated price will affect the overall profitability of the business. Tracking the reasons for the deductions may point to strategies to reduce these over time or to decide to switch to another processing plant.

Tracking yields from ponds will provide a basis from which to calculate the yield risk. If some ponds consistently produce lower yields than others over time, overall farm-wide yield risk can possibly be reduced by reworking those ponds or planting them in a different crop.

Similarly, recording total feed fed to each pond for each year will allow the manager to track FCRs, overall performance of various ponds, and how this performance varies from year to year.

PRACTICAL APPLICATION

The enterprise budget presented in Chapter 10 (Table 10.2) was used as the basis for a risk simulation analysis using the Microsoft Excel add-in program Crystal Ball (Decisioneering, Denver, Colorado). Table 10.2 shows net returns to operator's labor, management, and risk of $2,792 for the 256-acre catfish farm.

Stochastic distributions were entered into the spreadsheet for the following parameters: yield of catfish, price of catfish, feed prices, feeding rate, and

prices of fingerlings, labor, fuel (gas and diesel), electricity, repairs and maintenance, bird depredation, telephone expenses, office supplies, farm insurance, and legal/accounting expenses. Normal distributions were used for the yield of catfish, fish prices, feeding rate, and feed prices. Triangular distributions were used for prices of fingerlings, labor, and the other types of inputs. The simulations were run with 1,000 iterations.

Figure 16.6 shows the frequency of net returns to operator's labor, management, and risk that resulted from the simulation analysis.

Mean net returns were −$4,577 ± $8,683 for the 256-acre catfish farm. The profitability of making a profit above total costs was 51% while the probability of covering all variable costs was 71%. Table 16.2 shows the rank correlations of the contributions of the relative contributions of each stochastic parameter on overall risk. Yield clearly had the greatest effect on overall risk (0.89) and was followed by the price of catfish (0.27). The analysis demonstrates the care and attention that must be paid to achieving consistent yields on the farm, from both a management and research perspective.

OTHER APPLICATIONS IN AQUACULTURE

Zucker and Anderson (1999) used a dynamic stochastic model to develop recommendations for a newly

Table 16.2. Rank Correlation Coefficients for the Stochastic Parameters in the Risk Simulation Analysis of a 256-acre Catfish Farm.

Parameter	Rank correlation coefficient
Yield	0.89
Catfish price	0.27
Feed price	−0.21
Feeding rate	−0.19
Farm insurance	−0.07
Repairs and maintenance	0.06
Office supplies	0.04
Electricity	−0.03
Fingerling price	−0.03
Fuel (gas and diesel)	−0.03
Telephone	0.02
Salary	0.01
Bird depredation	0.01
Legal/accounting	−0.01
Second salary	−0.01

emerging aquaculture technology, that of land-based production of summer flounder. For a new startup aquaculture business, there are both production and marketing risks that can determine the success or failure of the new business. The model developed was used to evaluate a series of scenarios. The scenarios evaluated included: (1) the effects of improvements in the biological production that would result in 25% improvements in the growth rates at the lower end of the distribution and 5% improvement in the fastest growth rates expected; (2) effects of a low-interest loan rate of 2% on real estate and vehicle loans; (3) engineering improvements that would increase the production capacity of the system by 25%; and (4) change in regulations that would allow for fish smaller than 0.34 kg to be sold.

The least risky alternative, in terms of generating profits from the business, was to locate the business near a ready source of saltwater, sell to sushi chefs who prefer a medium-sized fish, and actively manage the business to adjust to market dynamics. The scenario that resulted in the greatest improvement was the policy option that would allow flounder farmers to sell a smaller fish, although all options considered showed some improvement.

SUMMARY

This chapter presents an overview of risk in aquaculture, discussing the various types of risk (production, marketing, and financial) and the difference between risk and uncertainty. Risk measurements based on calculating probabilities and probability density functions are presented. Literature on the use of spreadsheet-based risk simulations, stochastic dominance analysis, and risk programming model-based studies is summarized.

REVIEW QUESTIONS

1. What is risk and how is it different from uncertainty to economists?

2. List three major categories of risk and describe aquaculture examples of each.

3. What is commonly the relationship between risk and expected profits? Draw a graphical illustration of this relationship.

4. How is risk measured? Describe several ways to measure risk.

5. What is a cumulative distribution function and how is it interpreted?

6. Describe several financial indicators that can be used to interpret the level of risk for an aquaculture business. Use examples from aquaculture.

7. What is a spreadsheet-based risk simulation and what output does it provide?

8. Name and describe two software programs that can be used to develop a risk simulation with a spreadsheet program.

9. Describe some uses of risk programming mathematical programming models.

10. Describe several situations for which a risk analysis would be a more appropriate type of analysis than a deterministic analysis.

REFERENCES

Arnason, Ragnar. 1992. Optimal feeding schedules and harvesting time in aquaculture. *Marine Resource Economics* 7(1):15–35.

Dalton, Timothy J., Kate M. Waning, and Linda Kling. 2004. Risk efficient juvenile haddock production systems: an ex-ante recursive stochastic approach. *Aquaculture Economics and Management* 8(1/2): 41–59.

Engle, Carole R. and Upton Hatch. 1988. Economic assessment of alternative aquaculture aeration strategies. *Journal of the World Aquaculture Society* 19(3):85–96.

Gempesaw, Conrado M., Richard Bacon, and Ferdinand F. Wirth. 1992. Economies of pond size for hybrid striped bass growout. *Journal of the World Aquaculture Society* 23(1):38–48.

Heaps, Terry. 1993. The optimal feeding of farmed fish. *Marine Resource Economics* 8:89–99.

Hishamunda, Nathanael, Curtis M. Jolly, and Upton Hatch. 1998. Evaluating and managing risk in small-scale fish farming in a developing economy: an application to Rwanda. *Aquaculture Economics and Management* 2(1):31–38.

Hochman, P.S. Leung, L.R. Rowland, and J. Wyban. 1990. Optimal scheduling in shrimp mariculture: a stochastic growing inventory problem. *American Journal of Agricultural Economics* 72:382–393.

Jin, Di, Hauke Kite-Powell, and Porter Hoagland. 2005. Risk assessment in open-ocean aquaculture: a firm-level investment-production model. *Aquaculture Economics and Management* 9:369–387.

Kazmierczak, Richard F., and Patricia Soto. 2001. Stochastic economic variables and their effect on net returns to channel catfish production. *Aquaculture Economics and Management* 5(1/2):15–36.

Knight, Frank H. 1921. *Risk, Uncertainty, and Profit.* 1964 Edition. New York: Augustus M. Keley.

Levy, Haim. 2006. *Stochastic Dominance: Investment Decision Making Under Uncertainty.* New York: Springer Science Business Media.

Martinez, José A. and Juan Carlos Seijo. 2001. Economics of risk and uncertainty of alternative water exchange and aeration rates in semi-intensive shrimp culture systems. *Aquaculture Economics and Management* 5(3/4):129–145.

Medley, Paul B., Robert G. Nelson, Upton Hatch, David B. Rouse, and Gerard F. Pinto. 1994. Economic feasibility and risk analysis of Australian red claw crayfish *Cherax quadricarinatus* aquaculture in the southeastern United States. *Journal of the World Aquaculture Society* 25(1):135–146.

Pomerleau, Steeve and Carole R. Engle. 2003. Production of stocker-size channel catfish: effect of stocking density on production characteristics, costs, and economic risk. *North American Journal of Aquaculture* 65:112–119.

Seijo, Juan Carlos. 2004. Risk of exceeding bioeconomic limit reference points in shrimp aquaculture systems. *Aquaculture Economics and Management* 8(3/4):201–212.

Tveteras, Ragnar. 1999. Production risk and productivity growth: some findings for Norwegian salmon aquaculture. *Journal of Productivity Analysis* 12(2):161

Valderrama, Diego and Carole R. Engle. 2001. Risk analysis of shrimp farming in Honduras. *Aquaculture Economics and Management* 5(1/2):49–68.

Valderrama, Diego and Carole R. Engle. 2002. Economic optimization of shrimp farming in Honduras. *Journal of the World Aquaculture Society* 33(4): 398–409.

Zucker, David A. and James L. Anderson. 1999. A dynamic, stochastic model of a land-based summer flounder *Paralichthys dentatus* aquaculture firm. *Journal of the World Aquaculture Society* 30(2): 219–235.

17
Whole-Farm Modeling of Aquaculture

INTRODUCTION

Economic analysis of aquaculture frequently has focused on examining the economic consequences of specific changes that might be made on the farm based on a new experiment that has yielded what appear to be beneficial results. Enterprise budgets, partial budgets, and cost analyses have been commonly used. However, these types of analyses are limited in scope in that they examine only one enterprise (enterprise budget), one small change on the farm (partial budget), or costs associated with a specific production activity (cost analysis). In many cases, the analysis is developed on the basis of a single pond as the unit of analysis.

These types of analyses are useful as long as they are interpreted correctly. Incorporating some cost and return elements into basic production research does provide insights into the results and can suggest implications.

These types of analyses, however, are limited in that many variables from across the farm may affect whether or not the farmer can implement the new research results. The number of ponds, availability and skill level of the workforce, access to capital for operating or for additional investment, availability of adequate types and suitability of equipment, credit capacity of the farm business, and regulatory constraints may prevent new technologies or new research results from being adopted or from being feasible on farms. Enterprise budgets, partial budgets, and cost analyses do not allow these types of constraints to be considered in the analysis.

An enterprise is part of a greater farming system that involves sometimes complex interactions across the entire farm. Ignoring these interactions may lead to farm plans that are not feasible or are not realistic.

Whole-farm models incorporate the availability of resources, constraints, and their interactions across the entire farm. Biological, economic, and financial factors are integrated into one analysis. Whole-farm models can be used to compare the economics of adopting new technologies on a farm, to identify the optimal set of production practices for that particular farm, or to identify which types of resources are the most constraining to that particular farm business.

There are two basic broad categories of whole-farm modeling: whole-farm budgeting and mathematical programming models. Whole-farm budgeting is similar in some ways to developing an enterprise budget, with the difference that the costs and revenues are developed for the entire farm. Mathematical programming models are more flexible than a whole-farm budget and provide a comprehensive model that results in a more complex set of results that demonstrate how to optimize the use of the farm's resources.

Development of a whole-farm budget relies primarily on farm records of costs and returns. If a farmer maintains appropriate records and develops balance sheets and income statements each year, there is little difficulty in converting these numbers into a whole-farm budget. If the farmer wishes to consider changes for the next year, the effects of those changes can be estimated for an "end-of-year" whole-farm budget that shows the effect on farm profits. The primary disadvantage of whole-farm budgeting is that budgets are cumbersome mechanisms for examining a variety of levels associated with different technologies and changes. For example, a farmer looking at changing stocking densities on his/her farm would have to choose exactly which densities to use in which ponds prior to developing the whole-farm budget. If he/she then decides that additional densities need to be analyzed, an entirely new farm budget will need

Aquaculture Economics and Financing: Management and Analysis, Carole R. Engle, © 2010 Carole R. Engle.

to be developed. The degree of labor and financial constraints to the farm business must be set prior to developing the budget and new budgets must be developed for each different level of labor or financial constraint.

Mathematical programming models, on the other hand, allow for inclusion of a variety of levels of different changes being considered for the farm. If stocking density is the major change being considered, the model can include 5, 10, or more different densities. Moreover, the model can be developed to include different stocking densities of different sizes, species, or strains of fish. Different levels of availability of operating and investment capital, terms of lending, varying availability of degrees of skill of labor along with varying prices of inputs can be modeled in mathematical programming models. However, development of mathematical models requires a great deal of data. Each option included in the model requires complete data. If few data are available on the various alternatives and levels desired for the analysis, a whole-farm budget may be the only option available. It is better to develop a simpler model from reliable records, then to develop a more sophisticated model with too few data. Training and skill are required to develop useful mathematical programming models.

This chapter discusses both whole-farm budget procedures and mathematical programming model analyses in aquaculture. The various types of analyses are compared and contrasted.

WHOLE-FARM BUDGETING

A whole-farm plan outlines the production activities for the farm and identifies the resources needed to produce this set of activities. The whole-farm budget compiles the resource requirements and expected costs and returns for the combination of amounts of commodities selected as the best combination for the farm business.

The bottom line of the whole-farm budget is the projected net farm income. If various scenarios are considered, the net farm income can be compared to select the specific plan that results in the greatest net farm income.

The structure of a whole-farm budget is similar to that of an income statement. The various sources of revenue are itemized by type of crop to be produced. Variable and fixed costs for the entire farm are estimated. Total costs are subtracted from total revenue to calculate net farm income.

MECHANICS OF CONSTRUCTING A WHOLE-FARM BUDGET

Development of a whole-farm budget begins with an inventory of resources. These include the land available, buildings, wells, ponds, and other production facilities, a marketing or grading shed, equipment available, access to both investment and operating capital, labor, and management. It is important to specify details on the type, quality, and quantity of each type of resource. Pond sizes, number, volumes, and sizes of vats and tanks, all need to be specified. The ability to borrow additional capital must be assessed realistically. The lack of credit capacity can limit the adoption of new technologies and limit the ability of the farmer to make needed upgrades to more efficient equipment. Liquidity problems can arise if input prices increase rapidly and unexpectedly and the farmer has no capacity to acquire additional operating capital.

Labor resources must be assessed as well. The level of skill required for the types of enterprises selected for the farm business must be specified. Management skill must be assessed as well. Age, experience, management performance, special skills, and any weaknesses can be important factors.

When resources available to the farmer have been identified, the enterprises to be included on the farm must be selected. Decisions should include the species to be raised and whether to raise foodfish crops or intermediate crops of fingerlings or stockers. Enterprise budgets must be developed for each enterprise to be included on the farm. The enterprise budget specifies the resources required, costs and revenue, and gross and net margins for each unit of production.

The next step is to determine how much of each enterprise to raise on the farm. The key is to identify the best use of the resources available and which combination of enterprises will result in the highest profits given the set of resources available to the farm.

The whole-farm budget then estimates the expected revenue and expenses. Cash inflow, outflow, and liquidity should be evaluated along with the effects of various combinations of enterprises on profits and liquidity. Other options for changing the farm plan, such as adopting a new technology or expansion or contraction of the farm business should be considered. Finally, the exact amount of resources required for the optimal farm plan should be detailed.

Whole-farm budgets can be used to project outcomes for proposed changes on the farm. Net returns and liquidity can then be compared across the various combinations considered. Net cash flow can be projected for each combination considered by adding expected farm and off-farm revenue for each plan under consideration. Cash farm income from nonfarm work and investments are then added to total cash inflows. Cash farm expenses and cash outlays to replace capital assets and debt payments all constitute cash outflows. Total farm cash outflows are subtracted from cash inflows to obtain net cash flows for the budget for each whole-farm plan developed.

MATHEMATICAL PROGRAMMING

Mathematical programming is a computational method to allocate limited resources to maximize profit or minimize costs. It is a mathematical technique that uses a systematic procedure to find the best possible enterprise combination. It allocates resources to their best possible uses to achieve a given objective, given the limitations and restrictions on the use of each type of resource. It can be used at the farm, industry, or regional level. The word "programming" refers to planning and not developing computer code. However, computers are used to solve large mathematical programming models to find the "optimal" (best or most profitable) plan.

There are many uses of mathematical programming models. They can be used to (1) identify profit-maximizing combinations of scarce resources (labor, land, feed, fry) for production of a given species; (2) allocate land to particular crops; (3) identify the best allocation of scarce resources for production of different species; (4) develop plans for the most efficient use of labor and machinery; (5) identify the least-cost ingredients of a diet to meet specified nutritional requirements and formulate least-cost rations for livestock; (6) identify least-cost geographic sources from which fry and raw materials should be purchased; or (7) identify the best allocation of output among geographic markets.

FUNDAMENTAL CONCEPTS UNDERLYING MATHEMATICAL PROGRAMMING MODELS

Linear programming (LP) models essentially trace a production possibilities curve (for more details on the economic concept of a production possibilities curve, see (Baumol and Blinder 1999)) with linear segments. A production possibilities curve shows a series of points where combinations of production activities are feasible, given the quantities of resources available and the quantities required for each type of crop.

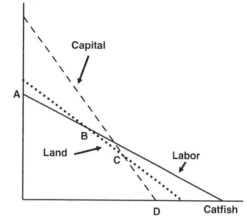

Figure 17.1. Graphical illustration of the concept of linear programming.

Figure 17.1 illustrates the fundamental concept that underlies LP with a simplified two-dimensional graph. Only two decision variables can be represented in such two-dimensional graphs. Hybrid striped bass and catfish production quantities are graphed on the axes. The lines represent the linear equations of requirements for the resource constraints of land, labor, and capital. The points A, B, C, and D define the area called the feasible region. Any combination of tilapia and catfish activities on or within this area is feasible given the resource restrictions or constraints. The optimal solution is found by identifying the line of equal gross margin that just touches one of the vertex points (A, B, C, or D).

The optimal plan that is generated from a LP model consists of the number of acres of ponds or the head of fish stocked needed to maximize profits. This combination of enterprises then needs to be used to develop the financial and economic statements that correspond to that optimal plan.

STRUCTURE OF BASIC LP MODEL

For whole-farm planning, a mathematical programming model typically maximizes an objective function subject to specified constraints. LP models (in which the equations in the model are linear) are most commonly used because they do not require as much data as some of the other types of models.

The LP tableau is a table that organizes the information for a LP model. Table 17.1 demonstrates

Table 17.1. Simple Model of Fish Farm in Thailand (in Thai Baht).

	Extensive polyculture	More intensive polyculture	Intensive tilapia monoculture	Catfish culture	Sell tilapia — Locally	Sell tilapia — Village	Sell tilapia — District	Sell Catfish	Constraint type	RHS
Objective function	−2,395	−2,531	−2,179,032	−52,402	30	40	30	27	MAX	
Pond balance	1	1	1	1					LE	4
Tilapia fry	0	0	25,000	0					LE	100,000
Catfish fingerlings	0	0	0	132,400					LE	150,000
Urea balance	0	1,332	1,719	859					LE	100,000
TSP balance	0	872	1,125	562					LE	100,000
Lime	0	766	988	494					LE	100,000
Animal manure	960	1,920	12,413	8,288					LE	100,000
Feed for fry	0	297.6	297.6	0					LE	100,000
Feed for growout	25.4	25.4	1,476.5	211,061					LE	500,000
Labor for growout	19	42.75	50	294					LE	100,000
Labor for harvesting	4	4	12	82,794					LE	100,000
Pond maintenance	550	2,550	2,550	2,550					LE	100,000
Marketing (Bt/kg) local	0	469	0	0					LE	100,000
Marketing tilapia farther	0	0	6,022	0					LE	100,000
Marketing catfish	0	0	0	82,794					LE	100,000
Operating capital	1,605	3,210	6,420	6,420					LE	2,500,000
Annual investment capital									LE	5,000,000
Tilapia production for local sales	−437.5	0	0	0	1				LE	0
Tilapia production for village sales	0	−1,875	0	0		1			LE	0
Tilapia production for district sales	0	0	−60,222	−6,022			1		LE	0
Catfish production	0	0	0	0				1	LE	0

LE, less than or equal to; RHS, right-hand side value.

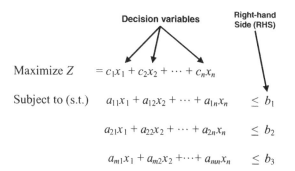

Figure 17.2. Mathematical form of a linear programming model.

a tableau for an LP model for an aquaculture business in Thailand.

LP models include a linear objective function that describes mathematically the objective of the farm plan to be developed and the decision variables or activities to be optimized (Figure 17.2). This equation consists of the sum of revenue-generating opportunities with the costs subtracted from the revenue generated. The objective function for farm plans usually maximizes net returns on the farm.

Decision variables, the variables whose values will be calculated by the model, are selected and defined. Each decision variable represents a particular farm enterprise or a particular level of production of a farm enterprise. An example could be the number of ponds to allocate to hybrid striped bass production in either monoculture or a specific polyculture combination. Each different production level for each enterprise is specified as a separate decision variable. These levels may represent various stocking densities, various levels of aeration usage, different sizes of ponds, or some other production or management system.

The resources that limit how much and what can be produced are modeled as constraints in the model. Constraints represent the limitations or restrictions on the number of units of an activity or variable that can be put into production, given the level of resources available. For example, farms have only a certain number of ponds available and the farm plan cannot allocate activities to more ponds than are available (unless the model allows for the possibility of an activity to build additional ponds). Typical constraints include land, labor, fingerlings, and capital. Credit capacity used in the plan cannot exceed the farm's credit or exceed the maximum on the line of credit. The combination

of activities selected cannot require more labor than the size of the workforce unless additional activities are added to allow additional labor to be hired. Constraints typically include the resources available, but can also include market constraints. For example, if the market for hybrid striped bass for the farm is only 100,000 lb/year, then it is important to specify in the model that no more than 100,000 lb of hybrid striped bass can be sold in a year.

For the model to be a LP model, the objective function and the constraints must be expressed as linear equations. These must also be deterministic and not contain random elements. The decision variables are continuous and non-negative (they must assume positive values).

There are certain assumptions that underlie LP. These include separability, proportionality, divisibility, and non-negativity. Separability implies that there is no interaction among the decision variables and allows the objective function to be written as the sum of several functions. For example, a model that includes purchases of hatchery-raised and wild postlarvae of shrimp is built upon the assumption that greater purchases of wild postlarvae do not reduce purchases of hatchery-raised shrimp. There are special techniques that can be used to relax this assumption.

The proportionality assumption implies that there are no economies of scale. If proportionality exists, the value of each constraint function is directly proportional to the value of each decision variable. For example, as larger quantities of postlarvae are produced, the unit price remains the same if proportionality exists. Thus, it would not be correct to assume that unit prices fall if larger quantities of shrimp postlarvae are produced. Special techniques have been developed to introduce economies of scale in LP models.

Divisibility allows for resources and decision variables to be used in fractional quantities. Techniques (integer programming) have been developed for decision variables that require integer values. For example, a farmer cannot stock 0.3 ponds; when he or she stocks fish into a pond, the entire pond is stocked.

Non-negativity assumptions require that all decision variables are either zero or positive. Decision variables and resource constraints are finite in LP models. Moreover, single-valued expectations are used because all parameters are known with certainty.

Data requirements for a mathematical programming model begin with an inventory of the resources available. Potential cropping alternatives must be identified. Then, data will be required on the technical coefficients of production. These coefficients are developed from the resource requirements for each unit of enterprise included in the model. Net returns are estimated for each cropping alternative or activity. Because the equations in a mathematical programming model are inequalities, the maximum quantity of each that is available to the farmer must be specified and recorded as the right-hand side value (RHS).

OBJECTIVE FUNCTION

The objective of a mathematical programming model is to either maximize or minimize something. In a farm planning model, the objective typically is to maximize net returns. The gross margin of each unit is multiplied by the number of units selected to include in the farm plan. These are summed to calculate the total net returns. Each enterprise (shrimp, tilapia fingerlings, tilapia foodfish, etc.) must be included as a term in the equation for the objective function.

Table 17.1 presents the tableau for a LP model designed to develop an optimal farm plan for a fish farm in Thailand. The farmer wishes to maximize net returns, thus the objective function is to maximize net returns. This objective function can be expressed as follows:

$$\text{Maximize } Z = 0X_1 + 0X_2 + 0X_3 + 0X_4 + 30X_5$$
$$+ 40X_6 + 30X_7 + 27X_8 \qquad (17.1)$$

ACTIVITIES AND DECISION VARIABLES

In Table 17.1, the fish farmer is considering raising either tilapia or catfish. Tilapia enterprises considered are either extensive or intensive polyculture or intensive monoculture. Moreover, tilapia can be sold locally at lower prices, in villages at slightly higher prices, or at district markets at even higher prices. Thus, the decision variables are to produce tilapia in either extensive or somewhat more intensive polyculture, in intensive monoculture, to raise catfish intensively, to sell catfish, or to sell tilapia in either local, village, or district markets. Mathematically, the decision variables can be expressed as:

$X_1 = $ tilapia production in extensive polyculture in a 1-ha pond
$X_2 = $ tilapia production in more intensive polyculture in a 1-ha pond
$X_3 = $ high-intensity tilapia monoculture in a 1-ha pond

$X_4 = $ catfish production in monoculture in a 1-ha pond
$X_5 = $ tilapia sales in local markets, in kg
$X_6 = $ tilapia sales in village markets, in kg
$X_7 = $ tilapia sales in district markets, in kg
$X_8 = $ catfish sales, in kg

CONSTRAINTS

There is always a limit to the quantity of resources available to the farmer. The value of this limit restricts what the farmer can and cannot do and different levels of this value can alter the profit-maximizing solution. The resource constraints in a LP model are developed as technical coefficients that equate to the amount of the particular resource needed to produce one unit of that enterprise. The equation then consists of the technical coefficients of the resources required for production of that enterprise multiplied by the number of ponds selected to be used for production of that enterprise. The sum of the quantities of resources used for each enterprise must be less than or equal to the total amount of that resource available. The total number available is specified as the RHS value for each resource and constitutes the constraint or limit on its use. Small programs can be solved by hand, but programs with large numbers of enterprises or with many restrictions require a computer to solve.

The first row of the tableau is the objective function and includes the gross margin coefficients for each activity. The following rows represent the resource constraints and balance rows. Each column represents a decision variable and the values in each row show how much of each resource is needed for each unit of the activity. Other columns represent sales activities and the RHS shows how much of that resource is available.

In the example provided in Table 17.1, the constraints include the availability of ponds, tilapia fry and catfish fingerlings, fertilizers (urea, triple superphosphate or TSP, lime, and animal manure), feed (for fry and for growout), labor (for growout and for harvesting), pond maintenance, marketing (tilapia locally or farther and catfish), capital (operating and investment, and fish production transfer rows). In this example, each production activity (extensive tilapia polyculture, more intensive tilapia polyculture, high-intensity tilapia production, and high-density catfish production) would each require at least one pond and would be added into the solution in units of a pond. For example, the first row in Table 17.1 is the pond constraint.

Each type of production system is modeled as a unit of one pond and all technical coefficient values are calculated on a per-pond basis. The numeral 1 for each production activity in the pond constraint row indicates that if selected in the optimal solution, then the corresponding amounts of other resources will be entered proportionally. The RHS value of 4 represents the four ponds available on the farm. This constraint would be expressed mathematically as:

$$1X_1 + 1X_2 + 1X_3 + 1X_4 + 0X_5 + 0X_6 \\ + 0X_7 + 0X_8 \leq 4 \qquad (17.2)$$

The next row is the tilapia fry resource. The intensive tilapia monoculture activity stocks 25,000 fry per pond and up to 100,000 fry (RHS value) are available for the entire farm. Similarly, the catfish monoculture activity requires 132,400 catfish fingerlings per pond. The next rows are the fertilizer resources of urea, TSP, lime, and animal manure. The number in each column represents the amount of fertilizer or lime needed for each pond in each alternative. The next row is the amount of feed required and then labor, and maintenance costs. As another example of how the constraints are expressed mathematically, the constraint of labor required for growout would be expressed mathematically as:

$$19X_1 + 42.75X_2 + 294X_3 + 50X_4 + 0X_5 + 0X_6 \\ + 0X_7 + 0X_8 \leq 100,000 \qquad (17.3)$$

Marketing constraints follow that indicate how many pounds of product can be sold in each type of market outlet. Operating capital requirements for each alternative and annual investment capital are next. The following rows (tilapia production for sale, tilapia production for village sales, and tilapia production for district sales) are transfer rows to move production of fish from production to sales.

The fish production transfer rows link the production activities to the sales activities. For example, the tilapia production for village sales constraint would be expressed mathematically as:

$$0X_1 - 1,875X_2 + 0X_3 + 0X_4 + 0X_5 + 1X_6 \\ + 0X_7 + 0X_8 \leq 0$$

Redundancy in constraints can increase the computing time. Redundancy occurs when a decision variable can satisfy more than one constraint. In computer models, redundancy does not have to be eliminated from the problem.

RIGHT-HAND SIDE

The constant terms that appear on the RHS of the constraint equations are referred to as the right-hand side values or RHS values. These values represent the amount of the resource available on the farm. RHS values must clearly be non-negative also.

SOLVING AN LP MODEL

In order to solve the inequality constraints in the model, the equations must first be converted to equalities. Thus, the Equation 17.1 becomes

$$1X_1 + 1X_2 + 1X_3 + 1X_4 + 0X_5 + 0X_6 \\ + 0X_7 + 0X_8 + s_1 = 4$$

The s_1 variables are referred to as slack variables. For less-than-or-equal-to inequalities, the slack variables appear with positive signs and are called positive slacks, or surplus variables. Negative slack variables are associated with greater-than-or-equal-to inequalities.

Originally developed for the solving of multisectoral economy-wide models, the general algebraic modeling system (GAMS) (Brooke et al. 1998) was adopted as a standard tool by economists who work in applied optimization. GAMS is a high-level language that uses concise algebraic statements easily read by both modelers and computers, are easily modified, and easily moved from one computer environment to another. It contains a series of solvers that use various algorithmic methods to approach a wide array of problems from linear and nonlinear, mixed integer, mixed integer nonlinear, and other types of problems.

Small mathematical programming models can be solved with Microsoft Excel. The practical application in this chapter presents how to set up and run small models in Excel.

An unbounded solution is one that has an infinite boundary, or no upper limit. Without an upper limit, the problem cannot be solved; the computer will continue to increase the number of units to be produced in an infinite manner. An unbounded solution indicates that the model was not developed properly. Models that have severe resource restrictions such that positive values cannot be found for the decision variables are referred to as infeasible.

INTERPRETING THE OUTPUT

The printed output includes the value of the objective function. Information provided on the rows includes status indicators of BS (row not a limiting factor), LL (solution lies on lower limit of constraint), or UL (solution lies on upper limit of the constraint). The values indicating the quantities selected for the solution and the amount of each resource used as well as slack values are indicated in the printed solution.

Dual values, including shadow prices and reduced costs, are also reported in the output. Shadow prices are the amount by which total gross margin would be increased if one more unit of that resource were available. For enterprises, the shadow price shows how much the gross margin would be reduced by forcing one unit of an enterprise into the plan. If resources are not used completely, the shadow price is zero. However, if the resource is used up completely, then the shadow price indicates whether additional units of that resource should be acquired. The decision maker should compare the cost of acquiring another unit of resource with the shadow price. If the shadow price is higher, its value is greater than its cost and it is worthwhile to acquire more. However, if the shadow price is less than its cost, then it is not profitable to acquire more.

The reduced costs indicate what the effect on gross margins would be by using another unit of an enterprise that was not included in the plan. Only enterprises that were not included in the optimal plan will have a reduced cost other than zero. The output also includes the input costs entered for each variable that entered the objective. The reduced cost indicates the amount by which the input cost would need to change for the optimal solution to be positive for that variable (reduced cost is zero for values selected for the optimal solution).

SENSITIVITY ANALYSIS, THRESHOLDS, BASIS CHANGES

The optimal solution, the combination of activities that maximizes profits, is known as the basis. The basis consists of the set of variables that enter into the optimal solution. Changes in the RHS values or changes in the price, cost, or other technical coefficients can result in changes in the basis. Sensitivity analyses that systematically vary key parameter values will demonstrate what price or cost levels will result in a change in the basis. Thus, the range of values over which the basis remains stable is identified. These types of sensitivity analyses are often referred to as parametric programming.

OTHER TYPES OF MATHEMATICAL PROGRAMMING MODELS

LP models have been the most commonly used types of mathematical programming models. However, there are a number of other types of mathematical programming models. Examples include integer, quadratic, goal programming, multiperiod, transshipment, and risk programming models (E-V, MOTAD, Target MOTAD), among others.

Integer programming refers to restricting certain parameter values to integer values as opposed to the continuous values that are a requirement of LP models. A goal programming model specifies an objective function as a target or goal instead of either maximizing or minimizing one specific objective. Thus, in goal programming, the objective is to minimize deviations from a target.

When parameters are not known with certainty, the assumption of single-valued expectations is relaxed and risk programming models are developed. Risk modeling in economics is based on the assumption that the probability distributions of the risky parameters are known with certainty. The risk model, then, represents the decision-maker's response to parameter risk (McCarl and Spreen 1997). The optimal plan developed may not necessarily be the best for each possible type of event, but is intended to establish a position that is robust across the entire set or distribution of parameter values.

One of the first types of risk programming models, the mean-variance, or E-V model was first developed by Markowitz (1959) to identify an optimal investment strategy for a portfolio. The variables indicate the amount of funds invested in each alternative investment option. Each is constrained by a total funds constraint, or the total amount of investment capital available. One assumption behind this approach is that investors do not place all their funds in the highest yield option; thus, a LP model would not be appropriate because linear models would put all the funds in the option that generates the highest yields.

In an E-V model, a risk aversion coefficient is multiplied by the variance of total income and subtracted from the expected income. The E-V model is based on the assumption that decision-makers trade expected income for lower levels of risk (reduced variance). The model can be solved for varying levels of risk aversion parameters. Each new solution reduces the variance of total returns, lowering risk, but also reducing net returns.

The E-V model is a quadratic programming problem that is more difficult to solve than a linear program. Software programs are available to solve it, but data requirements can still be difficult.

The difficulties involved in developing quadratic programming models led to the development of linear approximations to the E-V problem. The minimization of total absolute deviations, or MOTAD, model defines various states of nature related to the deviation from mean income. The model then characterizes the trade-offs between the expected income and the absolute deviation of income.

Roy (1952) introduced an approach referred to as a safety-first model. This approach assumes that decision markers will first select plans to assure a given safety level for income, typically set as a level of income that will cover production costs. Low levels of safety (high risk) correspond to the profit-maximizing LP solution. However, as the safety level increases, risk decreases, but the expected returns also decrease.

Tauer (1983) extended the safety-first model by allowing negative deviations from the safety level. A target income level is set and variables are included that indicate negative deviations from income. The maximum average income shortfall is set at the safety level.

In a Target MOTAD model, allowing for large deviations results in a profit-maximizing LP solution. However, as the allowable deviation is reduced, solutions are diversified. Thus, risk reductions are accompanied by larger deviations.

Sequences of decisions can be modeled as optimization problems with dynamic programming. Dynamic programming methods were developed by Bellman (1957) and Bellman and Dreyfus (1962). Dynamic programming models trace an optimal path from an initial state to another state. In dynamic programming, states are defined that indicate a particular point or configuration. Stages are the transitions from one state to another. Policies are the sets of actions that equate to a particular objective.

To solve dynamic programming models, recurrence relations are constructed that relate the various states. The recurrence relations are then used to calculate an optimal policy.

RECORD-KEEPING

Many of the types of records needed to develop a whole-farm model are those required also for enterprise budgets. A whole-farm model begins with a complete inventory of resources available to the farm, from land to types of ponds and facilities.

Information on the various alternative types of production options then must be collected. Keeping this information on hand over time provides a basis for future consideration. Data over time of prices of the products under consideration is used to project price trends and to identify any correlations that can be taken advantage of to reduce risk by diversifying production.

PRACTICAL APPLICATION

Table 17.2 illustrates the tableau for the 256-acre catfish farm whose owner is considering adding a hybrid striped bass enterprise to the farm. This example problem will be solved using the Excel solver program and Figure 17.2 illustrates how to lay out the tableau to be solved with the Excel solver.

The objective function is to maximize profits for the farm. The decision variables included production of catfish for growout or fingerlings or growout of hybrid striped bass for foodfish. Other decision variables included the possibility to purchase fingerlings or to grow fingerlings on farm. The constraints included a pond constraint and the necessary transfer rows to transfer fingerlings from either the fingerling purchase activity or the on-farm fingerling production activity to the growout activity, and to transfer hybrid striped bass produced from the hybrid striped bass production activity to the hybrid striped bass sales activity.

The first line in the Excel solver version of the tableau contains 0s. The first line is the one that Excel will change to find the solution that will indicate the number of ponds (production activities) or pounds of fish to be sold (sales activities) assigned to each decision variable in the optimal solution. The next row is the pond constraint. Each pond production variable (catfish foodfish growout, catfish fingerling production, or hybrid striped bass growout) is based on a unit of one 10-acre pond. Thus, a "1" is entered in the pond constraint row for each of these activities. The fingerling transfer row shows the number of catfish fingerlings needed for one 10-acre catfish pond (56,900). In the column of the on-farm fingerling production activity, the number of fingerlings produced in one 10-acre pond is recorded (−93,665). The possibility to purchase fingerlings is recorded as −1. The foodfish transfer row shows the number of pounds of catfish foodfish produced in one 10-acre pond (−45,000). The foodfish are transferred to the catfish sales activity with the one in the sell catfish column. The hybrid striped bass foodfish transfer row transfers

Table 17.2. Tableau for 256-acre Catfish Farm.

	Grow catfish from 5" fingerlings	On-farm fingerling production	Buy fingerlings	Grow HSB	Sell catfish	Sell HSB			RHS
Number of ponds to use	0	0	0	0	0	0			
Ponds	1	1	0	1	0	0	0	LE	25
Fingerling transfer	56,900	−93,665	−1	0	0	0	0	LE	0
Foodfish transfer	−45,000	0	0	0	1	0	0	LE	0
HSB foodfish transfer	0	0	0	−45,000	0	1	0	LE	0
Marketing constraints	0	0	0	0	0	1	0	LE	100,000
Objective function	−22,943	−16,192	−0.07	−46,307	0.7	1.50	0		

RHS, right-hand side value; HSB, hybrid striped bass.

the −45,000 lb of hybrid striped bass foodfish from the hybrid striped bass production to sell hybrid striped bass sales (1). Finally, the marketing constraint for hybrid striped bass limits the amount of hybrid striped bass foodfish to the size of the market for the product, with a "1" in that row.

The RHS values indicate the amount of each resource available. For the pond constraint, there are 25 10-acre ponds on the farm. The marketing constraint for hybrid striped bass is 100,000 lb. The objective function indicates both costs associated with each decision variable and the revenue to be obtained from each activity. The production costs for catfish foodfish production are approximately −$22,943 for each 10-acre pond. One 10-acre pond in catfish fingerling production costs −$16,192 per pond. Each fingerling purchased costs −$0.07 while one 10-acre pond of hybrid striped bass foodfish costs −$46,307 per pond. Each pound of catfish sold generates $0.70 of revenue while hybrid striped bass bring $1.50 per pound sold.

For the solver to function, a separate column is needed next to the RHS values. In this column, the "sumproduct" function is used to multiply the row of cells that will be the values to change to identify the optimal amount by each individual row of constraints. The sumproduct function must also be used to multiply the row of changing cell values by the values in the objective function.

After setting up the tableau, the solver can be used in Excel to find the profit-maximizing combination of activities. The solver is available as an add-in function under the "Data" tab in Excel 2007 and under

"Tools" on the Excel menu on older versions of Excel. Clicking on the solver brings up a dialogue box with several smaller boxes to fill in. The first box to fill in is the target cell. This is the cell that will record the net returns value for the farm that results in the optimal solution that shows the combination of activities that results from solving the LP model. This is the cell with the sumproduct function associated with the objective function. The next box asks for the cells that will have changing values and that will identify the number of each activity to be included. The next step is to identify the constraints. The cell with the sumproduct formula for each constraint row is indicated as the constraint. The symbol for less-than-or-equal-to or greater-than-or-equal-to or equal to is selected as appropriate and then the cell that contains the RHS value is indicated. Before running the model, the options button should be selected. In the options dialogue box, the "linear model" and "non-negative" boxes should be checked. After clicking "OK" to return to the previous dialogue box, it is time to click "solve." Excel will then indicate whether a feasible solution was found. If found, various report options can be selected.

Table 17.3 shows the optimal solution for the catfish-hybrid striped bass model. In the base solution, 23 ponds were put into catfish foodfish production and two ponds were put into hybrid striped bass growout production. Fingerlings were purchased and the farm sold 1,025,000 lb of foodfish catfish and 100,000 lb of hybrid striped bass, for a total net returns of $144,801.

The sensitivity report option shows a reduced cost of −$13,457 for on-farm fingerling production. This

Table 17.3. Base Solution for 256-acre Catfish Farm.

	Catfish production from 5" fingerlings	On-farm fingerling production	Buy fingerlings	Grow HSB	Sell catfish	Sell HSB		RHS
Number of ponds to use	23	0	1,296,056	2	1,025,000	100,000		
Ponds	1	1	0	1	0	0	LE	25
Fingerling transfer	56,900	−93,665	−1	0	0	0	LE	0
Foodfish transfer	−45,000	0	0	0	1	0	LE	0
HSB foodfish transfer	0	0	0	−45,000	0	1	LE	0
HSB marketing constraints	0	0	0	0	0	1	LE	100,000
Objective function	−22,943	−16,192	−0.07	−46,307	0.7	1.50	LE	151,281

RHS, right-hand side value; HSB, hybrid striped bass.

indicates that, if fingerlings had been produced on a pond on the farm that gross margins would have been reduced by $13,457. Shadow prices were $4,290 for ponds and $1.12 for hybrid striped bass sales. This means that gross margins would have increased by $4,290 if another pond had been available, $1.12 if an additional pound of hybrid striped bass could have been sold.

OTHER APPLICATIONS IN AQUACULTURE

Engle and Valderrama (2004) used an economic optimization model to evaluate several different practices that had been recommended as best management practices for shrimp farming in Central America. The model was used to identify several options that increased farm profits by improving production efficiencies while also reducing net quantities of nutrients discharged as effluents. The most effective practices were to reduce water exchange rates and the use of feed trays. These are technologies that can be combined and practiced by a wide array of farm sizes.

SUMMARY

This chapter presents methods to develop whole-farm planning models. Whole-farm modeling reveals potential conflicts in the use of resources on the farm and identifies a pattern of resource use that will maximize farm profits. Both whole-farm budgeting and mathematical programming can be used. Whole-farm budgeting is simpler to use, but each option to be considered requires a separate whole-farm budget. Mathematical programming, on the other hand, is a very flexible planning tool, but requires a great deal of skill to develop. Mathematical programming models include linear, goal, risk, and dynamic programming options.

REVIEW QUESTIONS

1. Why are whole-farm models of value?

2. Explain the relative advantages and disadvantages of whole-farm budgets and mathematical programming models.

3. Describe the process of developing a whole-farm budget.

4. List the major types of uses of mathematical programming models.

5. What are the specific features that make a mathematical programming model be a LP model?

6. Give an aquaculture example of a LP model that is unbounded.

7. Describe the major components of a LP model (objective function, decision variables, constraints, and RHS values).

8. Explain what information is provided by shadow prices and reduced costs in the output of a LP model.

9. What are goal and risk programming? Give several examples of how LP models can be converted into goal or risk programming models.

10. Describe what dynamic programming can be used for and how a dynamic programming model is developed.

REFERENCES

Baumol, William J. and Alan S. Blinder. 1999. *Microeconomics: Principles and Policy*. 7th ed. Mason, OH: Southwestern Publishing Company.

Bellman, Richard E. 1957. *Dynamic Programming*. Princeton: Princeton University Press.

Bellman, Richard E. and Stuart E. Dreyfus. 1962. *Applied Dynamic Programming*. Princeton: Princeton University Press.

Brooke, Anthony, David Kendrick, Alexander Mieraus, and Richard E. Rosenthal. 1998. *GAMS User Guide*. Release 2.50. San Francisco: The Scientific Press.

Engle, Carole and Diego Valderrama. 2004. Economic effects of implementing selected components of best management practices (BMPs) for semi-intensive shrimp farms in Honduras. *Aquaculture Economics and Management* 8:157–177.

Markowitz, Harry M. 1959. *Portfolio Selection*. New York: John Wiley and Sons.

McCarl, Bruce and Thomas Spreen. 1997. Applied mathematical programming using algebraic systems. Available at agrinet.tamu.edu.

Roy, A.D. 1952. Safety-first and the holding of assets. *Econometrica* 20:431–449.

Tauer, Loren W. 1983. Target MOTAD. *American Journal of Agricultural Economics* 65(3):606–610.

18
Managing Government Policies and Regulation in Aquaculture Businesses

INTRODUCTION

Aquaculture worldwide is developing at a rapid pace during a time of increased regulatory pressure. The increased focus and attention on aquaculture comes from a number of sources, but much of it has stemmed from the pressure from environmentalist and public advocacy groups. Many of these groups have targeted aquaculture. They see aquaculture emerging as an industry across the world and hope to influence its development and evolution because they expect aquaculture to play an ever-increasing role in the world's future food supply. These groups have hired lawyers and other personnel who mount well-orchestrated initiatives to pressure regulators to develop new policies.

Increasing numbers of countries have developed policies, plans, and regulations related to aquaculture. The fisheries and aquaculture sectors have been increasingly integrated into national policy documents. New regulations that result from these policies can lead to a variety of management changes that add new costs for aquaculture businesses. Moreover, the costs of reacting to the various types of allegations from public advocacy groups are not borne by the public agencies, but must be addressed by the businesses that are affected. These types of costs are not normally included in typical financial statements. Yet, if the owners and managers of aquaculture businesses do not engage in the public and regulatory discussions related to the development of new regulations, the end result may be new laws, ordinances, and regulations that increase costs or force farm-level changes that turn the business into an unprofitable or infeasible one.

Aquaculture Economics and Financing: Management and Analysis, Carole R. Engle, © 2010 Carole R. Engle.

The issues involved rarely are resolved permanently. While new issues may emerge as time goes on, many surface into prominence for a period of time, then are pushed into the background for a while only to resurface farther down the road. This chapter discusses these issues and their effects on the economics of aquaculture businesses.

COMPLYING WITH EXISTING REGULATIONS

All businesses must comply with existing regulations. These regulations include local ordinances, state and federal policies, and, increasingly, international regulations. Compliance with such a variety of permits can be costly in terms of the time required, hiring specialized personnel, and paying fees. Maintaining the permits entails additional costs to keep up with changes in regulations and permitting.

Local regulations are highly variable and specific to the particular area. Most aquaculture businesses were constructed originally on farmland. However, as urban areas expand, more and more aquaculture businesses find themselves within city limits. As zoning ordinances change, and restrictions related to zoning emerge, there is potential for conflicts to emerge with existing aquaculture farms. Zoning has also become relevant to mariculture businesses in coastal and offshore waters. Some parts of the world (Spain, United States) have experimented with mariculture parks or zones that restrict aquaculture activities to that area. Canada, Malaysia, Sri Lanka, and Australia have all developed zoning programs for aquaculture, primarily in coastal and marine areas.

Table 18.1 lists examples of the variety and prevalence of various types of state regulations in the United States related to aquaculture. Most states in the United

Table 18.1. State Regulations and Permits Required for Aquaculture-Related Activities.

Category	Description	Number of states
Regulations	Licensing and permitting for aquaculture	40 states and territories
	Possession of aquatic animals for aquaculture	30 states
	Water supply	26 states and territories
	Non-native species	22 states and territories
	Discharge control and protection of fish and wildlife	16 states and territories
	Processing	15 states
	Leasing of submerged public lands	13 states
	Discharges (other than federal NPDES permits)	11 states
	Use of coastal areas	2 states
	Depuration for shellfish	2 states
	Inspection	Arizona
Permits		24 states and territories
	Discharges (NPDES)	44 states and 1 territory
	Tideland use, including dredging	5 states
	Food quality sanitation and safety	3 states
	Other permits	7*

*Include aquaculture, certification, land use and development, gear or equipment, and disposal of mortalities.

States (40 states and territories) require some sort of permit or license to run an aquaculture business. Of these, 30 states and territories have regulations related specifically to the possession of aquatic species for aquaculture. For processing, 15 states have regulations related to processing the products from aquaculture and 13 states have regulations related to leasing submerged public lands. A few states have regulations related to inspection (1), shellfish depuration (2), conservation districts (1), and the use of coastal areas (2).

Twenty-four states and territories require various types of permits for aquaculture businesses (Table 18.1). Common permits required include food sanitation, quality, and safety, tideland use (including dredging), and discharges. Other less frequently used permits are for gear or equipment (1 state), disposal of mortalities (2 states), land use and development (1 state and 1 territory), animal damage permits (2 states), food quality sanitation and safety (3 states), shellfish certification guidelines (1 state), and an aquaculture permit (1 state).

There are a variety of federal regulations that aquaculture business owners must comply with (Table 18.2). The primary agencies involved with federal permits for aquaculture include U.S. Army Corps of Engineers, U.S. Department of Commerce, and the U.S. Fish and Wildlife Service. The U.S. Army Corps

of Engineers (USACE) regulates the construction of dams and levees in dredged and filled sites through Section 404 permits. The USACE is authorized to issue these permits under the Federal Water Pollution Control Act of 1972, as amended by the Clean Water Act of 1977, and Water Quality Act of 1987. The Federal Coastal Management Zone Act of 1972 regulates proposed federal activities that affect a state's coastal zone. Section 10 of the Rivers and Harbor Act of 1989 regulates the creation of obstructions not authorized by Congress to navigable waterways of the United States. The Federal Sanitation Standards for Fish Plants of the Department of Commerce provides for inspection of processing plants and facilities and grades aquaculture products for quality assurance. The U.S. Fish and Wildlife Service regulates activities that might affect endangered or threatened species or their habitat through the Endangered Species Act of 1983. The Lacey Act amendments of 1981 are also enforced by the U.S. Fish and Wildlife Service. Under the Lacey Act, it is unlawful to import, sell, acquire, or purchase fish, wildlife, or plants taken, possessed, transported or sold in violation of U.S. or Indian law or in interstate or foreign commerce. The Migratory Bird Act (enforced by the U.S. Fish and Wildlife Service) regulates the use of lethal control methods on migratory birds that cause aquaculture crop losses.

Table 18.2. Federal Regulations Related to Aquaculture.

Category	Specific title of regulatory authority	Agency
Construction of dams and levees for dredged and filled sites	Section 404 of the Federal Water Pollution Control Act of 1972, amended by Clean Water Act of 1977, and Water Quality Act of 1987	U.S. Army Corps of Engineers
Use of coastal zones	Federal Coastal Zone Management Act of 1972	
Use of navigable waters	Section 10 of the Rivers and Harbors Act of 1989	
Processing plants and grading standards	Federal Standard Sanitation Standards for Fish Plants	U.S. Department of Commerce
Activity that might affect endangered or threatened species or their habitat	Endangered Species Act of 1973	U.S. Fish and Wildlife Service
Interstate transportation of fish and wildlife	Lacey Act Amendments of 1981	U.S. Fish and Wildlife Service
Bird depredation	Migratory Bird Treaty Act	U.S. Fish and Wildlife Service

INFLUENCING THE OUTCOMES OF REGULATORY PROCESSES

Aquaculture has grown to a size that has attracted the attention of numerous nongovernmental organizations and regulatory bodies. Yet, many individuals in many countries including the United States have little knowledge or understanding of aquaculture. This is particularly problematic when aquaculture is such a diverse industry, producing nearly 500 different species of plants and animals in a wide array of different production systems (FAO 2000–2008). The likelihood of the individuals who develop new rules and regulations having some background or experience in aquaculture is very low. Regulators often are faced with high workloads, stringent deadlines, and have little time to research and study the industries that they are charged with regulating. At the same time, most regulatory authorities are charged to base their rule making on the best available science. How are regulators to find, understand, and integrate the best available science for every industry for which they are expected to develop new rules?

The only answer to this question is that it is in the best interest of the aquaculture industry to position itself to be the primary source of information for regulators, the media, and the public. Individual farmers frequently do not have the time to mount public relations campaigns on their own, much less to provide the level of detailed information that is needed by regulators. However, they can join and participate in the trade association that is most closely associated with their products.

Producer associations have played an increasingly important role in aquaculture development and regulation. These producer groups can be small, village-level groups, or international organizations. Participation in a trade association such as those listed in Table 18.3 is necessary for several reasons. Trade associations can have a major influence on regulatory decisions by insisting that there be an adequate scientific basis for the rule, that the rules developed be reasonable, rational, and not overly burdensome financially. Information presented by trade associations can counterbalance the negative media reports that pressure regulatory agencies to adopt new regulations for aquaculture.

The power and influence of a trade association is directly related to the size of its membership. The more stakeholders that an association represents, the more influence it has with which to negotiate. If an association has a sufficient dues-paying membership, it may be able to afford to hire lobbyists, pay travel expenses to have representatives at crucial meetings in different locations, and participate in key workshops and events.

Active involvement in a trade association entails costs associated with dues and attendance at association meetings. These constitute business costs in an era of increasing regulation. As a member of an association, one is in the position of serving as representative

Table 18.3. Examples of Aquaculture Trade Associations.

Type of trade association	Name of trade association
International	Global Aquaculture Alliance
	World Tilapia Association
Regional	Federation of European Aquaculture Producers
National	Aquaculture Association of Canada
	Arkansas Bait and Ornamental Fish Growers Association
	Association of Chilean Salmon Farmers
	Catfish Farmers of America (U.S.)
	Finnish Fish Farms Association
	Icelandic Fish Farmers and Sea Ranchers
	Irish Aquaculture Association
	National Aquaculture Association (U.S.)
	Norwegian Fish Farmers Association
	Striped Bass Growers Association
	U.S. Trout Farmers of America (U.S.)
Provincial/State	British Columbia Salmon Farmers Association

on various task forces and working groups that provide input to various rule-making efforts.

Every rule-making effort in the United States involves different periods of public comments. Both individuals and associations have opportunities to submit comments during these official periods. Coordinating key messages with trade associations is most effective.

In addition to participating in trade associations, individual businessmen can engage in the rule-making effort through the political process. Politics is the system within which a democratic society "converses" about its priorities, choices, and makes decisions. Elected officials must be kept informed of the consequences to their constituents of various policies under consideration. However, to have credibility, it is critical that any input or information provided to elected officials be well researched, credible, and defensible.

The first step is to get to know the appropriate elected officials. Inviting members of the political delegation to visit farms, contributing to campaigns, and attending fund-raisers are all ways of developing a relationship that provides a path of communication to ensure that one's elected officials have the best possible and most accurate information on the aquaculture industry.

EXAMPLE OF BENEFITS OF ENGAGING IN THE REGULATORY PROCESS: THE ENVIRONMENTAL PROTECTION AGENCY EFFLUENTS RULE OF 2004

On June 30, 2004, the Administrator of the Environmental Protection Agency (EPA) of the United States signed the final rule for aquaculture. This signature followed a 3-year rule-making period that was unique in several respects. Firstly, it was the final rule developed under a court-ordered consent decree, and it was developed with the engagement of a National Task Force that not only included both EPA officials and staff but also representatives of various sectors of the aquaculture industry as well as university and agency scientists (Westers, 2000).

The backdrop to this rule included an initial evaluation of aquaculture by EPA in the early 1970s (Whitman et al. 2002). At that time, EPA did not propose any regulations because the emphasis at that time was directed toward sectors that discharged effluents that contained toxic metals and organics. In 1973, EPA proposed and implemented National Pollutant Discharge Elimination System (NPDES) permit application rules for aquaculture. The NPDES permits are required for farmers that discharge at least 30 days/year (warmwater farms) or that produce less than 100,000 lb/year. For coldwater species, permits are required for farms that discharge more than 30 days/year and produce more than 20,000 lb/year.

The Clean Water Act provides for development of effluent limitation guidelines for point sources of pollution. In 1992, following a lawsuit by the Natural Resources Defense Council, a Consent Decree was signed by the court. The Consent Decree required the EPA to set a timetable for the development of effluent limitation guidelines for designated industries. In the original list of industries, Rule #12 was the industrial container cleaning industry. In late 1999, EPA asked the court to substitute aquaculture for the industrial container cleaning industry. The reasons given by EPA were: (1) the only relevant guidance was more than 20 years old; (2) the aquaculture industry had changed in terms of the species raised and the industrial processes employed; and (3) that states had identified aquaculture as a common cause of water quality impairment on the 305b and 303b lists.

Scrutiny of the 305b and 303d lists by aquaculture scientists raised a number of questions. For example, substances such as mercury, arsenic, pesticides, and metals were listed as the cause of impairment from aquaculture in Louisiana, California, and Ohio. Such substances are not used in aquaculture production. Moreover, a number of pesticides will kill fish, crustaceans, and other aquatic organisms and are not used on aquaculture farms. Moreover, a number of the major aquaculture-producing states were absent from the list all together. If aquaculture were a major source of impairment, the greatest impacts would be expected to be visible in the major production states. A follow-up survey by the North Carolina Department of Agriculture showed that the 305b and 303d lists provided by the states were compiled by listing all types of industries present in the watershed, without indication of which industry was responsible for the contamination. Moreover, the survey documented that states considered existing regulations adequate to handle any effluent problems associated with aquaculture.

A National Aquaculture Effluents Task Force was formed by the Joint Subcommittee on Aquaculture to support a nationally coordinated, systematic process to identify and report the best available and appropriate science and data related to discharges from aquaculture. The task force included representatives from various aquaculture trade associations. The involvement of the more than 180 industry representatives and scientists provided a thorough sounding board for EPA on a wide variety of technological alternatives that were discussed by EPA.

The EPA process for effluent limitation guidelines is based on establishing mandatory production processes for treating effluent discharges (Environmental Protection Agency 2006). Essentially, EPA searches for technological alternatives that reduce the levels of discharge without causing an excessive number of businesses to cease operations. EPA staff members have no training in aquaculture. They typically work through contractors who carry out the analyses required, but the contractors frequently do not have direct experience in the businesses under consideration. Moreover, the specific analyses prescribed for agency use may not cover some of the more relevant impacts and consequences of the rule.

As an example of the work done by the Task Force, Engle et al. (2005) showed that the three effluent treatment options under consideration (each option referred to a set of treatment options) would all cause small and medium-sized trout farms to go out of business. The analysis also showed that, while large farms would still have positive net returns under the options proposed, their option would cause the probability of being profitable to fall from 84% without any additional treatments to 10–11%, depending on the options selected. Moreover, the mathematical programming model developed demonstrated a high level of sensitivity to increased requirements for operating and investment capital. No additional effluent treatments were feasible for the options proposed, due to the additional investment capital required. The models showed that imposing new effluent treatment options forced farms to take raceways out of production to put them into effluent treatment. This was due to the investment capital required, limited access to capital by many trout farmers, and competing uses for land in trout farming areas that have driven land prices in these areas upward.

EMERGING POLICY AND REGULATORY ISSUES

There are a number of policy and regulatory issues that are emerging but are not yet well developed enough to be able to estimate the economic and financial costs involved. The following descriptions will present an overview of economic and financial considerations related to emerging policies related to trade, biosecurity, certification, aquatic nuisance species (ANS), and food safety issues.

TRADE ISSUES

Increasing globalization has had dramatic effects on aquaculture worldwide. The total volume of seafood traded worldwide was 108 million metric tons in 2005. A high percentage of the most frequently traded seafood products, such as salmon and shrimp, is from aquaculture.

The increase in trade volume has been accompanied by an increase in trade conflicts. The trade conflicts have resulted in a number of antidumping accusations and lawsuits. The World Trade Organization provides for antidumping suits to protect against sales of products at unfairly low prices. All countries have some sorts of tariffs that have been imposed on some other country. Many include agricultural products and most include some type of tariffs or regulations related to imported fish and fish products (World Trade Organization 2008).

Table 18.4. Examples of Trade Disputes in Aquaculture.

Species	Country		Date
	Petition initiated	Against	
Salmon	United States	Norway	1988
	United States	Chile	1997
	European Union	Norway	1999
Swimming crab	United States	Venezuela, Thailand, Indonesia, Mexico	2000
Crawfish	United States	China	1994
Basa/tra	United States	Vietnam	2002
Mussels	United States	Canada	2001
Shrimp	United States (shrimp fishermen)	Ecuador, Brazil, India, Thailand, China, Vietnam	2003

Many trade conflicts have involved salmon, particularly as the farm-raised salmon industry has grown worldwide (Table 18.4). Some of the earliest conflicts were between the United States and Norway in the 1980s. These were followed by conflicts between the United States and Chile in 1997, and the European Union and Norway in the early 1990s. Blue crab imports into the United States from Venezuela, Thailand, Indonesia, and Mexico were disputed in 2000. Increased imports of crawfish into the United States from China in 1994 prompted an antidumping petition and subsequent countervailing tariffs.

The most recent trade conflicts in aquaculture have involved the U.S. catfish industry and imports of basa/tra (*Pangasius* sp.) from Vietnam. The U.S. catfish industry filed an antidumping petition in 2002 and antidumping tariffs were imposed on Vietnam in 2003. After falling in 2002–2003, exports of basa/tra have increased again, although at a slower rate than their exports to other countries. The import data reflect growth in the number of countries exporting basa/tra to the United States. Whether this reflects production from those countries or transshipment (sales through a third country to avoid countervailing tariffs) is unknown.

The United States and Canada have experienced trade conflicts over mussels imported into the United States from Prince Edward Island. In 2003, an antidumping petition was filed by shrimp fishermen and processors against shrimp growers from Ecuador, Brazil, India, Thailand, China, and Vietnam.

Trade conflicts, particularly antidumping petitions, involve substantial sums of capital. The legal fees involved in filing these petitions often amount to mil-

lions of U.S. dollars. Many aquaculture industries are too small to be able to afford such expensive litigation. Some antidumping petitions have been filed by processors rather than growers and some of the costs involved have been paid by trade associations. Regardless of the source of capital, the costs become substantial and eventually, in one form or another, add cost to the farms and production units.

Biosecurity, as related to fish and other aquatic animals has been an issue of growing concern and interest. Increased globalization and trade have increased the potential to spread aquatic pathogens across the globe. Transmission of pathogens (such as whirling disease of trout and infectious salmon anemia virus of coldwater fish species) has been monitored and, in some cases, regulated, since the 1960s. However, less attention has been paid to the transmission of warm water fish pathogens, until recent years.

The Office International des Epizooties (OIE) (World Organization for Animal Health) regulates the spread of pathogens worldwide. The OIE Aquatic Code was first developed in 1995. The OIE requires that fish health inspections be done for the interstate transport of live fish and maintains a list of pathogens for which aquatic animals must be inspected. The health certificates issued must indicate the absence of these pathogens.

Recent outbreaks such as that of spring viremia of carp in the United States (2002) and, of viral hemorrhagic septicemia (VHS) in the Great Lakes and in Europe in more recent years have heightened awareness of aquatic biosecurity issues in the United States. The United States Department of Agriculture-Animal

Plant and Health Inspection Service (USDA-APHIS) is the regulatory authority responsible for control and surveillance of aquatic pathogens. With the VHS outbreak in the Great Lakes, USDA-APHIS is the regulatory authority responsible for control and surveillance of aquatic pathogens. With the VHS outbreak in the Great Lakes, USDA-APHIS first issued an Emergency Order and then an Interim Rule related to transportation of live fish. The rules prohibited the shipment of VHS-susceptible species of live fish from states where VHS had been detected, unless the fish had been inspected and found to be free of VHS. This rule continues to evolve over time.

To comply with this rule, farmers and live fish distributors must have fish inspected. There are provisions for both farm-level (requires a two-year history) and lot inspections (every shipment must be tested). The total annual expense to the farm business is much higher for lot inspections due to the much higher total number of inspections to be done. The cost of inspections includes fees to veterinarians to collect the samples and the costs for the diagnostic laboratories to conduct the analyses. The costs and logistics of complying with inspections must be included in planning, monitoring, and financial analysis of the farm business.

In addition to the USDA-APHIS federal regulation, individual states in the United States have adopted additional regulations to prevent the introduction of VHS into their states. These regulations vary considerably from state to state in terms of the species of fish to be tested, the numbers to be tested, the frequency of inspections, and the pathogens to be tested for. Aquaculture business owners who ship live fish in the United States must familiarize themselves with these regulations and position themselves to have immediate access to changes in these regulations.

Some segments of aquaculture have developed certification programs as a means to avoid delays that may occur due to changing regulations pertaining to biosecurity and the interstate shipment of live fish. One such example is the Arkansas Baitfish Certification Program. To be certified under this program the farm must have a two-year history of testing that shows the farm is free of the diseases identified in the program (spring viremia of carp, infectious pancreatic necrosis, and viral hemorrhagic septicemia, among others). Each farm is sampled twice a year by a licensed veterinarian and samples tested for a variety of pathogens by an APHIS-approved laboratory. Certificates are issued by the Arkansas State Department of Agriculture-State Plant Board on paper that cannot be copied. Inspectors

of the State Plant Board inspect each farm in the certification program twice a year to check for the presence of Aquatic Nuisance Species (ANS) that are prohibited in the program (zebra mussels, bighead carp, and silver carp, among others). The program includes a fee paid to the Arkansas Department of Agriculture in addition to the veterinarian and laboratory testing fees.

CERTIFICATION

There has been a dramatic increase in the number of certification programs for aquaculture in the last several years. The impetus for these programs stems primarily from concern on the part of major retailers like Wal-Mart, Tesco, and Carrefour to maintain or improve their reputation, competitiveness, and market share, in reaction to consumer concerns and to reduce risk of recalls, regulatory action, and poor quality. Recalls result in significant losses in terms of sales and in damages to a company's reputation. Lawsuits by environmentalist groups have resulted in an increase in use of certification programs by large retailers to demonstrate their commitment to environmental sustainability.

Growing reports of adulterations in seafood products from China and Vietnam (antibiotics, fungicides, and melamine) have raised issues for congressional and the Food and Drug Administration (FDA) action in the United States. The end result is likely to be increasing federal inspection and regulation.

Certification programs have been developed as a proactive response to these situations. Certification programs have focused on some of the key issues such as Hazard Analysis and Critical Control Point (HACCP), food and feed safety, food security, quality, environmental impact and sustainability, traceability, inspection, auditing, third-party verification, and certification by the Global Aquaculture Association (GAA), the World Wildlife Federation (WWF), Euro-Retailer Produce Working Group, the Irish Quality Eco-Mussel Standard, and others. Each program is designed differently. For this reason, the Food and Agriculture Organization of the United Nations (FAO) has initiated a process to develop standards for certification programs. Some, like the Irish Quality Eco-Mussel Standard is a single-species standard while the GAA/ACC is for multiple species. Most certification programs rely upon some type of third-party certification. Companies such as the Aquaculture Certification Council (used by GAA), SureFish, Inc. of Seattle, Washington, and others have been developed to provide this service. EUREPGAP began in 1997 by the Euro-Retailer Produce Working Group, driven mostly by British retailers. EUREPGAP evolved into GLOBALGAP in 2007 in response to the growing interest from producers and retailers worldwide. The GAA/ACC program includes a variety of components that range from food safety, sanitation, good manufacturing practices, environmental sustainability, effluents, community involvement, and traceability.

All certification programs entail cost to the producer, although the amount of the cost varies. What is not clear is to what extent this cost can be passed through to the consumer. There is no clear research that indicates that consumers generally are willing to pay higher prices for product that has been certified as environmentally sustainable. There are what appear to be small segments of consumers who will pay more but as of 2008, it does not appear to be a substantial consumer segment.

It is difficult to judge the point at which it is worth the cost to enter a certification program. Depending on the specific nature of the program, there is a risk to the producer in addition to the cost. The risk consists of the probability that a disease, ANS, or other problem may be found during a routine inspection. This risk needs to be weighed against potential advantages. If a primary sales customer begins to require participation in a certification program, a farmer or processor may have no choice. In other situations, it may be possible to turn the certification program to a marketing advantage. These trade-offs, particularly whether there is a way to use certification to expand into new markets or to develop a competitive edge over competitors in existing markets must be weighed against the costs and risks.

Exotic and non-native species have become a major issue affecting aquaculture. The concern is that non-native species will out-compete native species that will result in a priceless loss of biodiversity through homogenization. Once a species is lost, that loss is forever. In the United States, there are an estimated 185 non-native species of which 75 have become established. In addition, there are 316 species that are now found outside their native ranges. FAO lists 5,612 introduced species records worldwide in the database on Introductions of Aquatic Species (FAO 2000–2008).

Asian carps (primarily bighead and silver carp, but black carp and grass carp are included) have become a poster child for ANS in 2004. The jumping ability of silver carp has attracted much media and other attention. The populations of silver and bighead carp in the Mississippi River have expanded dramatically over the last decade. Silver and bighead carp were stocked in sewage treatment lagoons and waters in the 1970s from which escape was possible. EPA-funded research in the 1970s on aquaculture systems for wastewater treatment, in sewage treatment lagoons, some of which were directly connected to the wild. The Third Report to the Fish Farmers published by the U.S. Fish and Wildlife Service in 1984 stated, "The bighead carp is an excellent food animal and highly prized by Asians in the United States. It is well suited for culture in combinations with other fishes such as the grass carp, silver carp, and common carp." The first escape of a grass carp was from a U.S. Fish and Wildlife Service research facility.

Aquatic vegetation problems cost states in the United States from \$1 to \$10 billion each year. Of the various methods for controlling aquatic vegetation, the use of grass carp is the cheapest alternative, at \$45 to \$125/acre as compared to \$100 to \$11,000/acre for mechanical cutting, \$500 to \$2,400/acre for mechanical pulling, and \$1,100 to \$26,200/acre for dredging and rotovations (Greenfield et al. 2004).

Bighead carp have been raised for sale to ethnic grocery stores since the 1970s. Bighead carp sales have resulted in additional net revenue of \$180–371/acre for catfish farmers. This additional revenue has served to diversify production risk as a secondary crop during times of low catfish prices and have helped primarily smaller growers, especially small-scale growers, to stay in business and recover from adverse market risk.

Flood control measures have reduced wetland areas for migrating birds. Pelicans and double-crested cormorants have targeted commercial fishponds for feeding. Flood control efforts have reduced habitat that migrating birds traditionally used for resting areas, and birds look for alternative areas. While pelicans and cormorants damage farms by eating fish, pelicans have also spread an exotic trematode to catfish farms. Treatment costs of trematodes range from \$8.30/acre with black carp to \$156/acre with copper sulfate, and \$195/acre with hydrated lime. Without control, costs are \$363–1,073/acre.

What do fish farmers do to prevent escapes from farms? The majority of levee ponds have been constructed on land formerly in production of soybeans, and other row crops with no direct connection with open waters. The water management systems on fish farms also assist to prevent escapes. Fish farmers maintain water levels to allow for storage capacity for rainwater in ponds. Most farmers already practice water conservation to minimize pumping costs due to concerns over groundwater supplies. Proper site selection, pond construction, and management assist to minimize the chance of escape.

The regulatory authority over fisheries resources lies with the state fish and game agencies. In Arkansas, a fish farming permit is required, following inspection. Arkansas has an approved aquaculture species list. The approved list includes those species that can be cultivated without additional permits. This "clean list" includes species such as catfish, golden shiners, largemouth bass, and goldfish. The restricted species list provides an additional level of prevention to prevent escapement. Restricted species require a permit that details the location of the facility, measures taken to eliminate the possibility of escape, and numbers and species to be held. Permitted farmers must also construct barriers that prevent escape.

Species not listed on either list are evaluated on a case-by-case basis. The burden of proof to show adequate prevention of escape lies with the applicant. Biologists review the life history of the species under consideration.

FOOD SAFETY

Food safety concerns have continued to grow as new worldwide scares are communicated globally through the media. In 2008, concerns grew over the use of melamine to enhance measurable protein levels in products from China. The associated deaths of pets that ate feed made from melamine-enhanced ingredients and children in China who drank powdered milk laced with melamine further enhanced the level of concern.

In the United States food safety responsibility rests with the FDA. In the event of food safety scares, the FDA has the authority to shut down sales and processing of aquaculture products to protect the safety of food in the supply system. It also has the authority to require exporters to test food products shipped to the United States.

The European Food Safety Authority (EFSA) was developed in 2002 to provide independent scientific advice on food safety. It develops and publishes opinions on the basis of risk assessments of issues pertaining to food safety and works closely with national authorities (EFSA 2004). The risk assessments are prepared by scientific panels convened in areas that include: food additives, substances used in animal feeds, plant health and protection, genetically modified organisms (GMOs), dietetic products, biological hazards, contaminants in the food chain, and animal

welfare. Food safety legislation in the European Union addresses animal feeds, animal welfare, contaminants and residues, food additives, food supplements, organic products, and packaging. The EU's food and veterinary office in Dublin is charged with overseeing and monitoring food safety throughout the supply chain.

RELATIONSHIP WITH UNIVERSITIES/ SCIENTISTS/EXTENSION PERSONNEL

Universities have a wealth of scientific expertise that can be valuable when interacting with regulatory agencies during rule-making efforts and when involved in regulatory actions. While many university scientists are disengaged from the practical problems of industries, there are scientists who, with some encouragement, may be willing to provide summaries of the relevant knowledge base and interpret scientific data as it relates to specific regulatory issues. Others may be willing to conduct trials that may answer some specific questions that provide useful guidance in a rule-making effort.

Land-grant universities, in particular, are charged to identify stakeholders and direct research and extension resources toward solving their problems. Scientists with aquaculture experience and extension aquaculture scientists who are skilled at explaining research results to industry representatives, trade associations, and regulatory agency personnel are valuable resources. Many industry growers do not realize how much effort some university scientists will make on their behalf with some encouragement and recognition.

RECORD-KEEPING

Record-keeping requirements for managing the aquaculture business to ensure compliance with state and federal regulations will vary with each type of regulation. However, extension personnel and state and federal agencies frequently issue guidance and fact sheets on the records required to be in compliance with each type of regulation. The guidance documents should be kept on file and all reports submitted to regulatory authorities must be kept on file for future reference.

PRACTICAL APPLICATION

The U.S. EPA developed guidelines related to the discharge of effluents from aquaculture facilities. These are related only to flow-through, net pen, and recirculating aquaculture systems that discharge more than

30 days a year. Thus, there is no additional record-keeping required for the 256-acre catfish farm that has been used for the practical application throughout this book.

However, if this farm would consider switching to a cage operation in a lake, for example, then the new guidelines would become relevant, if the farm expected to produce more than 100,000 lb a year. The records required for flow-through, net pens, and recirculating aquaculture systems begin with records of fish production, feed usage, maintenance and repair logs, inspection logs, and records of employee training programs. Any spills of oil or chemicals require that the dates, places, and times be recorded, as well as the name of any inspectors who arrive to view the site. Additional information may be required to meet state guidelines that may be more stringent than federal guidelines.

OTHER APPLICATIONS IN AQUACULTURE

Engle et al. (2005) evaluated the economic feasibility of various options proposed by the U.S. EPA for treating effluents from flow-through trout farms. Surveys were conducted of trout farms in North Carolina and Idaho. Budget analyses showed that, without imposing additional treatment options, trout farming in these two states in the United States was generally profitable. The additional treatment options proposed caused medium-sized trout farms to become unprofitable. While the larger farm sizes analyzed were still profitable after imposing additional treatment options, the probability of achieving positive net returns decreased from 83 to 10.5%.

The additional treatment options created substantial amounts of financial risk because of the additional capital investment that would be required to adopt the new treatment options. Limited amounts of capital reserves on many trout farms exacerbated the levels of risk incurred and prevented the expansion that would be necessary to avoid substituting production units for treatment units. The overall result was to force trout farms to operate at smaller, less efficient scales of production.

SUMMARY

This chapter discussed the increasing role of regulations in aquaculture. Aquaculture businesses must contend with a wide array of different regulations. Compliance with regulations entails costs associated with sampling, inspection, and verification. Certifica-

tion and inspection programs have been developed for purposes of market security and to maintain confidence in the product. The inspection process subjects farms to the risk of detection and the subsequent economic and financial consequences.

Aquaculture growers must engage in the rule-making process to ensure that accurate information on the situation and of possible consequences is taken into consideration. Participation can be through trade associations, submitting comments during comment periods, and participation on task forces and panels. While there is cost involved, the alternative may be a far costlier set of rules.

Specific regulatory issues are discussed in the chapter beginning with the EPA's 2004 rule on aquaculture effluents. Trade issues, ANS, biosecurity, and food safety are discussed.

REVIEW QUESTIONS

1. What types of costs are incurred from new regulations developed for aquaculture? Give some specific examples.

2. Why is it important for growers to engage in the regulatory process?

3. What types of permits are required for aquaculture businesses?

4. What is the basis and authority to file an antidumping petition?

5. What has been the impetus for the proliferation of certification programs for aquaculture and what are their major characteristics?

6. Give several examples of the types of regulations that have affected aquaculture businesses.

7. What is the role of scientists in the policy debates that result in regulatory actions?

8. Describe several certification programs and discuss their role in the regulatory environment.

9. Describe how aquaculture businesses can engage in the regulatory process.

10. Contrast regulatory issues related to environmental sustainable with those related to food safety.

REFERENCES

Engle, Carole R., Steeve Pomerleau, Gary Fornshell, Jeffrey M. Hinshaw, Debra Sloan, and Skip Thompson. 2005. The economic impact of proposed effluent treatment options for production of trout *Oncorhynchus mykiss* in flow-through systems. *Aquacultural Engineering* 32:303–323.

Environmental Protection Agency. 2006. *Compliance Guide for the Concentrated Aquatic Animal Production Point Source Category*. Engineering and Analysis Division, Office of Science and Technology. EPA 821-B-05–001. U.S. Environmental Protection Agency. http://www.epa.gov/waterscience/guide/aquaculture. Accessed November 23, 2009.

FAO. 2000–2008. Database on Introductions of Aquatic Species. www.fao.org. Accessed November 23, 2009.

Greenfield, B.K., N. David, J. Hunt, M. Wittmann, and G. Siemering. 2004. *Review of Alternative Aquatic Pest Control Methods for California Waters*. Oakland, CA: San Francisco Estuary Institute. p. 109.

Welcomme, Robin L. 1988. *International Introductions of Inland Aquatic Species*. FAO Fish Technical paper 294, Rome, Italy. Food and Agriculture Organization of the United Nations, Rome, Italy.

Westers, Harry. 2000. *A White Paper or the Status and Concerns of Aquaculture Effluents in the North Central Region*. East Lansing: North Central Regional Aquaculture Center, Michigan: Michigan State University.

Whitman, Christine T., Tracy Mehan III, Geoffrey H. Grubbs, Sheila E. Frace, Marvin Rubin, Janet Goodwin, and Marta Jordan. 2002. *Development Document for Proposed Effluent Limitations Guidelines and Standards for the Concentrated Aquatic Animal Production Industry Point Source Categories*. Washington, DC: U.S. Environmental Protection Agency.

World Trade Organization. 2008. *World Tariff Profiles*. Geneva: World Trade Organization.

Bibliography

REFEREED JOURNALS

Aquaculture Economics and Management

This is the only journal that focuses strictly on economics issues related to aquaculture. Its scope includes: aquaculture production economics, farm management, processing, distribution, marketing, consumer behavior, pricing, government policy, international trade, and modeling. Additional information available at: www.iaaem.org.

Marine Resource Economics

This journal covers the broad area of marine resource economics. It includes aquaculture economics as one of a number of themes within its scope. Additional information available at: www.mre.cels.uri.edu.

Journal of the World Aquaculture Society

This journal focuses primarily on the biology, science, and methods underlying aquaculture, but also publishes results of economic and marketing analyses of aquaculture industries. Additional information available at: www3.interscience.wiley.com/journal/1222 17952/grouphome/home.html.

Aquaculture

This journal focuses primarily on the biology, science, and methods underlying aquaculture, but also publishes results of economic and marketing analyses of aquaculture industries. Additional information available at: www.elsevier.com/wps/find/journaldescription.cws_home/503302/description# description.

North American Journal of Aquaculture

This journal focuses primarily on the biology, science, and methods underlying aquaculture, but also pub-

lishes results of economic and marketing analyses of aquaculture industries. Additional information available at: www.afsjournals.org.

Journal of Applied Aquaculture

This journal focuses primarily on the biology, science, and methods underlying aquaculture, but also publishes results of economic and marketing analyses of aquaculture industries. Additional information available at: www.informaworld.com/smpp/title~content=1792306881.

American Journal of Agricultural Economics

This journal focuses on the economics of agriculture, natural resources and the environment, and rural and community development. Additional information available at www.wiley.com/bw/journal.asp?ref=0002-9092.

Journal of Agriculture and Applied Economics

This journal publishes creative and scholarly work in agricultural economics, including contributions on methodology and applications in business, extension, research, and teaching. Additional information available at: www.saea.org/jaae.

Journal of Food Distribution

Applied problem-oriented journal that emphasizes the flow of products and services through the food wholesale and retail distribution system. Additional information available at: http://fdr.ag.utk.edu/journal.

Journal of Food Policy

Multidisciplinary journal that focuses on policies for the food sector in developing, transition, and advanced economies including food production, trade, marketing, and consumption. Additional information available at: www.elsevier.com/wps/find/journaldescription.cws_home/30419/description# description.

Journal of Agricultural Economics

International journal that provides a forum for agricultural economics research, including statistics, marketing, business management, in agricultural, food and related industries, communities, and the environment. www.wiley.com/bw/journal.asp?ref=0021-857x.

BOOKS

Allen, P.G., L.W. Botsford, A.M. Shuur, and W.E. Johnston. 1984. *Bioeconomics of Aquaculture*. Elsevier: Elsevier Science Publishers.

This book provides an historical look at the economics of early attempts at intensive aquaculture, with an emphasis on bioeconomic modeling. Students of bioeconomic modeling would find the book of value.

Anderson, D.R., D.J. Sweeney, and T.A. Williams. 2004. *Quantitative Methods for Business*. Mason, OH: South-Western, a division of Thomson Learning.

This book covers a wide variety of quantitative analytical methods useful to the analysis of business situations that range from decision analysis to simulation and Markov processes. It includes five chapters on linear programming that present the method as well as applications that include transportation, assignment, transshipment, and integer programming models.

Barry, P.J., J.A. Hopkin, and C.B. Baker. 1983. *Financial Management in Agriculture*, 4th edn. Danville, IL: The Interstate Printers & Publishers

A classic textbook that applies the concepts of finance to the management decisions necessary on farms. This book works through sections on (1) financial analysis, planning, and control; (2) capital structure, liquidity, and risk management; (3) capital budgeting and long-term decision making; and (4) financial markets for agriculture.

Bjorndal, T. 1990. *The Economics of Salmon Aquaculture*. Oxford, UK: Blackwell Scientific Publications.

A useful book that covers salmon production and economics based on the technologies in use at the time. While there have been important changes in the salmon industry since this book was written, it paints a useful historical picture of the emergence of this important aquaculture industry worldwide.

Bjorndal, T., G.A. Knapp, and A. Lem. 2003. Salmon – a study of global supply and demand. FAO/GLOBEFISH Research Programme, Vol. 73, Rome, FAO. 151 pp.

This book provides a useful overview of global production of salmon and country-by-country breakdowns of the markets for fresh and frozen salmon. It includes a discussion of markets for other salmon products such as roe, and smoked, canned, and organic salmon products.

Brooke, A., D. Kendrick, A. Mieraus, and R.E. Rosenthal. 1998. *GAMS User Guide*. Release 2.50. San Francisco, CA: The Scientific Press.

This book provides detailed guidance and instructions on GAMS software. It is an essential reference for those using GAMS programs.

Chaston, I. 1988. *Managerial Effectiveness in Fisheries and Aquaculture*. Oxford, UK: Fishing News Books.

This book is out of print, but if a copy can be found, provides good background in effective management for those who aspire to be managers of a fisheries company such as a processing plant, wholesale company, or a corporate-style fisheries company.

Chaston, I. 1984. *Business Management in Fisheries and Aquaculture*. Oxford, UK: Fishing News Books.

This book is out of print, but if a copy can be found, there is some useful material on accounting practices, investment appraisal and financial analysis with an emphasis on corporate-level businesses.

Dantzig, G.B. 1963. *Linear Programming and Extensions*. Princeton, NJ: Princeton University Press.

This book is a classic text in linear programming methodologies written by an individual who was a major force in the development of the techniques. It covers topics that range from the general concept of linear programming and how to formulate a model, to various forms of classic linear programming models.

Engle, C.R. 2007. *Arkansas Catfish Production Budgets*. MP466, Cooperative Extension Program.Pine Bluff, AR: University of Arkansas.

This bulletin provides comprehensive enterprise budgets that were developed for five sizes of U.S. catfish farms. Values were developed based on a cost of production survey, not economic engineering techniques. Excel-based spreadsheet templates are available from www.uaex.edu/aqufi/extension/ aquaculture/aquaculture economics.

Engle, C.R., and I. Neira. 2005. *Tilapia Farm Business Management and Economics*. Corvallis, OR: Oregon State University.

This bulletin is a training manual for business planning and financial analysis with a particular emphasis on tilapia businesses. The default values used are form Kenya. Downloadable file available from http:pdacrsp.oregonstate.edu.

Engle, C.R., and D. Valderrama. 2001. Economics and management of shrimp farms training manual. In: M.C. Haws and C.E. Boyd (eds). *Methods for Improving Shrimp Culture in Central America*. Managua, Nicaragua: Editorial-imprenta, Universidad Centroamericana.(in English and Spanish). pp. 231–261.

This bulletin is a training manual for business planning and financial analysis with a particular emphasis on shrimp businesses. The examples used are based on data from Honduras and Nicaragua.

Engle, C.R., and K. Quagrainie. 2006. *The Aquaculture Marketing Handbook*. Ames, IA: Blackwell Publishing.

An introduction to aquaculture marketing for those interested in aquaculture and those new to the professional field. This book discusses fundamental principles of marketing and economics presented from a user-friendly, how-to perspective.

Gittinger, J.P. 1984. *Economic Analysis of Agricultural Projects*. Washington, DC: Economic Development Institute, The World Bank.

A classic reference on investment analysis with particular detail on the process of analyzing investments in major development initiatives. It covers the identification of costs and benefits, assigning values to costs and benefits, farm investments, and potential effects on government income and expenditures from various projects.

Hardaker, J.B., R.B.M. Huirne, and J.R. Anderson. 2002. *Coping with Risk in Agriculture*. New York: CAB International.

This book covers the risk and uncertainty faced by farmers from the perspective of the theory of probability and risk preference. Methods to analyze risky decisions and develop plans to manage risk on farms are presented with numerous practical examples.

Jolly, C.A., and H. Clonts. 1993. *Economics of Aquaculture*. New York: The Haworth Press.

This book is patterned after a general microeconomics textbook and is similar in format and content to that of Chang (1990). Theoretical aquaculture examples are used within the very broad scope of the book.

Kay, R.D., W.M. Edwards, and P.A. Duffy. 2008. *Farm Management*, 6th edn. New York: McGraw-Hill.

A well-known standard farm management text used in numerous undergraduate farm management courses in the U.S. An excellent introduction to the concepts and methods necessary to manage and monitor farms effectively.

Lee, W.F., M.D. Boehlje, A.G. Nelson, and W.G. Murray. 1988. *Agricultural Finance*, 8th edn. Ames, IA: Iowa State University Press.

This book is a standard text on agricultural finance. It applies the concepts of finance to the management decisions necessary on farms.

Leung, P-S., and C.R. Engle. 2006. *Shrimp Culture Economics, Market, and Trade*. Ames, IA: Blackwell Scientific.

This book includes a variety of primary literature related to the economics, markets, and trade of shrimp. It includes papers that provide an overview of the global market and trade issues, specific management strategies related to the economics of sustainable shrimp development, and the economics of shrimp farming.

Levy, H. 2006. *Stochastic Dominance: Investment Decision Making under Uncertainty*. New York: Springer Science Business Media.

This book provides a thorough treatment of topics related to investment decision-making under uncertainty. It covers various measures of risk, stochastic dominance decision rules and algorithms, and presents several empirical studies and applications.

Libbin, J.D., L.B. Catlett, and M.L. Jones. 1994. *Cash Flow Planning in Agriculture*. Ames, IA: Iowa State University Press.

This book is written in the form of a manual to provide easy-to-follow instructions on the use of cash flow budgets to improve management of farm businesses. The book emphasizes development and monitoring of complete farm plans.

Meredith, J., S. Shafer, and E. Turban. 2002. *Quantitative Business Modeling*. Mason, OH: South-Western Thomson Learning.

This book covers a variety of topics related to modeling business decisions that range from data collection and basic statistical models to implementation of model results in the business. It includes chapters on linear programming and decision analysis under risk and uncertainty, regression analysis and forecasting, and simulation.

Olson, K.D. 2004. *Farm Management Principles and Strategies*. Ames, IA: Iowa State University Press.

This is a fundamental text on farm management written for farm managers. It covers planning and budgeting, financial analysis, labor management, organization, and management of risk.

Palfreman, A. 1999. *Fish Business Management*. London: Fishing News Books.

This book reviews issues and strategies related to managing businesses built from capture fisheries.

Much of the book focuses on markets, exporting, and issues specific to fisheries in the European context.

Shang, Y.C. 1990. *Aquaculture Economic Analysis: An Introduction*. Baton Rouge, LA: The World Aquaculture Society.

This book presents a generalized overview of microeconomic concepts on production economics, feasibility analysis, and marketing, as these relate to aquaculture. It includes several chapters on farm-level management topics and finance, based on hypothetical case studies.

Shaw, S.A., and J.F. Muir. 1987. *Salmon: Economics and Marketing*. New South Wales, Australia: Croom Helm.

This book provides some historical economic values and analyses for the salmon industry. While out of date in terms of the current structure and economics, the historical perspective can be useful.

Smith, I.R., E.B. Torres, and E.O. Tan. 1985. *Philippine Tilapia Economics*. Manila, The Philippines: International center for Living Aquatic Resources Management.

This book provides an overview of the economics of raising tilapia in a variety of culture systems and levels in the Philippines in the 1980s. Hatcheries, cage culture, and land-based systems are analyzed.

Thornley, J.H.M., and J. France. 2007. *Mathematical Models in Agriculture*. Cambridge, MA: CABI.

This book covers a wide array of types of mathematical models that can be used to analyze various aspects of agriculture. It includes one chapter on mathematical programming that introduces the development, formulation, and solutions of mathematical programming, with examples of parametric, separable, integer, goal, and dynamic programming.

Warren, M. 1998. *Financial Management for Farmers and Rural Managers*, 4th edn. Malden, MA: Blackwell Science.

This is a practical book that describes basic management skills primarily targeted towards farm business managers. It provides computer spreadsheets for calculation of financial ratios and indicators.

Webliography

agecon2.tamu.edu/people/faculty/mccarl-bruce/books.htm

Downloadable book on linear programming. A comprehensive textbook with ample examples and applications (McCarl, B., and T. Spreen. 1997. Applied mathematical programming using algebraic systems. On agrinet.tamu.edu.).

aquanic.org

Aquaculture web site that includes discussion groups, species, systems, contacts, sites, publications, newsletters, educators, and a page by the Joint Subcommittee on Aquaculture that lists programs of the Agricultural Marketing Service and the Foreign Agricultural Service and the National Aquaculture Act.

europa.eu.int/scadplus/leg/en/lub/166002.htm

Details provisions of the common fisheries policy of the European Union.

fis.com/salmonchile

Site of the Association of Chilean Salmon Farmers, providing information about the Chilean aquaculture industry.

info.ag.uidaho.edu

Site of the University of Idaho that includes information on farm machinery costs.

ia.ita.doc.gov

Federal register notice that includes the regulations on antidumping and countervailing duty proceedings to conform to the Department of Commerce's regulations related to the Uruguay Round Agreements Act.

internationalecon.com/v1.0

Site contains an introductory course/text on international trade theory and policy (Suranovic, S.M. 1997–2004. International theory and policy analysis. The International Economics Study Center).

News.uns.purdue.edu

Site of Purdue University that includes information on farm machinery costs.

oregonstate.edu/dept/IIFET

Site of the International Institute of Fisheries Economics and Trade (IIFET). This organization is an international group of economists, government managers, private industry members, and other interested in the exchange of research and information on marine resource issues.

www.adcvd.com

A complete guide to U.S. antidumping and countervailing duty law.

www.ams.usda.gov/cool/

Agricultural Marketing Service of USDA web site that provides information about the country of origin labeling.

www.amstat.org/sections/srms/whatsurvey.html

Site of the American Statistical Association (ASA) that provides brochures on Survey Research, includes: "What is a Survey?" "How to Plan a Survey;" "How to Collect Survey Data;" "Judging the Quality of a Survey;" "How to Conduct Pretesting;" "What are Focus Groups?" "More About Mail Surveys;" "What is a Margin of Error?" "Designing a Questionnaire;" and "More About Telephone Surveys".

Aquaculture Economics and Financing: Management and Analysis, Carole R. Engle, © 2010 Carole R. Engle.

www.aquacultureassociation.ca

Site of the Aquaculture Association of Canada, with goals to foster an aquaculture industry in Canada, to promote study of aquaculture, to gather and disseminate technical and scientific information, and to encourage private industry and government agencies.

www.aquanic.org/publicat/usda_rac/racpubs.htm

The Regional Aquaculture Center pages are located on the AQUANIC site and include a large number of extension fact sheets on a wide variety of topics related to aquaculture, including marketing and economics fact sheets.

www.aquariumcouncil.org

Site of the Marine Aquarium Council (MAC), an international, not-for-profit organization that brings marine aquarium animal collectors, exporters, importers and retailers together with aquarium keepers, public aquariums, conservation organizations, and government agencies.

www.asabe.org

Site of the American Society of Agricultural and Biological Engineers, an educational and scientific organization dedicated to advancement of engineering applicable to agricultural, food, and biological systems.

www.atsea.org

Site for the At-sea Processors Association (APA), representing U.S.-flag catcher/processor vessels that participate in the groundfish fisheries of the Bering Sea.

wwww.bls.gov/blswage.htm

Site of the U.S. Bureau of Labor Statistics that provides databases on wage rates and other labor-related issues.

www.census.gov/epcd/ec97/industry/E311712.htm

Provides detailed national statistics for the fresh and frozen seafood processing industry from the Census, including number of firms, employees, payroll, and revenue by employment-size of the enterprise.

www.census.gov/epcd/susb/1999/us/US311712.htm

Provides statistics of U.S. fresh and frozen seafood processing including employment, size of enterprise, number of firms, number of plant establishments, number of paid employees and annual payroll.

www.census.gov/cir/www/mqc1pag2.html

Provides results from the Survey of Plant Capacity Utilization conducted jointly by the U.S. Census Bureau, the Federal Reserve Board (FRB), and the Defense Logistics Agency (DLA) and reports on the number of days and hours worked, estimated value of production at full production capability, and estimated value of production achievable under national emergency conditions.

www.cfsan.fda.gov

Page on the U.S. FDA web site that deals with seafood HACCP. It provides an overview of HACCP as it relates specifically to seafood and includes a summary of the provisions in the rule and full text of the final seafood HACCP rule.

www.cllie.plus.com/byrd.pdf

This site includes full text file of the manuscript:
Collie, D.R., H. Vandenbussche. 2004. Antidumping duties and the Byrd amendment.

www.commerce.gov

Official site of the U.S. Department of Commerce, providing information on the state of the U.S. economy, and includes export-related assistance and market information, export regulations, and summaries of trade statistics.

www.cites.org

Site of the Convention on International Trade in Endangered Species of Wild Fauna and Flora. Includes species and trade databases, registers, export quotas, reports, contacts, resolutions, and reports of the standing, animals, plants, and nomenclature committees.

www.daff.gov/au/fisheries/aquaculture/starting

Information on starting an aquaculture business, market access and trade, legal issues, and an aquaculture action agenda, from the Department of Agriculture, Fisheries and Forestry, Australia.

www.efr-central.com/

Site for the Efficient Foodservice Response (EFR) project, an industry-wide effort to improve efficiencies in the foodservice supply chain linking manufacturing plants to distribution warehouses to the retail end of the foodservice industry.

www.efsa.eu.int

Site of the European Food Safety Authority (EFSA), containing the latest opinions and reports of various scientific panels.

www.efsnetwork.com/

Site of the EFS Network, Inc. that provides supply chain solutions for the foodservice industry by combining collaborative workflow technology, hosted application modules, and robust data management services.

www.eia.doe.gov

United States Department of Energy web site that provides databases of energy prices, gas, diesel, and electric rates.

www.entrepreneur.com

Site that contains a market planning checklist, tools and services to enhance marketing success, marketing tips, business coaches, and business services.

www.epa.gov/waterscience/guide/aquaculture

Site of the U.S. Environmental Protection Agency guidance related to aquaculture effluents.

www.ers.usda.gov/Data/

The Economic Research Service (ERS) site provides online databases, spreadsheets, and web files on farm income, trade, food prices, food markets, diet and health, natural resources, and food consumption trends.

www.eurep.org

Site of the Euro-Retailer Produce Working Group (GAP–EUREP). EUREP is made up of leading European food retailers that publishes production standards for commodities entering the retail trade and requires third-party verification by an accredited certification body is required.

www.europa.eu.int

Provides an overview of European Community agencies including the European Food Safety Authority and contains the standards, logo, and certification program for organic products in the European Union.

www.europa.eu.int/eur-lex

Posts all legislative actions and full texts of regulations coded by number, information and notices of the European Union, including anti-dumping orders and publishes the Official Journal of the European Union.

www.eurunion.org/legislat/home.htm

Site of the European Commission's Health and Consumer Protection Directorate General that contains the full text of the White Paper on Food Safety describing major policy provisions for food safety in the European Union.

www.extension.iastate.edu/Publications

Site of the Iowa State Extension that includes information on farm machinery costs.

www.extension.missouri.edu

Site of the University of Missouri that provides information on farm machinery costs.

www.extension.umn.edu/distribution/business management

Site of the University of Minnesota that includes information on farm machinery costs.

www.fao.org

Site of the Food and Agriculture Organization of the United Nations that includes the most current global statistics available on aquaculture and fisheries as well as articles that summarize trends and summary statistics, and databases such as *FISHSTAT+ (*A set of fishery statistical databases downloadable to personal computers together with a data retrieval, graphical, and analytical software) and Fishery Data Collection in FAOSTAT of WAICEN (World Agricultural Information Center). Site includes the FAO Code of Conduct for Responsible Fisheries that lays the foundation for responsible management of aquaculture and fisheries stocks.

www.fao.org/DOCREP

Site that presents the codex Alimentarius of FAO, an international regulatory framework for fish safety and quality, including World Trade Organization (WTO) agreements on the Application of Sanitary and Phytosanitary Measures (SPS) and the FAO Codex Alimentarius.

www.fao.org/docrep/003/x7353e/x7353e03.htm

Site that includes information on the Uruguay Round Agreement on Agriculture.

www.fao.org/figis

Site that includes a fact sheet on HACCP.

www.fao.org/waicent/faoinfo/economic/ESC/esce/ cmr/cmrnot

Site that includes commodity notes, tables of apparent consumption, estimated value of fishery production by groups of species, trade flow by region, international exports by species and year, and the relative importance of trade in fishery products.

www.fda.gov

The U.S. Food and Drug Administration web site that includes the mission statement, summaries of what FDA regulates, and its history.

www.feap.info

Site of the Federation of European Aquaculture Producers, an international organization composed of the national aquaculture associations of European countries.

www.fedstats.gov

Provides statistical profiles of States, counties, cities, Congressional Districts, and Federal judicial districts; comparison of international, national, State, county, and local statistics; descriptions of the statistics on agriculture, demographics, economics, environment, health, natural resources and others.

www.foodconnex.com

Site of Foodconnex, an e-commerce platform and software hosted by Integrated Management Solutions, a leading provider of technology solutions to the Food Distribution and Processing Industries.

www.gaalliance.org

Site of the Global Aquaculture Alliance (GAA), an international non-profit trade association dedicated to advancing environmentally responsible aquaculture, including the GAA Individual Codes of Practice Food Safety, including third-party verification.

www.globefish.org/presentations/presentations.htm

Site of Globefish, a publications unit within FAO that publishes a wide variety of reports and analyses related to fish and seafood markets around the world, including global overviews, world market reports by species, specific market situation analyses, international trade, fishmeal, and trade barriers.

www.ifdaonline.org/index.html

Site of the International Foodservice Distributors Association (IFDA), a trade organization representing foodservice distributors throughout the United States, Canada, and internationally.

www.infofish.org

Site of Infofish, that publishes articles on capture fisheries and aquaculture, processing, packaging, storage, transport, and marketing; includes announcements of upcoming meetings and seafood shows.

www.iso.org

Site that summarizes the ISO programs, members, and offers copies of a variety of technical summaries and brochures of the more than 14,000 International Standards for business, government and society.

www.ifremer.fr/cofepeche/referencetexten/text/ marche.htm

Web site of Ifremer (French Research Institute for Exploitation of the Sea) that publishes market studies on pilot projects, pricing, sector studies, socioeconomic studies, and market appraisals in France, Europe, Africa, Asia, and Latin America.

www.macmap.org

Site of Market Access Map that covers customs tariffs (import duties) and other measures applied by importing countries.

www.members.tripod.com/Tanganyika

Site that includes contact information for ornamental fish trade companies around Lake Tanganyika, online magazines, books, and photos.

www.montereybayaquarium.org

Site of the Monterey Bay Aquarium, source of a pocket guide for fish consumers that judges how sustainable each type of fish is.

www.money.howstuffworks.com

Site that discusses how marketing plans work.

www.morebusiness.com

Site that includes templates for developing marketing plans, sample market plans, and includes software for business planning.

www.nal.usda/atmic/pubs/srb9303.htm

Site of the National Agricultural Library (ARS, USDA) that posts information under "Seafood Marketing Resources" on aquaculture, trade, databases, hearings, legislation, seafood shows, and lists of distributors, exporters, and importers.

www.nmfs.noaa.gov/aquaculture.htm

The National Oceanic and Atmospheric Administration (NOAA) site provides information on U.S. aquaculture, bycatch, grants, international interests, legislation, permits, and recreational fisheries, the Department of Commerce's aquaculture policy, National Aquaculture Act of 1980, NOAA Aquaculture Policy, Policy Paper on the Rationale For a New Initiative in Marine Aquaculture, Department of Agriculture's National Aquatic Animal Health Plan, the Environmental Protection Agency's final aquaculture effluents rule, and a draft Code of Conduct for Responsible Aquaculture Development in the U.S.

www.nmfs.noaa.gov/trade/DOCAQpolicy.htm

Main front page of the U.S. Department of Commerce that outlines the mission and vision statements and objectives, for U.S. aquaculture.

www.ornamentalfish.org

Site of the Ornamental Aquatic Trade Association Worldwide that includes marketing and trade statistics for the ornamental fish trade, including a Code of Conduct for businesses, water quality criteria, and a customer charter.

www.ornamental-fish-int.org

Site of the Ornamental Fish International (OFI) that includes a Code of Ethics.

www.paloalto.com

Site that contains sample market plans and includes tutorials on how to write a marketing plan.

www.rurdev.usda.gov/rbs/coops/csdir.htm

Site of the Rural Development Agency of the U.S. Department of Agriculture that provides information on cooperative programs, data, charts, publications, and funding opportunities for research.

www.salmonfarmers.org

Site of the British Columbia Salmon Farmers Association, the association that is the voice of British Columbia's farmed salmon industry.

www.seafoodbusiness.com/archives/02feb/news_trade

Site of Seafood Business that provides a summary of the out-of-court settlement between Great Eastern Mussel Farms of Maine and mussel producers from Prince Edward Island, Canada.

www.srac.org/publications

This web site provides downloadable publications on a wide variety of aquaculture topics, including enterprise budgets for a variety of species and production systems, business planning and marketing information.

www.st.nmfs.gov/st1/market_news/index.html

Site of Fishery Market News of the U.S. National Marine Fisheries Service (NMFS) that reports data on shrimp and finfish landings, imports, exports, storage, and prices in various fish markets and auctions as well as data on fishmeal and fish oil production.

www.thenaa.org

Site of the National Aquaculture Association, a U.S. producer-based association dedicated to establishment of national programs that further the common interest of membership. Pages include: environmental stewardship, policies and resolutions, current issues, industry events.

www.tns-sofres.com

Site of SECODIP, a source of consumer panel survey data for France.

www.trademap.org

Site that provides trade statistics for international business development.

www.usace.army.mil

Provides information on policies of the U.S. Army Corps of Engineers on policies and permits for work done within their area of jurisdiction, including construction in floodplain areas.

www.uaex.edu/aqfi

Site of the University of Arkansas at Pine Bluff that provides extension information on technical assistance to the aquaculture industry, including economic, biological, and technical information. This is a comprehensive web site on U.S. warmwater aquaculture for the University of Arkansas at Pine Bluff Aquaculture/Fisheries Center, and includes a variety of

Excel-based spreadsheets to develop cost estimates, project fish growth and sales, and to develop cash flow-based management plans.

www.uaex.edu/OtherAreas

Site of the University of Arkansas that includes information on farm machinery costs.

www.uccnet.org

Site for the Uniform Code Council's (UCC) subsidiary UCCnet™, providing tools to synchronize item information and the transfer of information in a business-to-business environment.

www.unep-wcmc.org/marine/GMAD

Site of the United Nations Environment Programme-World Conservation Monitoring Centre along with the Marine Aquarium Council, including a database on 2,399 species from 45 representative wholesale exporters and importers.

www.usda.gov/agency/oce/waob/index.htm

Site of the World Outlook Board of USDA and serves as the focal point for economic intelligence on the outlook for U.S. and world agriculture with forecasts of supply and demand for major commodities at the world level, and for livestock products and refined sugar at the U.S. level.

www.usda.gov/nass/

The National Agricultural Statistics Service (NASS) site provides statistical information on aquaculture that includes, publications, charts and maps, historical data, statistical research, and Census of Agriculture.

www.usda.mannlib.cornell.edu/usda/

Contains nearly 300 reports and datasets from the economics agencies of the U.S. Department of Agriculture, covering U.S. and international agriculture and related topics. Aquaculture falls under Specialty Agriculture and Aquaculture Outlook (by ERS), Catfish Processing: Dataset (by NASS), Catfish Processing: Report (by NASS), Catfish Production (by NASS), and Trout Production (by NASS).

www.uscatfish.com

Site of the Catfish Farmers of America (CFA), that provides information on farmers, processors, feed mills, researchers, and suppliers.

www.usitc.gov

Official site of the International Trade Commission. Includes information on antidumping and countervailing duty orders for product group, country, and data, daily and weekly reports, and tariff schedules.

www.ustfa.org

Site of the U.S. Trout Farmers Association that provides the Trout Producer Quality Assurance Program.

www.was.org

Site of the World Aquaculture Society, an international non-profit society founded in 1970 with the objective of improving communication and information exchange within the diverse global aquaculture community.

www.wisc.edu/uwcc/

Site of the University of Wisconsin Center for Cooperatives (UWCC) website that provides information on all aspects of cooperatives including business principles, organizing cooperatives, cooperative financing, cooperative structure, cooperative management, leadership and governance, and related topics for both agricultural and consumer cooperatives.

www.worldwildlifefund.org

Site of the World Wildlife fund, the world's leading conservation organization.

www.wto.org

Site of the World Trade Organization, the only global international organization dealing with the roles of trade among nations, including a training package, videos, list of members, publications, calendar of events, news releases, committee reports, and international trade statistics.

Glossary

Accounts payable: An expense that has been incurred but not yet paid.

Accounts receivable: Income that has been earned but for which no cash payment has been received.

Acid test ratio: A measure of liquidity that is calculated from the balance sheet. The targeted minimum value is one. It is calculated as the sum of cash, marketable securities, and accounts receivable divided by current liabilities.

Additional costs (partial budget): Subcategory of a partial budget that itemizes any increase in expenditure that would result from making the relatively small change on the farm. Included under the major category of Costs.

Additional revenue (partial budget): Subcategory of a partial budget that itemizes any increase in revenue that would result from making the relatively small change on the farm. Included under the major category of Benefits.

Agricultural Credit Bank: One of the seven banks (the others are regional farm credit banks) that compose the Farm Credit System (FCS) of the Farm Credit Administration. FCS is a federally chartered network of cooperatives that lends to agricultural producers, rural homeowners, farm-related businesses, and agricultural, aquatic, and public utility cooperatives. Created by the Farm Credit Act of 1971.

American Society of Agricultural and Biological Engineers: An educational and scientific organization dedicated to the advancement of engineering applicable to agricultural, food, and biological systems. Founded in 1907.

Amortization schedule: A table detailing each periodic payment on a loan that has been distributed into smaller payments.

Annual capital recovery charge: Method to determine the annual cost of a fixed input that also accounts for the opportunity cost of the capital tied up in the fixed input.

Annual percentage rate (APR): True annual rate at which interest is charged on a loan.

Antidumping duties: Levies on products that are deemed to be imported at less than fair market value.

Antidumping petition: A legal procedure filed by domestic firms with the Department of Commerce (U.S.) and the International Trade Commission (in the United States) to seek redress against sales of imported products at prices below their fair market value.

Antidumping tariffs: Duties imposed on goods imported from companies found to be selling at prices below their fair market value at a percentage rate calculated to counteract the margin of difference between the sales price and the fair market price.

Aquatic nuisance species: An organism that threatens the diversity or abundance of native species or the ecological stability of infested waters, or the commercial, agricultural, aquaculture, or recreational activities dependent on such waters.

Assets: Physical or financial property that has value and is owned by a business or individual.

Assets, current: Assets normally used up or sold within a year.

Assets, depreciable: Assets that have a definable useful life such that the value of the asset declines with its use.

Assets, noncurrent: An asset that will normally be owned or used up over a period longer than a year.

Aquaculture Economics and Financing: Management and Analysis, Carole R. Engle, © 2010 Carole R. Engle.

Assets, total (balance sheet): The sum of the current and noncurrent assets on the balance sheet.

Assets, total current: Assets that are in the form of cash or that can be converted into cash in the next 12 months. Inventory is included in current assets.

Assets, total noncurrent: The sum of the assets that will normally be owned or used up over a period longer than a year.

Asset-generating loan: Loan that provides a source of collateral used by the lender to secure the loan in the event of a default of loan repayment.

Asset turnover ratio: Measure of how efficiently farm assets are being used to generate revenue.

Balance sheet: A financial report summarizing the assets, liabilities, and equity of a business at a point in time.

Balloon payments: A loan that does not fully amortize payback of all principal over the term of the note; a balance is left at maturity. The final payment is called a balloon payment because of its large size.

Bank examiners: An individual who reviews the operations of banks, including the bank's lending policies, guidelines, and practices. Employed by state and federal banking regulatory agencies.

Bayesian: Method in the fields of probability and statistics that considers the probability of the model considering all possible parameter values.

Beginning cash balance (cash flow budget): The amount of cash available to the business at the beginning of the planning period.

Benefits (partial budget): Major category of a partial budget. Sum of the additional revenue and reduced costs.

Book value: The original cost of an asset minus the total accumulated depreciation expense taken to date.

Breakeven price: The selling price for which total income will just equal total expenses for a given level of production. Calculated from the enterprise budget.

Breakeven price above variable costs: The selling price for which total income will just equal total variable costs for a given level of production. Calculated from the enterprise budget.

Breakeven price above total costs: The selling price for which total income will just equal total costs for a given level of production. Calculated from the enterprise budget.

Breakeven yield: The yield level at which total income will just equal total expenses at a given selling price. Calculated from the enterprise budget.

Breakeven yield above variable costs: The selling price for which total income will just equal total variable costs for a given level of production. Calculated from the enterprise budget.

Breakeven yield above total costs: The selling price for which total income will just equal total costs for a given level of production. Calculated from the enterprise budget.

Business plan: Formal statement of a set of business goals and the plan for reaching those goals.

Call options: Gives the holder the right to buy a particular number of shares of a designated common stock at a specified price.

Capital: A collection of physical and financial assets that have a market value.

Capital assets: An asset expected to last through more than one production cycle that can be used to produce other saleable assets or services.

Capital budget: Analysis that determines the profitability of a long-term investment.

Capital budgeting: Process of identifying, evaluating, and implementing a firm's investment opportunities.

Capital intensive: Enterprise that uses proportionately greater amounts of capital as compared to other factors of production.

Capital recovery factor: Annualized equivalent value of the initial investment cost of a capital asset.

Capital replacement and term debt repayment margin: Money remaining after all operating expenses, taxes, family living costs, and scheduled debt payments have been made. Cash generated by the farm business available for financing capital replacement such as machinery and equipment.

Capital reserves: Resource created by the accumulated capital surplus (not revenue surplus) of a firm.

Capper-Volstead Act: P.L. 67–146 is the Cooperative Marketing Association Act adopted by the U.S. Congress on February 18, 1922. It gave associations of persons producing agricultural products certain exemptions from antitrust laws.

Cash: Money or its equivalent (as a check) paid for goods or services at the time of purchase or delivery.

Cash available (cash flow budget): Quantity of money for given time period. Includes beginning cash and receipts of all types of other revenue.

Cash balance (cash flow budget): Line at bottom of cash flow budget that indicates the amount of cash available after all expenses and debt-servicing payments have been made. The ending cash balance becomes the beginning cash balance of the next time period.

Cash flow: Actual transfers of cash into or from the firm. Cash generated by the firm and paid to creditors and shareholders.

Cash flow budget: A projection of the expected cash inflows and cash outflows for a business over a period.

Cash flow coverage ratio (cash flow budget): Measure of the firm's ability to service debt (both interest and principal payments). Calculated as excess available cash divided by the sum of total interest paid plus payments on intermediate and long-term debt.

Cash flow deviation report: Tabulation of the differences between the projected cash flow and the actual cash flow.

Cash flow risk (cash flow budget): Fluctuations in the flow of cash inflow and outflow.

Cash flow statement: Financial tabulation that shows the annual flow and timing of cash coming into and out of a business.

Cash inflow (cash flow budget): Cash receipts by budgeting period. Includes beginning cash (from ending cash balance at end of previous planning period) and receipts from sale of any product during that planning period.

Cash outflow (cash flow budget): Cash expenses by budgeting period. Includes all variable costs and debt-servicing payments that include both principal and interest payments.

Cash position: The amount of cash that a farm has available to it at a given point in time.

Clean Water Act: The principal federal law related to protection of the quality of surface waters in the U.S. Created by the 1972 amendment to the Federal Water Pollution Control Act (known as the Clean Water Act or CWA). Section 402 of the CWA specifically required the Environmental Protection Agency (EPA) to develop and implement the National Pollution Discharge Elimination System (NPDES) program.

Cobb-Douglas function: A standard production function which is applied to describe how much output two inputs into a production process make. Exhibits constant returns to scale.

Coefficient of variation: Measures risk per unit of return. It is calculated as the standard deviation divided by the mean.

Compounding: Involves finding the future value of money invested today. Determines how money will grow over time.

Confidence interval: In statistics, a particular kind of interval estimate of a population parameter. Instead of estimating the parameter by a single value, an interval is given that is likely to include the parameter. Used to indicate the reliability of an estimate.

Constraints (mathematical programming): Restrictions or limitations imposed on a mathematical programming problem.

Continuous data: Data which has a potentially infinite number and divisibility of attributes.

Correlations: A statistical concept that relates movements in one set of variables to movements in another.

Costs (partial budget): Major category of a partial budget. Sum of the reduced revenue and additional costs.

Credit: Capacity or ability to borrow money.

Credit capacity: Maximum dollar amount that lenders would approve in loan funds to a particular individual or farm. Total amount of capital that could be borrowed by an individual.

Credit reserves: Source of liquidity that arises from the liability side of the firm's activities. Borrower's expectation of additional funds that lenders may be willing to loan to a firm or to an individual borrower to finance transactions and investment opportunities.

Credit scoring: Process of evaluating an individual's credit worthiness and likelihood of repayment of financial obligations.

Credit evaluation: Use of a statistical model based on applicant attributes to assess whether a loan automatically meets minimum credit standards. The model assigns values to potential borrowers' attributes, with the sum of the values compared to a threshold.

Credit worthiness: Lender's evaluation of the ability of an individual to repay borrowed capital.

Crystal Ball®: An add-in program to Microsoft® Excel that allows the user to add distributions into spreadsheet cells to replace point values.

Cultivated Clam Pilot Insurance Program: First federal U.S. crop insurance program for aquaculture. Began in 2000 in four Atlantic coast states.

Cumulative distribution function: Provides the probability that a value drawn from a distribution will be less than or equal to some specified value.

Current ratio: Total current assets divided by total current liabilities. The current ratio is used to measure short-term solvency of a firm.

Debt capital: Funds that constitute a financial obligation on which interest and other fees have to be paid.

Debt/asset ratio: Calculated as total debt (sum of current and long-term liabilities) divided by total assets. It measures the proportion of assets financed by borrowers.

Debt/equity ratio: Calculated as the total debt divided by the stockholders' equity.

Debt structure ratio: Measure of the capacity of the business to service short-term, intermediate-term, and long-term debt.

Debt servicing ratio: Measure of the sum of interest and principal payments as a proportion of the total cash available. Indicator of liquidity with respect to loan repayments.

Debt-servicing: Payment of debts according to a specified schedule.

Debtor turnover period: Efficiency ratio that examines the bills owed to the business. A high debtor period requires follow up to ensure swifter payment.

Decision criteria: Set of rules that can be used to evaluate the information available and make a decision even though the end results are not known with certainty.

Decision tree: A graphical representation of alternative sequential decisions and the possible outcomes of those decisions.

Decision variables (mathematical programming): A variable that represents an input that can be controlled, in mathematical programming models.

Depreciation: An annual, noncash expense to recognize the amount by which an asset loses value due to use, age, and obsolescence. It spreads the original cost over the asset's useful life.

Depreciation expense ratio: Measure of economic efficiency. Total depreciation expense divided by gross revenue.

Differentiated product: Good similar to others that differs in one or more ways, often in a minor attribute, such as a different color, shape, or size.

Discount rate: The interest rate used to find the present value of an amount to be paid or received in the future.

Discounting: Process of determining the present value, or the value as of today, of a future cash flow.

Discrete data: Numeric digits with no intermediate amounts possible.

Diseconomies of size: Production relation in which the average total cost per unit of output increases as more output is produced.

Distributions, beta: Family of continuous probability distributions on the interval between 0 and 1.

Distributions, gamma: Two-parameter probability distributions with a scale parameter.

Distributions, Weibull: Continuous probability distribution similar to gamma function. Can be used to model biological growth.

Diversification: Production of two or more commodities for which production levels and/or prices are not closely correlated.

Dual price (mathematical programming): Improvement in the value of the optimal solution per unit increase in the right hand side of a constraint.

Dumping: Selling products at prices below the cost of production and below normal domestic prices.

Dynamic optimization (mathematical programming): Identification of the best set of management solutions over time for various possible states.

Dynamic programming (mathematical programming): Type of mathematical programming model that allows for solutions to problems that have multiple stages using inductive principles. Defines optimal directions and payoffs for each stage depending upon the status at the beginning of the stage for various possible states.

Economic analysis: Analysis of how people, individually and in groups, allocate scarce resources among competing uses to maximize satisfaction over time.

Economic engineering: Process that uses data from university reports, manufacturers, and other secondary sources to prepare a budget instead of developing the budget based on historical farm records.

Economies of scale: Condition in which average per-unit costs decrease as the size of business increases; decreasing average costs with increasing output levels.

Economy of size: Larger companies can operate at relatively lower costs by having cost advantages.

Elasticity of demand: Degree of responsiveness of quantity demanded to a given change in price.

Enterprise budget: A projection of all the costs and returns for a single enterprise.

Equal principal payments: Loan amortization schedule in which the amount of principal paid in each periodic payment is constant but the total payment declines over the life of the loan.

Equal total payment loans: Loan amortization schedule in which the periodic payments are constant.

Equipment loan: Funds borrowed to purchase equipment.

Equity: Amount by which the value of total assets exceeds total liabilities. Amount of the owner's capital invested in the business.

Equity capital: Funds obtained from the owner(s), partners, and investors of the business.

Equity/asset ratio: Ratio of owner's equity to total assets. Measure of solvency.

E-V model (mathematical programming): Type of mathematical programming model in which the farmer's preferences for sets of management options are assumed to be based on the expected income and the associated variance of the income. See also Mean-Variance Model.

External opportunities: Component of a SWOT (Strengths, Weaknesses, Opportunities, and Threats) analysis in which potential benefits may arise from outside the immediate business context of the company.

External threats: Component of a SWOT (Strengths, Weaknesses, Opportunities, and Threats) analysis in which potentially adverse events may affect the business from outside the immediate business context.

Fair Labor Standards Act: U.S. federal law that establishes a minimum wage, guaranteed time and a half for overtime in certain jobs, and prohibits child labor.

Farm credit banks: Regional lending associations that provide funds and services to the Federal Land Bank Associations, Federal Land Credit Associations, and Production Credit Associations. Members of the Farm Credit System.

Farm credit system: Borrower-owned cooperative established by the authority of the U.S. Congress that makes loans to farmers and ranchers.

Farm Financial Standards Council: A committee of agricultural financial experts that developed a set of guidelines for uniform financial reporting and analysis of farm businesses.

Farm Service Agency (FSA): USDA agency that now includes what previously was known as the Farmers Home Administration. Also includes former Agricultural Stabilization and Conservation Service.

Federal Deposit Insurance Corporation (FDIC): Federal agency that insures deposits in member banks for up to $100,000.

Federal Insurance Contributions Act (FICA): Federal law that created a retirement and disability program commonly called Social Security.

Fee fishing: Business designed to generate revenue from fees paid by anglers to fish in a private pond or lake. Fees can be charged by the day, the hour, or the number or weight of fish caught. Also referred to as "pay lakes."

Feed conversion ratio: A measure of feed efficiency. Calculated as the pounds of feed fed divided by the pounds of weight gained by the fish during the same time period.

Feed yield ratio: Farm-wide measure of feed efficiency. Measured as the total feed fed divided by the weight of fish sold from the farm.

Financial contingency plan: Organized set of actions designed to be prepared to react quickly to adverse conditions that may negatively affect the farm business.

Financial leverage ratio: Measure of how well the business uses debt versus equity capital. Calculated by dividing the Return on Equity by the Return on Assets.

Financial analysis: Evaluation of a farm's financial health, its financial position and performance. Includes its profitability, solvency, liquidity, repayment capacity, and financial efficiency.

Financial performance: Evaluation of a farm's profitability, solvency, liquidity, repayment capacity, and efficiencies.

Financial position: Financial resources controlled by a farm and the claims against those resources.

Financial risk: Includes the cost and availability of debt capital, the ability to meet cash flow needs in a timely manner, ability to maintain and grow equity, and the increasing chance of losing equity by larger levels of borrowing against the same net worth.

Financial statement: Often used as another term for balance sheet but is a general term for other documents relating to the financial condition of a business such as an income statement, statement of cash flows.

Firm growth: Increase in the net worth and physical plant of a company.

Fishermen's Collective Marketing Act of 1934: Federal law that allows fishermen to jointly harvest, market, and price their product without violating antitrust laws.

Flow of funds analysis: Table that traces the flow of cash through the business during the year and monitors changes in assets and liabilities.

Food and Agriculture Organization: Agency of the United Nations that is responsible for issues related to worldwide food and agriculture production.

Fixed costs: Costs that will not change in the short run even if no production takes place.

Forward contracting: Contract between a buyer and seller that fixes the price of a commodity before it is delivered.

Forward price contracts: Contractual obligations for a buyer to pay a specified price for a specified volume of production from a farmer.

Full-Time Equivalents (FTEs): Standardized unit for reporting quantities of labor used that allows for adding part-time and full-time labor into one measure.

Future value (FV): Value that a payment or set of payments will have at some time in the future, when interest is compounded.

Futures contracts: Standardized, legally binding agreements to either deliver or receive a certain quantity and grade of a specific commodity during a designated delivery period.

General algebraic modeling system (GAMS): Software used commonly to run mathematical programming models.

General partnership: Form of unincorporated business with two or more co-owners who are known as general partners and who take active roles in the management and obligations of the business.

Hazard Analysis of Critical Control Points (HACCP): Food safety program that: (1) analyzes hazards; (2) identifies critical control points; (3) estab-

lishes preventive measures with critical limits for each control points; (4) establishes monitoring procedures for each critical control point; (5) establishes corrective actions; (6) develops record-keeping systems, and (7) establishes verification procedures.

Hedging: Strategy for reducing the risk of a decline in prices by selling a commodity futures contract in advance of when the actual commodity is sold.

Horizontal integration: Combining firms at the same level of the marketing chain (with similar marketing functions) join together to pursue a new marketing opportunity.

Illiquid assets: Assets that are expected to be used in the business for many years and are not easily converted to cash.

Income above variable costs (enterprise budget): Measure of receipts that remain after charging all variable costs in an enterprise budget.

Income statement: Table that itemizes revenues received by source and expenses for a specific period of time. Also known as a profit and loss statement.

Income elasticity: Measure of the response of the quantity demanded to changes in income.

Installment loan: Amortized loan. Loan that has periodic interest and principal payments.

Internal strengths: Component of a SWOT (Strengths, Weaknesses, Opportunities, and Threats) analysis in which opportunities arise from specific skills and talents available from within the company.

Internal weaknesses: Component of a SWOT (Strengths, Weaknesses, Opportunities, and Threats) analysis in which areas of deficiency within the company result in negative and problematic conditions.

Interest: Amount paid to a lender for the use of borrowed money.

Interest coverage ratio: Measure of solvency that accounts for the relative claims on the returns to farm assets. Calculated by dividing the firm's return to assets by the amount of interest charges for a specific time period.

Interest expense ratio: Measure of economic efficiency. Calculated by dividing farm interest expense by gross revenue. Indicates degree of dependence on borrowed capital.

Interest on operating capital: Charge for use of funds expended to cover the variable costs of the business.

Interest on investment: Charge for use of funds expended annually for assets and resources that are used over a period of time greater than a year, i.e., land, equipment, buildings.

Interest rate: Percentage of amount due that is charged for use of money borrowed.

Internal rate of return (IRR): Discount rate that sets the net present value of the investment to zero.

Inventory management: Decision making related to the number, type, and value of assets that are owned at a point in time.

Investment analysis: Analysis of the profitability of a dollar invested in a specific business venture over the course of its business life.

Investment capital: Funds used in the process of adding assets to a business.

Labor-capital substitution: Of the four factors of production (land, labor, capital, and management), labor and capital can substitute for each other in varying proportions. For example, an earthen pond can be dug entirely by hand if enough people and time are available, or it can be constructed in a much shorter period of time with many fewer people by using a tractor and pan and scraper.

Labor efficiency: Measure of the output, cost, or income from an enterprise or farm per person-year. Examples include the value of the farm production per employee, the labor cost per crop acre, the crop acres per person, etc.

Labor schedule: Table that shows the quantity and type of labor required for each activity on the farm by time period (week or month).

Lacey Act: U.S. federal law administered by the Department of the Interior under which it is unlawful for any person to import, export, transport, sell, receive, acquire, or purchase any fish or wildlife or plant taken, possessed, transported, or sold in violation of any law, treaty, or regulation of the United States or in violation of any Indian tribal law whether in interstate or foreign commerce. Used to restrict interstate transport of nuisance species listed under the Lacey Act.

Leasing: Act of entering into an agreement that allows a person to use and/or possess someone else's property in exchange for a rental payment.

Leverage: Practice of using credit to increase the total capital managed beyond the amount of owner equity.

Liabilities: Financial obligations (debt) that must be paid at some future time

Liabilities, current: Liabilities normally paid within a year. Obligations expected to require cash payment within one year.

Liabilities, noncurrent: Financial obligations that will normally be paid over a period longer than a year.

Liabilities, total current: Total of all financial obligations to be paid within a year.

Liabilities, total: Sum of total current liabilities and total noncurrent liabilities.

Liabilities, total noncurrent: Sum of all financial obligations that would be paid over a period longer than a year.

Liability insurance: A type of insurance designed to offer specific protection against third-party claims, someone who suffers a loss who is not a party to the insurance contract.

Limited partnership: Form of business in which more than one person has ownership, but some (the limited partners) do not participate in management and have liability limited to the amount of their investment.

Line of credit: Arrangement by which a lender transfers loan funds to a borrower as they are needed, up to a maximum amount.

Linear programming: Mathematical model with a linear objective function, a set of linear constraints, and non-negative variables.

Liquid assets: Assets that can be converted readily to cash.

Liquid reserves: Assets held that are not obligated and that can be readily converted to cash in the event of unanticipated adverse circumstances.

Liquidity: The ability of a business to meet its cash financial obligations as they come due.

Livehaulers: Individuals who own a fish hauling truck and contract to transport fish.

Loan amortization: Table that spreads the repayment of a loan out over the specified number of payments for the length of the loan and indicates the amount of each payment that is applied to the principal and that of the interest payment.

Loan limits: Maximum amounts that a lender will loan to an individual or a business.

Loans: Sum of money borrowed from a bank on a contractual basis that determines the length of the lending period, the interest rate, and the number of payments to be made.

Long position: Situation in which the trader owns the security in question and will profit if the price of the security goes up.

Long-run average cost curve: Graphical representation of the relationship between the cost per unit of production and the total quantity of production. Tends to fall with increasing quantity produced if economies of scale are present and increases if diseconomies of scale exist. Developed by combining a series of short-run average cost curves and then developing a curve that envelops the short-run cost curves.

Long-term capital: Funds used to acquire fixed assets that will be used for a number of years in the business.

Long-term debt to equity: Ratio of long-term debt to the equity in the business to assess the effect of the level of long-term debt to the level of financial risk of the business.

Management: Making and implementing decisions that allocate limited resources in ways to achieve an organization's goals.

Management intensive: Enterprise that uses proportionately greater amounts of management expertise and time as compared to other factors of production and to other types of enterprises.

Market power: Ability of a firm to alter the market price of a good or service.

Market risk: Variability and fluctuations in market prices, market requirements, and market access.

Marketing cooperatives: Special type of corporation in which the members who contribute capital enjoy limited liability. Remaining net income is distributed to members typically based on the amount of business of each member. Members must perceive that working together will provide more benefits than operating independently.

Marketing plan: Document that outlines and describes the current situation of the industry and firm, the marketing goals and objectives of the business, and describes the series of actions necessary to accomplish the goals and objectives of the plan.

Mathematical programming: Method of determining the best use of resources to achieve either a profit-maximizing or a cost-minimization objective.

Maturity period: The amount that will be received at the time a security is redeemed.

Maximin: Decision criterion that selects the action that, after identifying the minimum return of each possible action, has the largest minimum return. Chooses the maximum of the minimums.

Maximum expected returns: Decision criterion for risk management that is based on selecting the option with the largest total weighted potential returns.

Mean-variance model (mathematical programming): Type of mathematical programming model in which the farmer's preferences for sets of management options are assumed to be based on the expected income and the associated variance of the income. See also E-V model.

Median: Middle value of an ordered group of observations.

Migratory Bird Act: Federal statute that protects migratory birds. The original 1918 statute implemented a 1916 Treaty between the U.S. and Great Britain (for Canada) for the protection of migratory birds. Prohibits hunting, capturing, killing, or possessing migratory birds.

Minimax: Decision criterion that selects the action that, after identifying the maximum regret of each possible action, has the smallest maximum regret. Chooses the minimum of the maximums.

Minneapolis Grain Exchange: Regional market place organized to promote fair trade and prevent abuses in the market for wheat, corn, and soybeans. First organized as a cash market and then futures and options markets.

Mode: Most frequent value in a set of observations.

Monte Carlo simulations: Technique that involves using random numbers and probability to solve problems.

MOTAD (mathematical programming): Minimization of the Total Absolute Deviations model. Incorporates risk into the model and is most relevant when the variance of the farm income is modeled with sample data.

Multiperiod programming (mathematical programming): Type of mathematical programming model that incorporates several planning periods, such as multiple years or multiple quarters.

Multiple-peril insurance: Insurance program that provides comprehensive protection against weather-related causes of loss and other unavoidable perils to crop yield.

Net benefit (partial budget): Bottom line of a partial budget. Calculated by subtracting the value of Costs from that of Benefits. If positive, the change analyzed is considered to be economically worthwhile.

Net capital ratio: Measure of solvency. Calculated by dividing total assets by total liabilities.

Net farm income: The difference between total revenue and total expenses, including gain or loss on the sale of all capital assets. Also, the return to owner equity, unpaid labor, and management.

Net farm income from operations: The difference between total revenue and total expenses, not including gain or loss on the sale of certain capital assets.

Net present value: Sum of the present values of future after-tax net cash flows minus the initial investment.

Net returns (enterprise budget): Measure of the profitability of an enterprise. Bottom line on an enterprise budget.

Net returns to operator's labor and management (enterprise budget): Measure of the profits earned by an enterprise from use of all resources other than the operator's labor and management.

Net returns to risk (enterprise budget): Measure of the profits earned by an enterprise from use of all resources other than risk.

Net worth: The difference between the value of the assets owned by a business and the value of its liabilities. Also called owner equity.

Niche market: An area in which there is little competition for a specific type of product.

Non-negativity (mathematical programming): Condition that requires that decision variables take on a zero or positive value.

Normal distributions: Continuous probability distribution that describes data that clusters around a mean or average. The graph of the associated probability density function is bell-shaped, with a peak at the mean.

NPDES permit: National Pollution Discharge Elimination System. Federal law enforced by state environmental management agencies. Requires permits to discharge if the quantity of effluent or its concentration exceeds certain specified levels.

Objective function (mathematical programming): The mathematical equation that defines the variables to be maximized or minimized.

Office International de Epizooties: Now the World Organization for Animal Health. International governmental organization founded to: (1) guarantee the transparency of animal disease status worldwide; (2) collect, analyze, and disseminate veterinary scientific information; (3) provide expertise and promote international solidarity for the control of animal diseases; and (4) guarantee the sanitary safety of world trade by developing sanitary rules for international trade in animals and animal products.

Operating capital: Funds used to cover the variable costs of the business.

Operating expense ratio: Measure of economic efficiency. Calculated by dividing total operating expenses by gross revenue.

Operating line of credit: Type of loan in which the borrower negotiates to receive funds up to an approved maximum amount. Income received is used to pay first the accumulated interest on the loan, then is applied to the outstanding principal. There is no fixed repayment schedule.

Operating profit margin ratio (OPMR): Value represented by net farm income from operations, plus interest expense, minus opportunity cost of operator labor and management, expressed as a percentage of gross revenue.

Opportunity costs: The income that could be received by employing a resource in its most profitable alternative use.

Operating loans: Funds borrowed to cover variable costs of operating the farm business.

Optimization: Name of a group of tools to solve managerial problems in which the decision maker must allocate limited resources among various activities to identify the best outcome.

Organizational structure: Type of business ownership such as sole proprietorship, partnership, limited liability company, etc.

Outlier: One or more observations in a dataset that are so extreme in value that their inclusion is questioned for sampling error.

Owner equity: The difference between the total value of the assets of a business and the total value of its liabilities; also called net worth.

Parameter: Quantity that describes or characterizes a population. Typically estimated by random sampling of the population. Estimates referred to as statistics.

Partial budget: An estimate of the changes in income and expenses that would result from carrying out a proposed change in the current farm plan.

Partnership: Form of business organization in which two or more partners have joined to operate a business.

Pay lakes: Business designed to generate revenue from fees paid by anglers to fish in a private pond or lake. Fees can be charged by the day, the hour, or the number or weight of fish caught. Also known as "fee fishing."

Payback period: Number of years required to recover the initial cost of the investment.

Payment (lending): Periodic expense of a loan in which a portion of the principal and the accrued interest are returned to the lender.

Payoff matrix: Table of potential returns or payoffs that could be obtained if certain actions are taken and certain events occur.

Present Value (PV): Current value of a set of payments to be received or paid out over a period.

Price risk: Fluctuations in market prices that can create adverse economic consequences for a farm business.

Pricing: Strategy to select the value at which to offer a product in the market place.

Principal: Amount borrowed, or the part of the original loan that has not yet been repaid.

Probability density function: Relationship that describes the density of probability at each point in the sample space. Also known as a probability distribution function.

Product-space map: Diagram of dual continuums of price and quality across which products can be positioned to assist in the identification of the most effective pricing strategy.

Product positioning: Process of selecting the price-quality point in the market that matches consumer expectations.

Product life cycle: Patterns of sales of a product from the point when it is first introduced into the market until it reaches a point of decline in sales revenue.

Production efficiency: Quantity of production per unit of a factor of production, i.e., quantity produced per acre, quantity produced per dollar of investment capital.

Production risk: Fluctuations in the level of production that occur due to weather, pests, diseases, technol-

ogy, machinery efficiency and reliability, and quality of inputs.

Promotion: Efforts made to communicate the desirable attributes of a product to potential buyers. Includes advertising and public relations initiatives by the business.

Products with existing demand: A good that is produced that is currently available in the market.

Profit, accounting: Gross revenue minus total expenses where both values are computed using standard accounting principles and practices.

Profit, economic: Accounting profit less opportunity costs on all unpaid resources.

Profit and loss statement: Table that itemizes revenues received by source and the expenses for a specific period of time. Also known as the income statement.

Profitability: The degree or extent to which the value of the income derived from a set of resources exceeds their cost.

Proportionality (mathematical programming): Linear property that requires that the value of each variable is directly proportional to its use.

Proprietorship: Ownership of a business.

Put option: Contract that gives the buyer the right to sell a futures contract for an agricultural commodity at a specified price.

Quadratic programming (mathematical programming): Type of mathematical programming model that includes quadratic equations to represent the variance, or risk of various parameters.

Rate of return on farm assets (ROA): Percentage value represented by net farm income from operations, plus interest expense, minus the opportunity cost of operator labor and management.

Rate of return on farm equity (ROE): Percentage value of net return generated by the business before gains or losses on capital assets are realized, but after the value of unpaid labor and management is subtracted.

Real estate loan: Funds borrowed for the purpose of acquiring assets such as land and buildings.

Recommended management practices: Set of strategies and decisions that follow guidelines of established authorities in the particular subject matter.

Reduced costs (partial budget): Subcategory of a partial budget that itemizes any decreases in cost that would result from making a relatively small change on the farm. Included under the major category of Benefits.

Reduced costs (linear programming): Amount by which an objective function coefficient would have to improve (increase for a maximization problem, decrease for a minimization problem) before it would be possible for the corresponding variable to assume a positive value in the optimal solution.

Reduced revenue (partial budget): Subcategory of a partial budget that itemizes any decrease in revenue that would result from making the relatively small change on the farm. Included under the major category of Costs.

Redundancy (mathematical programming): One or more constraints in a linear programming model that do not affect the feasible region.

Regrets matrix: Table of potential regrets that result from having chosen a certain action instead of any other alternative action. Calculated from a payoff matrix.

Repayment capacity: Ability to cover cash outflow from cash inflows over a period of time.

Retained earnings: Net income generated by a farm business used to increase owner equity rather than being withdrawn to pay for living expenses, taxes, or dividends.

Return to labor (income statement): That portion of net farm income (profit measured on the income statement) that can be attributed to the labor used in the business.

Return to management: Net return generated by a business after all expenses have been paid and the opportunity costs for owner's equity and unpaid labor have been subtracted.

Return to labor and management (income statement): That portion of net farm income (profit measured on the income statement) that can be attributed to the combined labor and management used in the business.

Revenue: Economic gain resulting from the production of goods and services, including receipts from the sale of commodities, other cash payments, increases in inventories, and accounts receivable.

Revenue, total: Income received from sale of the total physical product; same as total value product.

Right-hand side value (mathematical programming): Capacities or availability of the various resources, usually expressed as an upper or lower limit. Express minimum requirements for a greater-than-or-equal to constraint.

Risk: A situation in which more than one possible outcome exists, some of which may be unfavorable.

@RISK®: An add-in program to Corel LOTUS-1-2-3 that allows the user to add distributions into spreadsheet cells to replace point values.

Risk programming model (mathematical programming): Type of mathematical programming model that incorporates risk-averse behavior.

Risk management: Systematic application of management policies, procedures, and practices to the tasks of identifying, analyzing, assessing, treating, and monitoring risk.

Risk and returns comparison: Evaluation of the returns possible at various levels of risk; alternatively, the risk levels associated with various levels of returns (profits).

Safety-first rule: Decision criterion that first eliminates all possible actions that violate the safety-first rule and then chooses from the remaining actions that which has the largest expected return.

Security collateral: Asset pledged as support for a loan application. If the borrower defaults on the loan, the asset becomes property of the lender.

Self-liquidating loan: Loan that will be repaid from the sale of assets originally purchased with the loan funds.

Sensitivity analyses: A procedure for assessing the riskiness of a decision by using several possible price and/or production outcomes to budget the results, and then comparing them.

Separability (mathematical programming): Property that specifies that various levels of a variable are independent and not indivisible.

Shadow pricing: Value obtained from a linear programming solution that shows the amount by which total gross margin would be increased if one more unit of a limiting input were available.

Share leases: Agreement in which a percentage of the physical yield is returned to the landowner in return for allowing the tenant to use the land.

Short position (hedging): Promise to sell a certain quantity of a good for a particular price in the future.

Short-term capital: Funds used in a business over the near term, typically a period of one year.

Simple rate of return (investment analysis): Measure of profit from a business investment. Calculated as the average net returns divided by the value of the investment and multiplied by 100.

Simulation model: Descriptive technique for conducting experiments with a system by checking the performance of different configurations or scenarios of the system. Often based on a computerized mathematical model of a management system operating for an extended period of time. Shortcoming is that it does not identify the best alteration.

Single-payment loan: Loan that is structured in a manner that both the principal and the total interest charged are repaid in one payment.

Small Business Administration: Agency of the U.S. federal government charged with advocacy for small-scale businesses.

Sole proprietorship: Business owned by one owner. All profits, losses, and liabilities accrue to the owner.

Solvency: Ability to pay off all debts at a certain point in time.

Specialty markets: Market developed for a product that is highly differentiated and occupies a higher-priced position of this segment in the market.

Statement of cash flow: Summary of the actual cash inflows and cash outflows experienced by a business during an accounting period.

Stochastic cost function: Relationship of determinants of costs which incorporates probabilities associated with fluctuations in their values.

Stochastic dominance: Method developed that uses partial information on the investor's preferences or random variables to order the available investment options from best to worst.

Stock turnover ratio: Calculates how many times the business has sold the value of its stocks during the year. Calculated by dividing the value of goods sold by the value of its stock.

Strategic goals: Targets established for strengthening the farm's overall position and competitive vitality.

Subchapter C Corporation: Corporation in the United States that is taxed separately from its owners. The corporation is taxed as a separate business entity.

Subchapter S Corporation: Corporation with between 1 and 100 shareholders that passes through net income or losses to shareholders.

Sunk cost: Cost that is not affected by the decision made. Cost that will be incurred regardless what values the decision variables assume.

Supply chain management: Managing the flow of resources, final products, and information among input suppliers, producers, resellers, and final consumers.

Tableau (mathematical programming): Array of coefficients in which the columns represent alternative production or marketing activities and the rows represent the limitations on those activities.

Target MOTAD (mathematical programming): Type of MOTAD risk model in which a target income is set and the model solves for the farm plan that deviates the least from this target.

Tariffs: Tax placed by a government on internationally traded goods, generally imports. Also called customs duties.

Technical risk: Fluctuations in the production or yield of the crop that cause adverse economic and financial outcomes.

Term debt and capital lease coverage ratio: Measure of the debt repayment capacity of the business. Ratio of the cash available for term debt payments for the past year, divided by the total of the term debt payments due in the next year, including principal and interest on amortized loans plus capital lease payments.

Terms of lending: Constitute the length of time of the loan, the interest rate, the number of payments, and the amount of each payment.

Total costs (enterprise budget): The sum of total fixed cost and total variable cost.

Total fixed costs (enterprise budget): The sum of all fixed costs.

Total variable costs (enterprise budget): The sum of all variable costs.

Transshipment (mathematical programming): Type of mathematical programming problem that allows for intermediate destinations in optimizing transportation of products from the locations where produced to those where purchased and consumed.

Triangular distributions: Estimating probabilities using only the most likely, lowest, and highest values.

Turnover ratio: Measure of the number of times a company's inventory is replaced during a given time period.

Unbounded solution: Situation in which the values in the solution can take on infinitely large values (in a profit-maximization model) and infinitely small values (in a cost-minimization model).

Uncertainty: Knowing some of the possible outcomes but not the probabilities.

Undercapitalization: Level of funds that is inadequate to meet all funding needs of the business.

Unpaid family labor: Quantity of uncompensated labor provided by family members.

Unsecured loan: Loan for which the borrower does not give the lender the right to possess certain assets if the repayment terms are not met. No collateral.

Valuation: Method to estimate the value of business assets.

Valuation, cost basis: Method that values assets by the original cost. Used for land, feed, and other supplies and purchased feeder livestock.

Valuation, market basis: Method that values assets by using its current market price. Used for stocks, bonds, and feeder livestock, items that will be sold in near future.

Variable costs (enterprise budget): Charges that reflect resources used that depend on and vary with the volume of production.

Variance (risk analysis): Measure of the dispersion or variability in the random variable.

Vertical integration: When a firm operates at more than one level of a series of levels leading from raw materials to the final consumer in the business chain.

Veterans Health Administration: Agency of the U.S. Department of Veterans Affairs. Provides health benefits to veterans of the U.S. Armed Forces.

Whole-farm modeling: Method to analyze the entire farm in one economic and financial model. Can include budgeting or mathematical programming methods.

Working capital: Difference in value between current assets and current liabilities. Measure of liquidity.

Years of useful life: Number of years used to fully depreciate a depreciable asset.

Yield risk: Variability in the yield of aquaculture crops that can result in adverse economic and financial outcomes.

Index

Acid test ratio 51, 52, 241
Actual Percentage Rate (APR) 85
Additional costs (partial budget) 127, 192, 241
Additional revenue (partial budget) 127, 241
American Society of Agricultural Engineers 84, 236, 241
American Statistical Association 235
Amortization factor 85
Amortization schedule 176, 177, 179, 180, 181, 241
Annual Capital Recovery Charge 84, 241
Annual cost and returns 28, 125, 189
Antidumping 223, 224
   Petitions 224, 241
   Tariffs 223, 224, 241, 252
Aquaculture Association of Canada 222, 236
Aquaculture Certification Council 225
Aquaculture insurance 100
Aquatic Nuisance Species (ANS) 223, 225, 241
Arkansas Bait and Ornamental Fish Growers Association 222
Arkansas Baitfish Certification Program 225
Arkansas State Department of Agriculture 225
Assets 54, 131, 241
   Current assets 53, 132, 241
   Depreciable assets 134, 241
   Non current assets 53, 132, 241
   Total assets 54, 242
   Total current assets 135, 242
   Total non current assets 135, 242
Asset-generating loan 173, 242
Association of Chilean Salmon Farmers 222
Australia 103, 219
Australasia 100
Atlantic cod 203

Baitfish 225
Balance sheet 30, 31, 34, 35, 64, 65, 67, 85, 131, 135, 242
Banks 177, 182
   Agricultural Credit Bank 177, 241
   Farm Credit Banks 177, 245
   Farm Credit System 59, 60, 177

Bank examiners 59, 65, 242
Bayesian 202, 242
Bighead carp 225, 226
Black carp 226
Brazil 224
Break-even prices 121, 124, 242
Break-even probability 95, 96
British Columbia Salmon Farmers Association 222
Budget 118
Business plan 4, 23, 242
Business startup 3

Call options 98, 242
Canada 103, 219, 224
Capital 170, 242
   Asset 81, 159, 242
   Budget 159, 242
   Budgeting 84
   Costs 81
   Debt 170
   Equity 43, 44, 46, 48
   Good 159
   Investment capital 4, 123
   Long-term capital 179
   Operating capital 4
   Reserves 173, 242
Capital intensive 4, 242
Capital Recovery Factor 84, 242
Capital replacement and term debt repayment margin 52, 71, 242
Capper-Volstead Act 19, 242
Carp 129
Carrefour 17, 225
Cash 242
   Beginning cash balance 147, 242
   Cash available 146, 243
   Cash balance 146, 243
   Cash inflow 143, 145, 243
   Cash outflow 143, 146, 243
   Cash position 147, 243
   Deviation report 74, 243

Cash flow 31, 67, 143, 233, 243
    Budget 31, 32, 34, 36–37, 38–39, 75–76, 77–78,
        86, 143, 149–150, 151–152, 154–155, 156–157,
        243
    Coverage ratio 50, 86, 243
    Risk 53, 86, 243
    Statement 143, 145, 147, 243, 252
Catfish
    Basa 224
    Channel 24, 27, 98, 148, 166, 167, 193, 203, 204, 209,
        215, 228, 232
    Clarias 194
    Striped 129
    Walking 129
Catfish Farmers of America 222
Central America 218
Certification 225
Chile 100, 103, 224
China 224, 225, 227
Clean Water Act 220, 221, 222, 243
Cobb-Douglas function 203, 243
Coefficient of variation 198, 243
Collateral 174
Compounding 161, 243
Confidence interval 198, 243
Constraints 211, 212, 243
Continuous data 199, 200, 243
Corporation 6
    Subchapter C 6, 252
    Subchapter S 6, 252
Correlations 97
Costs 119
    Capital 85
    Fixed 119, 190, 246, 252, 253
    Investment 187
    Production 43, 121, 188
    Variable 119, 189, 252, 253
Crawfish 194, 201, 224
Credit 173, 243
    Capacity 61, 86, 174, 208, 211, 243
    Determinants 61
    Evaluation 61, 243
    Reserves 61, 99, 243
    Scoring 61, 62, 243
    Sources 59, 177
    Worthiness 62, 243
Crystal Ball™ 101, 201, 202, 244
Cultivated Clam Pilot Insurance Program 100,
    244
Cumulative distribution function 199, 243
Current ratio 49, 50, 54, 101, 244

Debt/asset ratio 47, 48, 54, 86, 244
Debt/equity ratio 47, 48, 244
Debt servicing ratio 50, 86, 244
Debt structure ratio 47, 48, 49, 244
Debtor turnover period 51, 244

Decision
    Criteria 95, 244
    Rule 95
    Tree 93, 94, 244
    Variables 209, 211, 244
Demand 13
    Characteristics 13
    Elastic 13
    Inelastic 13
Denmark 103
Depreciation 84, 120, 124, 147, 191, 244
    Double declining balance method 120
    Expense ratio 51, 244
    Schedule 134
    Straight-line method 120
Differentiated product 14, 244
Discount 162, 244
    Discounting 162, 164, 166, 244
    Discount rate 161, 163, 164, 166, 167, 172, 244
Diseconomies of scale 82
Distributions
    Beta 200, 244
    Cumulative 200
    Gamma 200, 244
    Normal 200, 249
    Triangular 204, 252
    Weibull 200, 244
Diversification 97, 244
Divisibility 211
Dual values 214, 244
Dynamic programming 215, 244

Economic analysis 115, 244
Economic engineering 121, 186, 244
Economic performance 41
Economies of scale 81, 245
Ecuador 224
Elasticity 13, 245
Endangered Species Act 220, 221
England 100, 103
Enterprise budget 28, 29, 35, 117, 185, 245
Environmental Protection Agency 222, 237
Equipment cost 124, 188
Equity 58, 245
Equity/asset ratio 47, 48, 245
Equity capital 85, 245
European Food Safety Authority 227, 237
European Union 224, 235
Euro-Retailer Produce Working Group (EUROPGAP)
    225, 237
E-V model 214, 215, 245
External
    Opportunities 25, 245
    Threats 25, 245

Fair Labor Standards Act 111, 245
Farm Credit System 59, 60, 177, 245

Farm Financial Standards Council 71, 136, 137, 245
Farm Services Agency (FSA) 60, 245
Farmers Home Administration 177
Federal Coastal Management Zone Act of 1972 220, 221
Federal Deposit Insurance Corporation 59, 245
Federal Insurance Contributions Act (FICA) 111, 245
Federal Water Pollution Control Act 220, 221
Federation of European Aquaculture Producers 222
Fee fishing 17, 245
Feed conversion ratio 42, 189, 245
Feed yield ratio 189, 245
Financial analysis 28, 35, 115, 131, 245
Financial contingency plan 72, 245
Financial efficiency 51
Financial leverage ratio 45, 46, 245
Financial performance 41, 245
Financial plan 24
Financial position 131, 245
Financial risk 4, 245
Financing 7–8, 57
Finland 103
Finnish Fish Farms Association 222
Firm growth 83, 246
Fishermen's Collective Marketing Act 19, 246
Fishy 134
Flow of funds analysis 52, 246
Florida 187
Food and Agriculture Organization of the United Nations 225, 237, 246
Food safety 227
Forward contracting 98, 246
Forward price contracts 98, 246
France 103
Full-time Equivalents (F.T.E.'s) 30, 107, 246
Future Value 160, 161, 246
Futures contracts 98, 246

General Algebraic Modeling System (GAMS) 213, 232, 246
Germany 103
Global Aquaculture Alliance 222, 238
GLOBALGAP 226
Globefish 238
Goal programming 214
Grameen Bank 63
Great Lakes 93, 224, 225
Greece 103
Grouper 65, 193
Growth ratios 53

Haddock 194, 201
Hawaii 140
Hazard Analysis of Critical Control Points (HACCP) 108, 225, 236, 246
Hedging 98, 246
Herring 21
Holland 103

Honduras 102, 103, 202, 203
Horizontal integration 19, 246
Hybrid striped bass 209, 215, 216

Iceland 103
Icelandic Fish Farmers and Sea Ranchers 222
Illiquid assets 133, 246
Immigration Reform and Control Act 111
Income statement 31, 35, 131, 135, 138, 140, 141, 246
India 100, 224
Indonesia 224
Integer programming 214
Interest 246
    Interest rate 161, 247
    Compound interest 161
    Coverage ratio 47, 49, 246
    Expense ratio 51, 246
Internal
    Internal strengths 25, 246
    Internal weaknesses 25, 246
Internal rate of return 86, 160, 165
International Institute of Fisheries Economics and Trade 235
Inventory management 97, 247
Inventory valuation 134
Investment analysis 159, 247
Iowa State University 237
Ireland 10
Irish Aquaculture Association 222
Irish Quality Eco-Mussel Standard 221
Italy 103

Japan 100, 103
Job description 109

Kanga 194

Labor 105
    Costs 108
    Efficiency 108, 109, 247
    Management 108
    Quality 107
    Quantity 106, 107
    Relations 111
    Requirements 112
    Schedule 106, 247
Labor-capital substitution 108, 247
Lacey Act 220, 221, 247
Leasing 86, 247
Lending 173
Leverage 68, 247
Liabilities 31, 41, 54, 131, 247
    Current liabilities 31, 54, 133, 247
    Non current liabilities 31, 54, 133, 247
    Total current liabilities 31, 133, 247
    Total liabilities 31, 133, 247
    Total non current liabilities 31, 133, 247

Liability insurance 99, 247
Linear programming 209, 232, 235, 247
Liquid assets 51, 52, 247
Liquid reserves 49, 247
Liquidity 41, 49, 67, 132, 247
Lloyd's of London 100
Loans 247
   Amortization 176, 247
   Balloon Payments 177
   Equal Principal Payments 176
   Equal Total Payments 176
   Equipment 174
   Installment 176
   Line of credit 174, 175
   Maturity period 173
   Maximum capacity 176
   Operating 174
   Proposal 34
   Real estate 174
   Repayment 175
   Self-liquidating 173
   Single-payment 175
   Terms of lending 179
   Unsecured loan 175
Long position 98, 248
Long-run average cost curve 83, 248
Long-term contracts 97
Long-term debt to equity 99

Malaysia 219
Management 5, 248
Management-intensive 5, 248
Marine Harvest 19
Market
   Power 17, 248
   Risk 248
   Values 122
Marketing 13–21, 233
   Challenges 5
   Cooperatives 19, 99, 248
   Strategy 16, 20
Marketing plan 6, 16, 24, 26, 248
Mathematical programming 207, 209, 223, 248
Maximin 95, 248
Maximum expected value 95, 96
Mean-expected value 214, 215
Mean-variance Model 214, 215, 248
Median 198, 248
Mexico 103, 111, 202, 224
Migratory Bird Act 220, 221, 248
Minimax 95, 248
Minneapolis Grain Exchange 98, 248
Mode 198, 248
Monte Carlo simulations 201, 248
Monterrey Bay Aquarium 238
Most likely outcome 95, 96

MOTAD 214, 215, 248
Multiperiod programming 214, 215, 248
Multiple-peril insurance 100, 248
Mussels 224

National Aquaculture Association 222, 239
National Aquaculture Effluents Task Force 223
Natural Resources Defense Council 222
Net capital ratio 47, 48, 249
Net farm income 31, 44, 45, 136, 208, 249
   Net farm income from operations 31, 249
Net Present Value 84, 86, 160, 164, 165, 249
Net returns 30, 122, 191, 193, 249
   Above cash costs 124, 191
   Above opportunity costs of female labor, management, risk 122, 124, 191, 249
   Above risk 122, 124, 191, 249
Net worth 31, 47, 48, 54, 86, 132, 133, 249
New products 27
New Zealand 103
Niche market 27, 249
Non-negativity 211, 249
Norway 103, 203, 224
Norwegian Fish Farmers Association 222
NPDES permits 222, 249

Objective function 212, 249
Occupational Safety and Health Act (OSHA) 111
Office International des Epizooties 224, 249
Open-ocean aquaculture 203
Operating expense ratio 51, 249
Operating line of credit 174, 249
Operating Profit Margin (OPMR) 44, 45, 249
Opportunity cost 84, 249
Organizational structure 3, 6, 249
Outlier 189, 249
Owner equity 132, 133, 135, 249

Pacific threadfin 153
Pangasius 21, 224
Partial budget 110, 111, 112, 117, 126, 128, 185, 192, 250
Partnership 6, 246, 250
   General 6, 246
   Limited 6, 247
Payoff matrix 95, 250
Payment 161, 250
Payback period 160, 163, 250
Penaeus
   merguiensis 99
   monodon 99
   occidentalis 99
   schmitti 99
   setiferus 99
   vannamei 99
Percentage owed 51

Percentage owned 52
Permits 8
Portugal 103
Practical application
    Balance sheet 135
    Business plan 34
    Cash flow budget 148
    Cash vs. accrual accounting 139
    Enterprise budget 123
    Financing 63
    Income statement 138
    Investment analysis 166
    Lending 179
    Managing capital assets 88
    Managing cash flow 74
    Managing labor 111
    Managing risk 103
    Misuse of enterprise budgets 193
    Monitoring performance 53
    Marketing 19
    Partial budget 128
    Regulations 228
    Risk analysis 204
    Starting an aquaculture business 10
    Whole-farm modeling 215
Prawns 129
Present Value 160, 161, 170–171
Pricing 16, 27
Prince Edward Island 224
Principal 174, 175, 176, 177, 179, 250
Probability density function 199, 250
Product life cycle 15, 250
Product positioning 14, 27, 250
Product-space map 14, 250
Production efficiency 41
Promotion 13, 15
Profit and loss statement 131
Profit margins 43
Profitability 43, 67, 250
Profitability measures 122
Proportionality 211, 250
Proprietorship 6, 250
Purdue University 235
Put options 98, 250

Quadratic programming 214, 250

Rate of return 44
    Internal 160, 165, 167, 247
    Simple 160, 162, 164, 165
    On Farm Assets (ROA) 44, 45, 101, 250
    On Farm Equity (ROE) 45, 46, 101, 250
Ratios
    Acid test 51, 52
    Cash flow coverage 50
    Current 48, 50

Debt/asset 47, 48
Debt/equity 47, 48
Debt servicing 50
Debt structure 47, 49
Debtor turnover 51
Depreciation expense 51
Equity/asset 47, 48
Growth 53
Interest coverage 47, 49
Interest expense 51
Net capital 47
Operating expense 51
Stock turnover 43
Percentage owned 52
Turnover 51, 242
Record-keeping 9, 19, 34, 63, 74, 88, 102, 111, 128, 134,
        137, 147, 166, 177, 192, 203, 215, 227
Red claw 201
Reduced costs (partial budget) 127, 250
Reduced costs (mathematical programming) 214, 251
Reduced revenue (partial budget) 127
Regrets matrix 95, 251
Regulations 219
Repayment capacity 52, 251
Retained earnings 85, 251
Return on investment (ROI) 192
Return to labor and management 45, 46, 251
Return to labor 45, 46, 251
Return to management 45, 47, 251
Right-hand side value 212, 213, 251
Risk 93, 197, 233, 251
    Analysis 197
    Financial 4, 197
    Income 197
    Management 101
    Marketing 197
    Price 197
    Production 197
    Programming 202, 214
    Software 101
    Yield 197, 198
@RISK 101, 201
Risk and returns comparison 95, 96, 251
Risk-programming model 202, 203, 205, 251
Risk-rated management 101
Rivers and Harbor Act of 1989 220, 221
Rwanda 113, 202

Safety first 95, 96, 215, 251
Salmon 100, 119, 203, 224, 232, 234, 235
Salvage value 84
Scallops 79, 91, 194
Security 174, 251
Self-liquidating loans 60, 62, 99, 251
Sensitivity analyses 99, 101, 126, 192, 214, 251
Separability 211, 251

Shadow prices 214, 251
Share leases 97, 251
Short position 98, 251
Short-run average cost curve 83
Shrimp 7, 10, 68, 83, 88, 94, 95, 98, 99, 102, 103, 118,
    194, 202, 203, 218, 224, 223
  Giant tiger 99
Silver carp 225, 226
Simple rate or return 160, 164
Simulation model 201, 252
Small Business Administration 59
Small-scale growers 17, 63
Snakehead 129
Social Security (See Federal Insurance Contributions
    Act) 111
Sole proprietorship 6, 252
Solver 215, 216
Solvency 41, 47, 62
South Korea 103
Southern Regional Aquaculture Center (SRAC) 239
Spain 103, 219
Specialty markets 13, 14, 15, 252
Speculators 98
Spring Viremia of Carp 224
Sri Lanka 219
Statement of cash flow 69, 70
Strategic goals 24
Stochastic cost function 201, 252
Stochastic dominance 201, 233, 252
Stock turnover ratio 43, 252
Strike price 98
Striped Bass Growers Association 222
Summer flounder 97, 202, 205
Sunk cost 187, 252
Supply chain management 27, 252
SureFish 225
Sweden 103
Swimming crab 224

Tableau 209, 214, 215, 252
Target MOTAD 203, 214, 215, 252
Taura Syndrome Virus 102
Techniques to handle risk 96
Term Debt and Capital Lease Coverage Ratio 52, 71, 252
Tesco 17, 225
Thailand 224
Tilapia 21, 35, 113, 129, 194, 212, 232, 234
Time value of money 160
Tools for managing cash flow risk 100
Tools for managing financial risk 99
Tools for market risk 97
Tra 224
Transshipment 214, 252
Trade issues 223

Trout 203, 223, 228
Turkey 103
Turnover ratio 51, 253

Unbounded solution 213, 253
Uncertainty 197, 253
United Kingdom 103
United States of America 103, 111, 186, 203, 219, 224
University of Arkansas at Pine Bluff 239
University of Idaho 235
University of Minnesota 237
University of Missouri 237
Unpaid family labor 190, 191, 192, 193, 253
U.S. Army Corps of Engineers 220, 221
U.S. Bureau of Labor 236
U.S. Department of Agriculture 100, 177, 186, 203
    USDA-APHIS 235
    USDA-Regional Aquaculture Centers 236
    USDA-Agricultural Marketing Service 235
    USDA-National Agricultural Statistics Service 239
U.S. Department of Energy 121, 237
U.S. Department of the Interior
    Department of Commerce 220, 221, 236
U.S. Department of Labor 121
U.S. Economic Research Service 237
U.S. Fish and Wildlife Service 220, 221, 226
U.S. Food and Drug Administration 225, 227, 236, 238
U.S. Trout Farmers of America 222

Valuation of assets 134, 253
    Book value 134
    Cost basis 134, 253
    Market basis 134, 253
Variance 198, 199, 253
Venezuela 224
Vertical integration 19, 253
Veterans Administration 59, 253
Vietnam 224, 225
Viral Hemorrhagic Septicemia 93, 224

Wal-Mart 17, 225
Water Quality Act of 1987 220, 221
White Spot Syndrome Virus 102
Whole-farm budgeting 207
Whole-farm modeling 207, 253
Working capital 50, 253
World Aquaculture Society 240
World Tilapia Association 222
World Trade Organization 223, 240
World Wildlife Federation 225

Years of useful life 253

Zebra mussels 225